The Experimental Fire

synthesis

A series in the history of chemistry, broadly construed,
edited by Carin Berkowitz, Angela N. H. Creager, John E. Lesch,
Lawrence M. Principe, Alan Rocke, and E. C. Spary,
in partnership with the Science History Institute.

The
Experimental
FIRE

Inventing English Alchemy, 1300–1700

Jennifer M. Rampling

The University of Chicago Press

Chicago and London

The University of Chicago Press, Chicago 60637
The University of Chicago Press, Ltd., London
© 2020 by The University of Chicago
Published 2020
Paperback edition 2023
Printed in the United States of America

32 31 30 29 28 27 26 25 24 23   1 2 3 4 5

ISBN-13: 978-0-226-71070-9 (cloth)
ISBN-13: 978-0-226-82654-7 (paper)
ISBN-13: 978-0-226-71084-6 (e-book)
DOI: https://doi.org/10.7208/chicago/9780226710846.001.0001

Library of Congress Cataloging-in-Publication Data

Names: Rampling, Jennifer M., author.
Title: The experimental fire : inventing English alchemy, 1300–1700 /
    Jennifer M. Rampling.
Other titles: Synthesis (University of Chicago Press)
Description: Chicago : University of Chicago Press, 2020. | Series:
    Synthesis | Includes bibliographical references and index.
Identifiers: LCCN 2020018398 | ISBN 9780226710709 (cloth) | ISBN
    9780226710846 (ebook)
Subjects: LCSH: Alchemy—England—History. | Chemistry—
    England—Experiments—History.
Classification: LCC QD18.G7 R36 2020 | DDC 540.1/1209420903—dc23
LC record available at https://lccn.loc.gov/2020018398

In loving memory of Dr. Molly Rampling (1913–2008),
my grandmother

Contents

Figures

Abbreviations

Ashmole	Oxford, Bodleian Library, MS Ashmole
BCC	Jean-Jacques Manget, ed., *Bibliotheca Chemica Curiosa*, 2 vols. (Geneva: Chouet, 1702)
CCC	Oxford, Corpus Christi College Library
CCCC	Cambridge, Corpus Christi College Library
CRC	Jennifer M. Rampling, "The Catalogue of the Ripley Corpus: Alchemical Writings Attributed to George Ripley (d. *ca.* 1490)," *Ambix* 57 (2010): 125–201
Getty	Los Angeles, Getty Research Institute, MS
Harley	London, British Library, MS Harley
HMES	Lynn Thorndike, *A History of Magic and Experimental Science*, 8 vols. (New York: Columbia University Press, 1923–58)
ODNB	*Oxford Dictionary of National Biography* (Oxford: Oxford University Press, 2004; online ed., 2007)
OOC	George Ripley, *Opera omnia chemica*, ed. Ludwig Combach (Kassel, 1649)
Singer	Dorothea Waley Singer and Annie Anderson, *Catalogue of Latin and Vernacular Alchemical Manuscripts in Great Britain and Ireland Dating from before the XVI Century*, 3 vols. (Brussels: Maurice Lamertin, 1928, 1930, 1931)
Sloane	London, British Library, MS Sloane
TC	Lazarus Zetzner, *Theatrum chemicum*, 6 vols. (Ursel and Strasburg, 1602–61)

TCB   Elias Ashmole, ed., *Theatrum Chemicum Britannicum* (London, 1652)

Testamentum Michela Pereira and Barbara Spaggiari, eds., *Il Testamentum alchemico attribuito a Raimondo Lullo: Edizione del testo latino e catalano dal manoscritto Oxford, Corpus Christi College, 244* (Florence: SISMEL, 1999). Excerpts are denoted by book and page reference (e.g., 1:172).

Trinity   Cambridge, Trinity College Library, MS

Conventions

Since many of the texts cited in this book have never been published in print form, I rely throughout on transcriptions from manuscripts, where words are often abbreviated and spelling is inconsistent. In such cases, I use italics to denote the expansion of abbreviated text. Text between "\ /" indicates subsequent additions or amendments to the manuscript. Where additional text is required to determine the sense of a passage, this is placed within square brackets, as in the representation of alchemical symbols (e.g., [mercury]). When manuscripts are cited in references, "||" denotes page endings. Original spelling and capitalization have been retained, including the frequent use of "v" for "u" (and vice versa) and of "j" for "i." Thorn (þ) is replaced by "th" in square brackets. Where appropriate I have modernized punctuation by substituting commas for periods and dashes.

In early modern England, 25 March marked the first day of the new year. Dates between 1 January and 24 March are therefore indicated in the format "5 March 1573/4."

The names of famous alchemists are preserved in their usual anglophone forms (Raymond Lull for Ramon Llull, Arnald of Villanova for Arnau de Vilanova). All translations are my own unless otherwise stated.

Acknowledgments

As is usually the case with alchemical literature, this book rests upon the work of previous authorities, whom (in a departure from many medieval compilations) I hereby gratefully cite. My academic career would never have left port without the enthusiasm and erudition of Peter Forshaw and Stephen Clucas, who mentored my MA studies at Birkbeck, University of London, encouraged me to consider doctoral research, and supplied valuable advice and camaraderie throughout the PhD. Once embarked upon, the project remained on an even keel thanks to the calm good sense, humor, and unflagging patience of my supervisor, Lauren Kassell. Her wisdom and generous support underwrites both the book and many subsequent scholarly ventures. Peter Murray Jones trimmed sail with learned advice on many topics, and Hasok Chang provided endless encouragement during my post-doctoral fellowship in Cambridge. My colleagues in Princeton's History Department, and especially in the Program for the History of Science, have provided the most delightful of all scholarly harbors.

I am grateful to those who generously took the time to read the book at various stages of its gestation. The entire manuscript was read by Lawrence Principe, Stephen Clucas, Paula Findlen, Peter Forshaw, and Tony Grafton; needless to say, the final result has benefited enormously from their perspicacious comments and critical precision. In addition, individual chapters were read by Peter Jones, Sébastien Moureau, William Newman, Sophie Roux, and Dmitri Levitin, as well as the massed expertise of Princeton's Program Seminar in the History of Science, and Lorraine Daston's Department II

Colloquium at the Max Planck Institute for the History of Science. To all of these readers, my deepest thanks.

The book has taken shape over the course of a series of wonderful visiting fellowships, at (in date order) the Chemical Heritage Foundation, Philadelphia; Scaliger Institute, Leiden University; National Hellenic Research Foundation, Athens; Centre Simão Mathias for Studies in History of Science (CESIMA), Pontifical Catholic University of São Paulo; Beinecke Rare Book & Manuscript Library, Yale University; Clare Hall, Cambridge; Centre for Research in the Arts, Social Science, and Humanities (CRASSH), University of Cambridge; MPIWG (Department II), Berlin; and All Souls College, Oxford. I am immensely grateful to all these institutions for offering me house room, reading matter, and much more. The project has also required me to consult a large number of primary sources in archives worldwide, and I am grateful beyond measure to the librarians and archivists of these institutions for their hospitality and kind assistance. I would thank in particular Julian Reid at Corpus Christi College Library, Oxford; Jonathan Smith and Sandy Paul at Trinity College Library, Cambridge; and Frau Ruiz at Leipzig Universitätsbibliothek.

My doctorate was funded by the Darwin Trust of Edinburgh Martin Pollock Scholarship, and my postdoctoral research by a Wellcome Trust Postdoctoral Fellowship. In Princeton, my work has been supported by grants from the David A. Gardner '69 Magic Project, the University Committee on Research in the Humanities and Social Sciences, and the Office of the Dean for Research. Further funding for archival research was awarded at various times by the British Society for the History of Science, Cambridge European Trust, Clare College, Cambridge's Department of History and Philosophy of Science, Richard III Society, Royal Historical Society, Society for the History of Alchemy and Chemistry, Society for Renaissance Studies, and J. B. Trend Fund (Cambridge). I should like to express my thanks to all these bodies and institutions for making my research materially possible.

Throughout my studies and subsequent career I have benefited from the expertise, comradeship, and hospitality of many scholars. These include Ana Maria Alfonso-Goldfarb, Robert Anderson, Debby Banham, Marco Beretta, Donna Bilak, Harmut Broszinski, Charles Burnett, Antonio Clericuzio, Angela Creager, Chiara Crisciani, Surekha Davies, Jenny Downes, Seb Falk, Marcia Ferraz, Hjalmar Fors, Daniel Garber, Margaret Garber, Roger Gaskell, Michael Gordin, Molly Greene, John Haldon, Anne Hardy, Katherine Harloe, Felicity Henderson, Hiro Hirai, James Hyslop, Nick Jardine, William Chester Jordan, Didier Kahn, Vera Keller, Elisabeth Leedham-Green, Matteo Martelli, Erika Milam, Iris Montero Sobrevilla, Bruce Moran, Nicolette Mout, Signe Nipper Nielsen, Tara Nummedal, Kasper van Ommen, Cesare Pastorino, Michela Pereira, Will Poole, Katya Pravilova, Valentina Pugliano, Nicky Reeves, Helmut Reimitz, Anna Marie Roos, Simon Schaffer, Daniela Sechel, Katie Taylor, Pierre Teissier, Brigitte van Tiggelen, Anke Timmermann, Koen Vermeir, Keith Wailoo, and Tessa Webber. If I have omitted anyone, I apologize. In addition, Peter Wothers at Cambridge and Craig Arnold at Princeton generously furnished me with laboratory space for alchemical experiments, and Lawrence Principe provided important practical demonstrations of "Ripleian" alchemy. Rafał Prinke and Ivo Purš drew my attention to several important continental manuscripts. Special thanks to Elizabeth, Paul, Hasok, and Gretchen, who helped keep me going through the whole thing, and to Robin Sutton, who was there from the start.

I thank my editor, Karen Darling, and the Board of Synthesis at the University of Chicago Press for supporting the book in word and deed, Tristan Bates and Caterina MacLean for their help in nursing it into production, and Marian Rogers for her meticulous copyediting of the manuscript.

Finally, my deepest gratitude goes to those whose support is the canvas for all my endeavors: my parents, John and Susan Rampling; my brothers, Adam and Jonathan; and my grandparents.

Introduction

What Is Mercury?

One thyng, one Glasse, one Furnace and no mo.[1]

On 20 July 1577, the gentleman alchemist Samuel Norton completed the preface to a treatise addressed to his sovereign, Elizabeth I, Queen of England. The *Key of Alchemy* offered Elizabeth a taste of the extraordinary physical transformations wrought by chemistry. Who would not be amazed, asked Norton, to see hard iron turned into soft water, or glass made to withstand the blow of a hammer? To watch flowing quicksilver form "a stedfaste masse," and fixed steel "flye away in smoke"? As if these astounding metallurgical effects were not enough, his science also taught how metals and minerals could be used to heal the human body: "Copper to becom medicinable, gould and silver to be potable, tynne to remove great sicknesses, and lead in vertue exceedinge all, to haue almost the swettnes of sugare in taste." Using alchemical techniques, even minerals and deadly poisons could become perfect medicines—transformations that, Norton assured the queen, "will lightly be done, and are not of great difficultye."[2]

Yet in this remarkable list there is an interesting omission. Nowhere did Norton mention transmutation: the alchemists' dream of perfecting a technique for transforming base metals into silver and gold. His medieval authorities often referred to the agent of transmutation as the "philosophers' stone"

---

1. George Ripley, *Compound of Alchemy*, in *Theatrum Chemicum Britannicum: Containing Severall Poeticall Pieces of Our Famous English Philosophers, Who Have Written the Hermetique Mysteries in Their Owne Ancient Language. Faithfully Collected into One Volume with Annotations Thereon*, ed. Elias Ashmole (London: J. Grismond for Nathanial Brooke, 1652), 107–93, on 159; hereafter *TCB*.

2. Oxford, Bodleian Library, Ashmole 1421, fol. 169r-v.

I

(*lapis philosophorum*), a superperfected form of matter made using alchemical techniques.[3] This "stone" is typically introduced in the singular, implying
that the whole practice of alchemy tends toward this one, universal end. Yet,
rather than lingering over a single, unique stone, Norton offered a variety of
alchemical products, including several with medicinal applications. In addition to mineral, vegetable, and animal stones, the *Key* described an elixir of
life, a multipurpose "mixed" stone, and a "transparent" stone used for making precious gems.

Norton did not claim any novelty for his many-stranded approach. On the
contrary, the Somerset practitioner was keen to state his alchemical credentials by positioning himself within a lineage of England's great adepts. His
great-grandfather, he claimed, was the fifteenth-century Bristol alchemist
Thomas Norton (d. 1513), author of a famous poem, the *Ordinal of Alchemy*
(1477). Samuel's *Key* could also claim descent from another fifteenth-century
master: its recipes had been extracted from a book compiled by the great
English alchemist George Ripley, canon of Bridlington (fl. 1470s). Throughout the *Key*, Norton drew repeatedly on the authority of medieval English
adepts, noting that, for their services in clarifying the obscurities of the
alchemical art, no one deserved more honor than his own countrymen.[4]

Norton's treatise is emblematic of the alchemical preoccupations of the
late sixteenth century, a period characterized by powerful optimism about
the potential of the art. Writers were inspired by the transformative capa-
bilities of chemical operations, yet also driven by a pressing need for practical solutions to economic, political, and medical problems. Across Europe,
princes invested funds and credit in alchemical projects, medical practitioners appropriated alchemical techniques, and poets drew on alchemical language to express both material and metaphysical ideals. At the same time,
alchemy was increasingly the butt of satire and polemic, as critics dwelled
on the tricks and moral failings of those who professed to have knowledge of
transmutation. A reputational chasm opened between "philosophers," who
had truly mastered the secrets of alchemy, and others who had not, or who

---

3. The use of the term "stone" for the transmuting agent originates in Arabic alchemy, where
ḥajar (stone) denoted the matter used to make the elixir, regardless of whether that matter was
animal, vegetable, or mineral in nature. The term was translated directly into Latin as *lapis*. Sébastien Moreau, "*Elixir Atque Fermentum*: New Investigations about the Link between Pseudo-
Avicenna's Alchemical *De anima* and Roger Bacon; Alchemical and Medical Doctrines," *Traditio:
Studies in Ancient and Medieval Thought, History, and Religion* 68 (2013): 277–323, on 288–89.

4. Ashmole 1421, fol. 172v.

merely claimed to have done so—variously decried as fools, puffers, frauds, or simply "alchemists."[5]

It was in this environment of mingled optimism and skepticism that alchemical practitioners turned to the past in search of authoritative support for their current endeavors. In England, that usually meant looking across the English Channel to the lands of continental Europe: the source of influential alchemical texts and translations during the Middle Ages, and, in the sixteenth century, the site of continuing innovation in mining, metallurgy, chemical medicine, and the manufacture of chemical products, which English practitioners were eager to imitate. However, as the sixteenth century progressed and the Reformation reshaped English cultural life, Tudor alchemists became increasingly preoccupied with their medieval legacy. Competing with foreign practitioners for readers and patrons, they drew attention to their own Englishness. Past adepts, real and imagined—from Merlin and Saint Dunstan to Roger Bacon and John Dastin—were invoked in alchemical patronage proposals, the style of their alleged works imitated, and their accomplishments reenacted (so their early modern disciples claimed) through countless experiments. More recent writers like George Ripley and Thomas Norton in turn acquired a reputation for successful practice, and were enshrined in the pantheon of English alchemy as exemplars for new generations of hopeful adepts. Even Samuel Norton, the devoted Elizabethan interpreter of Ripley and Norton, eventually gained a lesser place in this pantheon, as his writings passed the torch of English alchemy down to his own seventeenth-century readers. Posterity thus achieved what Samuel was unable to accomplish during his own lifetime, by reinventing him as an alchemical philosopher—a new link in the golden chain that stretched back into antiquity.

READING LIKE AN ALCHEMIST

Samuel Norton was not the first to search for links between experimental practice and his own nation's history. European knowledge of the natural world expanded dramatically throughout the sixteenth and seventeenth centuries, a period still regularly characterized, albeit in increasingly broad

5. On the persona of the alchemical fraud, or *Betrüger*, in early modern Europe, especially in the German lands, see Tara Nummedal, *Alchemy and Authority in the Holy Roman Empire* (Chicago: University of Chicago Press, 2007). On associations with currency crime, see Jotham Parsons, *Making Money in Sixteenth-Century France: Currency, Culture, and the State* (Ithaca, NY: Cornell University Press, 2014), 223–31.

terms, as a scientific revolution. However, while early modern natural philosophers often emphasized what was new in their work, they were also deeply concerned to recover what was old. This engagement with the past, catalyzed by the rediscovery of ancient texts and artifacts, transcended disciplinary fields and, to an extent, territorial boundaries.[6] It was also, inevitably, value-laden. Whether gathering antique inscriptions, imitating classical artworks, or scouring medieval documents for evidence of early church practices, early modern knowledge-seekers were motivated by contemporary concerns, imposing their own political, religious, and scholarly preoccupations on frequently obscure or fragmentary source material. When these sources were missing or corrupt, ingenious readers might even attempt to fill the gaps by reconstructing "lost" content, in whole or in part.[7] One outcome was the invention of new traditions in the name of the old: from rewriting liturgy in the wake of the Reformation to seeking philosophical and scriptural precedents for new visions of the structure of matter.[8]

In this book, I trace how this fusion of authority and invention contributed to the development of a particular body of natural knowledge—alchemy—in the context of one national tradition. Over the last half century, historians of science and medicine have revealed the important role played by alchemy in shaping early modern scientific ideas and practices, as an experimental enterprise that was also grounded in sophisticated theories of nature. Historians of books and reading have also shown how readers studied past texts to shed light on problems they faced in their own time. But how, exactly, did book learning interact with practical experience? Did alchemical practitioners deliberately innovate, or did they rather view their experimental work as a form of historical reconstruction—an attempt to recover the lost practices of their medieval forebears?

In attempting to answer those questions, I have chosen to restrict my own reconstructive efforts to a specific place and time—the insular kingdom of

6. On natural philosophers' employment of humanist methods, including the study of ancient and medieval texts and philosophies, see, inter alia, Anthony Grafton, *Defenders of the Text: The Traditions of Humanism in an Age of Science, 1450–1800* (Cambridge, MA: Harvard University Press, 1991); on the English context in particular, Dmitri Levitin, *Ancient Wisdom in the Age of the New Science: Histories of Philosophy in England, c. 1640–1700* (Cambridge: Cambridge University Press, 2015).

7. Conjectural emendation, long employed in scriptural exegesis, provided one such technique; see Anthony Grafton, *Joseph Scaliger: A Study in the History of Classical Scholarship*, vol. 1, *Textual Criticism and Exegesis* (Oxford: Clarendon Press, 1983), 12–14.

8. On ecclesiastical traditions, see Anthony Grafton, "Church History in Early Modern Europe: Tradition and Innovation," in *Sacred History: Uses of the Christian Past in the Renaissance World*, ed. Katherine Van Liere et al. (Oxford: Oxford University Press, 2012), 3–26; on matter theory, Levitin, *Ancient Wisdom*, chap. 5.

England, from the beginning of the fourteenth century to the end of the seventeenth.[9] While limiting my scope geographically I seek to extend it temporally, and, in so doing, to chart how alchemists crafted a new kind of chemical practice, grounded in English history, over a significant chronological span. In England, this extended period witnessed the arrival of plague, the dissolution of the monasteries, the advent of Paracelsianism, and the rise of antiquarianism and experimental science: all of which affected how alchemical books were read, and to what ends. It is only by following texts and practices over time, and in granular detail, that we can grasp the cumulative impact of incremental changes in the science itself.

Alchemy offers promising fuel for this investigation precisely because its objects, although intimately concerned with the workings of nature, have no clear analogue in the modern sciences. No longer considered a fruitful topic of scientific study, alchemy in its premodern heyday nonetheless underpinned many activities, and offered answers to many questions, that are still considered germane to the chemical sciences today. Alchemy is not, however, the same as modern "chemistry," and most historians would agree that our understanding of its past can only be impoverished by attempts to read it solely in light of present-day definitions, standards, and expectations.[10] Yet our very willingness to take alchemy on its own historical terms is fostered by the assumption that its ideas and practices are no longer relevant to the science of our day—or, more bluntly, that they do not "work."

Early modern alchemists lacked that assumption. The recovery of alchemical knowledge invoked a special kind of antiquarian sensibility, one that was concerned not just with the form of practices in the past, but also with their effectiveness in the present. When sixteenth- and seventeenth-century alchemists opened their books, or assembled their materials for practice, they engaged with the medieval corpus as a tradition that, although temporally distant, was nonetheless living—and that promised incalcul-

9. By focusing on England rather than the British Isles more generally, I thereby regretfully exclude alchemy as practiced in Scotland, Wales, and Ireland. Alchemy was of course practiced elsewhere in the Isles, and attracted great interest at the Scottish royal court; see, for instance, the case of John Damian summarized in John Read, "Alchemy under James IV of Scotland," *Ambix* 2 (1938): 60–67.

10. The danger of driving a terminological wedge between "alchemy" and "chemistry" has been addressed by William Newman and Lawrence Principe, who propose the general use of "chymistry" as a solution: William R. Newman and Lawrence M. Principe, "Alchemy vs. Chemistry: The Etymological Origins of a Historiographic Mistake," *Early Science and Medicine* 3 (1998): 32–65. In this book I typically follow my historical actors in using "alchemy" and, more commonly still, "philosophy." To avoid anachronistic comparisons, I generally use "natural philosophy" rather than "science" when discussing the formal study of the natural world; on occasions when I do employ "science," I intend its broader, early modern sense of learned knowledge (*scientia*).

able material benefits, as well as unparalleled insight into the workings of nature.[11] In this context, medieval books provided vital sources of theoretical insight and practical instruction.[12] Even at the vanguard of developments in seventeenth-century chemistry, natural philosophers like Robert Boyle and Isaac Newton studied the fifteenth-century writings of George Ripley with attention, interest, and expectation of useful results.[13]

Like all living systems, medieval alchemy was also subject to change. Early modern readers knew that the task of extracting workable knowledge from these sources was no sinecure, and, like the editors of ancient texts, they sought to fill in the gaps. They studied, tested, and reinterpreted their authorities, using the most ingenious trials that reason and experience could suggest, often in ways unanticipated by the original writers. To translate is to interpret: accordingly, the very process of reconstructing past processes inevitably (and often unwittingly) transformed their content—and hence their practical outcomes—in a cyclical process that I call "practical exegesis."[14]

In this book I trace how this cycle of reinvention revolved in England over the space of four centuries, and how it resulted in alchemical change. During this period, successive generations of English alchemists transformed the theory and practice of their art: unpicking the clues of their forebears,

11. On the emerging concern with the past among early modern readers of English alchemica, see George R. Keiser, "Preserving the Heritage: Middle English Verse Treatises in Early Modern Manuscripts," in *Mystical Metal of Gold: Essays on Alchemy and Renaissance Culture*, ed. Stanton J. Linden (New York: AMS, 2007), 189–214; Lauren Kassell, "Reading for the Philosophers' Stone," in *Books and the Sciences in History*, ed. Marina Frasca-Spada and Nick Jardine (Cambridge: Cambridge University Press, 2000), 132–50. On English antiquarianism more generally, T. D. Kendrick, *British Antiquity* (New York: Barnes & Noble, 1950); Mary McKisack, *Medieval History in the Tudor Age* (Oxford: Clarendon Press, 1971); Graham Parry, *The Trophies of Time: English Antiquarians of the Seventeenth Century* (Oxford: Oxford University Press, 1995); Thomas Betteridge, *Tudor Histories of the English Reformations, 1530–83* (Aldershot: Ashgate, 1999); Angus Vine, *In Defiance of Time: Antiquarian Writing in Early Modern England* (Oxford: Oxford University Press, 2010).

12. See, for instance, the sources in Timothy Graham and Andrew G. Watson, eds., *The Recovery of the Past in Early Elizabethan England: Documents by John Bale and John Joscelyn from the Circle of Matthew Parker* (Cambridge: Cambridge University Press, 1998). On medieval manuscripts in domestic contexts, Margaret Connolly, *Sixteenth-Century Readers, Fifteenth-Century Books: Continuities of Reading in the English Reformation* (Cambridge: Cambridge University Press, 2019).

13. On Ripley's seventeenth-century reception, see chap. 9, below. On Boyle and his sources, see Lawrence M. Principe, *The Aspiring Adept: Robert Boyle and His Alchemical Quest* (Princeton: Princeton University Press, 1998); William R. Newman and Lawrence M. Principe, *Alchemy Tried in the Fire: Starkey, Boyle, and the Fate of Helmontian Chymistry* (Chicago: University of Chicago Press, 2002). On Newton, see Newman, *Newton the Alchemist: Science, Enigma, and the Quest for Nature's "Secret Fire"* (Princeton: Princeton University Press, 2018).

14. I introduce this term in Jennifer M. Rampling, "Transmuting Sericon: Alchemy as 'Practical Exegesis' in Early Modern England," *Osiris* 29 (2014): 19–34; see also chap. 2, below.

attempting to follow their instructions, and eventually feeding their own practical findings back into the textual record in the form of new treatises, recipes, and annotations. The cycle relied on a twofold process of reconstruction: not just the replication of practices, but the recovery of meaning hidden within texts. The densely encoded and frequently laconic guidance bequeathed by past philosophers to their hopeful descendants required a raft of special interpretative techniques, which challenged early modern readers just as they continue to perplex modern scholars. The history of practice is thus intimately related to the history of reading. To retrieve the original sense of a text—and hence to reconstruct, insofar as it is possible, the original practice—requires that we, too, learn to read like alchemists; or, even more specifically, like alchemical philosophers.

Throughout the book, I use the notion of the "alchemical philosopher" as a very particular instantiation of the natural philosopher: a reader-practitioner whose interests are neither wholly scholarly nor wholly grounded in craft, but who is presumed to have acquired special insight into the making of the philosophers' stone. While many alchemical writers self-identified as philosophers, the term was also bestowed as an accolade by later readers who recognized that success in the art trumped any formal educational qualifications. It therefore encompasses a remarkable range of historical actors: from university-trained scholars of European eminence, like Roger Bacon (ca. 1214–1292?) and John Dee (1527–1609), to men with mercantile or artisanal backgrounds, like the clothworker Thomas Peter (fl. 1520s–1530s) and unlicensed medical practitioner Thomas Charnock (1524/6–1581). Those who identified as alchemical philosophers also tended to view their knowledge as a route to social and economic advancement—thus, despite a wide disparity in their backgrounds, education, and connections, both Dee and Charnock aspired to become Elizabeth I's own "philosopher."[15] Accordingly, alchemical philosophy is often closely linked to patronage, although there was not always consensus over who counted as an adept: as we shall see, one man's philosopher was another man's fraud.[16]

15. Dee famously conceived of himself as a "Christian Aristotle" in search of royal patronage of the kind offered by Aristotle's own pupil, Alexander the Great—a trope discussed by Nicolas H. Clulee, *John Dee's Natural Philosophy: Between Science and Religion* (Oxford: Routledge, 1988), 189–99; Paula Findlen, *Possessing Nature: Museums, Collecting, and Scientific Culture in Early Modern Italy* (Berkeley: University of California Press, 1994), 352–65. Charnock sets out his aspirations in his *Booke Dedicated vnto the Queenes Maiestie*, British Library, MS Lansdowne 703, fol. 45v, discussed in chap. 6, below.

16. The classic study of alchemical courtly patronage is Bruce T. Moran, *The Alchemical World of the German Court: Occult Philosophy and Chemical Medicine in the Circle of Moritz of Hessen*

This hybrid status of alchemy raises the question of how its practitioners first came to view their enterprise as philosophical. Although alchemy was already viewed as a subject of philosophical provenance in Greco-Roman Egypt and the Islamic lands, in twelfth-century Latin Europe it was still a newcomer by the standards of other fields of knowledge.[17] Accordingly, its early proponents sought to establish its prestige by positioning it as *scientia* (learned knowledge), and hence proper to the study of natural philosophy, rather than as *ars* (craft knowledge). The discipline of scholastic natural philosophy—named for the schools where it first took shape—was itself a medieval invention, concerned with the content of Aristotle's natural books.[18] Its goal was to generate certain knowledge through the derivation of universal principles from particulars: a form of knowledge building distinct from artisanal or "mechanical" practices of the kind implicated in much alchemical activity.[19] By arguing that their work was similarly grounded in general, natural principles, proponents of alchemy claimed that it was as much a "science" as other branches of learned knowledge, and hence worthy

(1572–1632) (Stuttgart: Franz Steiner Verlag, 1991). Other important studies include R. J. W. Evans, *Rudolf II and His World: A Study in Intellectual History 1576–1612* (Oxford, 1973; repr., London: Thames & Hudson, 1997); Pamela H. Smith, *The Business of Alchemy: Science and Culture in the Holy Roman Empire* (Princeton: Princeton University Press, 1994); Nummedal, *Alchemy and Authority*; David C. Goodman, *Power and Penury: Government, Technology, and Science in Philip II's Spain* (Cambridge: Cambridge University Press, 1988); Alfredo Perifano, *L'alchimie à la cour de Côme Ier de Médicis: Culture scientifique et système politique* (Paris: Honoré Champion, 1997); Nils Lenke, Nicolas Roudet, and Hereward Tilton, "Michael Maier—Nine Newly Discovered Letters," *Ambix* 61 (2014): 1–47. Jonathan Hughes has written two speculative studies of royal interest in alchemy in medieval England, to be treated with caution: Jonathan Hughes, *Arthurian Myths and and Alchemy: The Kingship of Edward IV* (Stroud: Sutton Publishing, 2002); Hughes, *The Rise of Alchemy in Fourteenth-Century England: Plantagenet Kings and the Search for the Philosopher's Stone* (London: Continuum, 2012).

17. The arrival of alchemy was an outcome of the great Arabic-to-Latin translation movement of the twelfth and thirteenth centuries; see the references on p. 32, note 29, below. For an overview of alchemy's earlier history, see Lawrence M. Principe, *The Secrets of Alchemy* (Chicago: University of Chicago Press, 2013), chaps. 1–3. On alchemy as a *novitas* in Latin Europe, see Robert Halleux, *Les textes alchimiques* (Turnhout: Brepols, 1979), 70–72.

18. On the incorporation of Aristotle's *libri naturales* into the medieval curriculum, see Edward Grant, *The Foundations of Modern Science in the Middle Ages: Their Religious, Institutional, and Intellectual Contexts* (Cambridge: Cambridge University Press, 1996); Grant, *God and Reason in the Middle Ages* (Cambridge: Cambridge University Press, 2009). The significant role played by the mendicant orders in shaping the identity of medieval natural philosophy is examined (if somewhat provocatively) in Roger French and Andrew Cunningham, *Before Science: The Invention of the Friars' Natural Philosophy* (Aldershot: Ashgate, 1996).

19. On the relationship between art and nature in scholastic natural philosophy, and its consequences for the status of alchemy as *scientia*, see William R. Newman, "Technology and Alchemical Debate in the Late Middle Ages," *Isis* 80 (1989): 423–45; Newman, *Promethean Ambitions: Alchemy and the Quest to Perfect Nature* (Chicago: University of Chicago Press, 2004).

to be counted as philosophy. The English philosopher Roger Bacon went so far as to propose alchemy as the foundation of science and medicine, since it teaches how all things are generated from the elements.[20]

Despite these attempts, alchemy failed to secure a foothold in the medieval university curriculum, although its practitioners did not abandon their philosophical aspirations. By the fifteenth century, even less well-educated practitioners had learned to present their work in the form of "philosophical" treatises that expounded the theory of alchemy alongside its practice. This positioning did not convince critics like the naturalist Conrad Gessner (1516–1565). While admitting that the objects of alchemy (such as metals) were proper to natural philosophy, Gessner assigned it to the mechanical rather than the liberal arts on the grounds that it was practiced by ignorant and illiterate men.[21] In the face of such criticism, many alchemists made it their object to convince readers and patrons that they were, despite any deficiencies in formal education, highly literate within the specific context of alchemical philosophy. One way of doing so was to reproduce the distinctive methods and topoi of earlier authorities in their own alchemical writings. Such stratagems preserved the status of alchemy as a privileged form of knowledge, while allowing practitioners to retain their individual authority—and to keep their secrets.[22]

Such strategies place alchemists in an analogous position to that of other highly skilled artisans in early modern Europe who chose to redefine them-

20. Roger Bacon, *Opus tertium*, in *Opera quaedam hactenus inedita Rogeri Baconis*, fasc. 1, ed. J. S. Brewer (London: Longman, Green, Longman, and Roberts, 1859), 3–310, on 39–40; translated in William R. Newman, "The Alchemy of Roger Bacon and the *Tres Epistolae* Attributed to Him," in *Comprendre et maîtriser la nature au moyen age: Mélanges d'histoire des sciences offerts à Guy Beaujouan* (Geneva: Librarie Droz, 1994), 461–79, on 461–62.

21. Conrad Gessner, *Bibliotheca universalis, sive catalogus omnium scriptorum locupletissimus, in tribus linguis, Latin, Graeca, & Hebraica: extantium & non extantium veterum & recentiorum . . .* (Zurich: Christophorus Froschouerus, 1545) and *Pandectarum sive Partitionum universalium libri XXI* (Zurich: Christophorus Froschouerus, 1548); cited in Jean-Marc Mandosio, "L'alchimie dans les classifications des sciences et des arts à la Renaissance," in *Alchimie et philosophie à la Renaissance*, ed. Jean-Claude Margolin and Sylvain Matton (Paris: Vrin, 1993), 11–41, on 15–16.

22. On the intellectual and economic value of secret knowledge, and the various methods of preserving it (and, paradoxically, of publishing it) in medieval and early modern science, see especially Pamela O. Long, *Openness, Secrecy, Authorship: Technical Arts and the Culture of Knowledge from Antiquity to the Renaissance* (Baltimore: Johns Hopkins University Press, 2001); William Eamon, *Science and the Secrets of Nature: Books of Secrets in Medieval and Early Modern Culture* (Princeton: Princeton University Press, 1994); Elaine Leong and Alisha Rankin, eds., *Secrets and Knowledge in Medicine and Science, 1500–1800* (Farnham: Ashgate, 2011), 47–66. On alchemical traditions of secrecy, see Barbara Obrist, "Alchemy and Secret in the Latin Middle Ages," in *D'un principe philosophique à un genre littéraire: Les secrets; Actes du colloque de la Newberry Library de Chicago, 11–14 Septembre 2002*, ed. D. de Courcelles (Paris: Champion, 2005), 57–78; Principe, *Secrets of Alchemy*, esp. chap. 6.

selves as something more than manual workers. Painters and architects emphasized their own mastery of subject matter and materials, turning to classical models like Vitruvius in order to raise the status of their practice in the eyes of their patrons.[23] The flow of knowledge was not unidirectional: when patrons took note of the utilitarian applications of ancient knowledge, humanist scholars also profited from relating ancient knowledge to the practical problems of their own day.[24]

Yet alchemy differs from most fields of knowledge in the deliberate inaccessibility of its language, which requires aspirants to read widely and carefully in order to extract practical sense from the textual record. Its philosophically oriented treatises serve as guides to more than chemical operations alone: they also function as manuals of reading practice, educating their readers in the proper modes of communicating alchemical knowledge.[25] Understanding this function helps to explain the idiosyncratic form of many alchemical treatises, but also shows how they were meant to be read, and hence how we, too, must attempt to read them. For instance, students of alchemy are frequently warned to be suspicious of literal readings, to instead approach their texts on multiple levels in a manner reminiscent of medieval techniques of scriptural exegesis, delving into metaphorical and analogical interpretations of even outwardly straightforward terms, such as "mercury."

In such an exegetical minefield, changing or misconstruing a single word might alter the outcome of the work. Among the church fathers, Irenaeus had famously warned his own scribes to take care when transcribing his

---

23. There is a vast literature on the self-presentation of Renaissance painters; for an overview, see Francis Ames-Lewis, *The Intellectual Life of the Early Renaissance Artist* (New Haven: Yale University Press, 2000); Bram Kempers, *Painting, Power, and Patronage: The Rise of the Professional Artist in the Italian Renaissance*, trans. Beverley Jackson (London: Penguin, 1984).

24. Pamela O. Long, *Artisan/Practitioners and the Rise of the New Sciences, 1400–1600* (Corvallis: Oregon State University Press, 2011). Pamela Smith proposes that artisans from the late Middle Ages were successful in promoting their own "vernacular epistemology" as a counterpoint to text-based knowledge, based on their experience of working materials; Pamela H. Smith, *The Body of the Artisan: Art and Experience in the Scientific Revolution* (Chicago: University of Chicago Press, 2004); see also the collected essays in Sven Dupré, ed., *Laboratories of Art: Alchemy and Art Technology from Antiquity to the 18th Century* (Cham: Springer, 2014).

25. On the use of philosophical texts as manuals for expounding alchemical reading techniques, see Jennifer M. Rampling, "Reading Alchemically: Early Modern Guides to 'Philosophical' Practices," in "Learning by the Book: Manuals and Handbooks in the History of Knowledge," ed. Angela Creager, Elaine Leong, and Matthias Grote, *BJHS Themes* 5 (forthcoming). On the interpretative techniques employed by some prominent seventeenth-century English alchemists, see Newman and Principe, *Alchemy Tried in the Fire*, 174–88; Newman, *Newton the Alchemist*, chap. 2. Some other contentious fields, including natural magic and Kabbalah, called for similar interpretative expertise, as did the discipline of law; see Ian Maclean, *Interpretation and Meaning in the Renaissance: The Case of Law* (Cambridge: Cambridge University Press, 1992).

writings: an exhortation that still carried weight among alchemical writers a millennium later.[26] After all, when copying from heavily contracted medieval sources, a slip of the pen or skip of the eye is all it takes to transmute "vitriolum," or vitriol (a class of metal sulphates used to make mineral acids), into "vitrum," or glass: an error presenting obvious hazards for unwary readers. As Thomas Norton warned in the "Prohemium" to his famous poem, the *Ordinal of Alchemy*,

> And changing of som oone sillable
> May make this boke vnprofitable.[27]

Despite the frequency of such admonitions, in practice it was almost impossible to avoid altering a text, knowingly or otherwise. Reading is inherently a historical process, because readers living at different times and in different places did not approach their texts in the same way. Their interpretations of alchemical texts—and, consequently, their practices—were shaped by their own experience of substances and materials, and by the distinctive social, intellectual, and religious contexts within which they worked. These conditions must be borne in mind as we learn to mind the gaps between what alchemical treatises say, and how they were actually read.

RECOVERING ALCHEMICAL PRACTICE

When the Reformation wrought transmutations in every sphere of English life, alchemy was not excluded. From the 1530s, the libraries of religious houses, replete with alchemical books written or owned by former brethren, were dispersed. Those that survived the dissolution offer tantalizing glimpses of a lost world of monastic practice, littered with the names of priests, monks, friars, and canons both regular and secular, who pledged their credit on a bewildering array of chemical theories and practices. Given this bounty, it is surprising how little we know about the state of monastic alchemy in England prior to the Reformation.[28] The writings of named alchemists like

26. As related in Eusebius, *Historia ecclesiastica*; cited in Anthony Grafton and Megan Williams, *Christianity and the Transformation of the Book: Origen, Eusebius, and the Library of Caesarea* (Boston: Harvard University Press, 2008), 187. For similar concerns in medieval Europe, see Daniel Hobbins, *Authorship and Publicity before Print: Jean Gerson and the Transformation of Late Medieval Learning* (Philadelphia: University of Pennsylvania Press, 2013), 165–68.

27. *Thomas Norton's The Ordinal of Alchemy*, ed. John Reidy (Oxford: Early English Text Society, 1975), 10 (ll. 73–74); hereafter *Ordinal*.

28. Monastic alchemy still awaits systematic treatment. Although Sophie Page focuses primarily on magic rather than alchemy, her work provides useful context for English alchemy as well;

John Sawtrey of Thorney (fl. ca. 1400) and George Ripley of Bridlington provide precious, contextualizing landmarks in a sea of anonymous and pseudepigraphic texts whose provenance and dating have proved as difficult to fix as mercury itself. However, if we are to map the entire ocean we cannot rely on these islets alone, written by "alchemical philosophers" whose rhetoric, if not their practice, presents their activities as solitary, secret, and consistent with a unified, learned tradition. It is only when we brave the surrounding waters that we discover the true variety of approaches and ingredients employed by English alchemists: approaches preserved in hundreds of manuscripts, only a handful of which have received systematic study.

The sheer difficulty of charting this territory becomes obvious as soon as we search for a place to begin. Alchemical treatises often outline a detailed succession of chemical processes; but, as in any other serial procedure, knowing where to start is vital to success—one cannot ascend the ladder unless the first step is sturdily in place. Yet in alchemical writing, the final stages are often described with far greater consistency than the first step—namely, the selection of the starting materials, or *prima materia*. The identity of the elusive first matter is, in many alchemical texts, both the most closely guarded secret and the most intently sought.

For instance, alchemical philosophers often claimed that their work was founded upon one, single prime matter, requiring the addition of no other ingredient. For authority on this point, readers could turn to the most revered alchemical authorities—such as the *Emerald Tablet*, reputedly engraved on a precious stone by Hermes Trismegistus, the legendary founder of alchemy, which describes the marvelous working of "one thing" (*miracula rei unius*) whose father is the Sun, and mother the Moon.[29] The influential *Secretum secretorum* (Secret of Secrets), supposedly comprised of Aristotle's secret teachings to Alexander the Great, further emphasized the ubiquity of this matter, which is "founde in euery place, in euery time, in euery man."[30]

Sophie Page, *Magic in the Cloister: Pious Motives, Illicit Interests, and Occult Approaches to the Medieval Universe* (University Park: Pennsylvania State University Press, 2013). For a brief, general overview, see W. Theisen, "The Attraction of Alchemy for Monks and Friars in the 13th–14th Centuries," *American Benedictine Review* 46 (1995): 239–51. On the practice of alchemy by friars, see the collected essays in Andrew Campbell, Lorenza Gianfrancesco, and Neil Tarrant, eds., "Alchemy and the Mendicant Orders of Late Medieval and Early Modern Europe," *Ambix* 65 (2018); and chap. 2, note 3, below.

29. Hermes Trismegistus, *Tabula Smaragdina*, in J. Manget, *Bibliotheca Chemica Curiosa* (Geneva, 1702), 1:381; hereafter *BCC*.

30. Translation based on Ashmole 396 (fifteenth century), in *Secretum Secretorum: Nine English Versions*, ed. Mahmoud Manzalaoui (Oxford: Oxford University Press, 1977), 67.

Medieval alchemists often took such riddles to refer to mercury, or quick-silver: *mercurius* or *argentum vivum* in Latin, "argent vive" in Middle English. Mercury was an object of fascination to alchemical practitioners, both for its peculiar physical properties and for its role in medieval theories of metal-lic generation. According to the sulphur-mercury theory, two primordial vapors—a dry, earthy "sulphur" and cool, moist "mercury"—combine in varying proportions within the earth to create the various metals: *prima materia* in the most general sense. These two material principles do not cor-respond to elemental quicksilver and brimstone, but instead provide the fun-damental constituents of all metals.[31]

Quicksilver had particular value for medieval writers, who sought to ele-vate alchemy's status as *scientia*. In Aristotelian natural philosophy, like must stem from like: thus a pear tree can bear pears, but not figs, and a lioness can produce lion cubs, but not a donkey. Alchemical theorists extended the analogy to the mineral kingdom, arguing that a transmuting agent capable of generating gold and silver should also derive from a metallic body: typi-cally, from a purified and subtilized form of mercury. By assuming that mer-cury already contained its own, inner "sulphur," proponents of this approach could claim that additional sulphur was not required in the work, justify-ing the choice of mercury as their single, prime ingredient. This view, which underpins much late medieval transmutation theory, has been dubbed "mer-cury alone" by Lynn Thorndike, and, more recently, "mercurialist" by Wil-liam Newman and Lawrence Principe.[32]

Yet the language of "one thing" posed problems in practice. Premised on the generation of metals, the mercurialist approach was more appropriate as a justification for gold-making (chrysopoeia) and silver-making (argyro-poeia) than for other chemical applications, particularly medicinal reme-dies. Strictly interpreted, this philosophy eliminated a wide range of poten-tial ingredients from all the kingdoms of nature, including such chemically interesting substances as herbs, blood, urine, eggs, and a wide variety of salts

31. The theory, based on Arabic adaptations of Aristotle's *Meteorology*, is examined by John A. Norris, "The Mineral Exhalation Theory of Metallogenesis in Pre-Modern Mineral Science," *Ambix* 53 (2006): 43–65. On some aspects of its medieval reception, see Newman, "Technology and Alchemical Debate"; Newman, *Atoms and Alchemy: Chymistry and the Experimental Origins of the Scientific Revolution* (Chicago: University of Chicago Press, 2006), chap. 1.

32. Lynn Thorndike, *A History of Magic and Experimental Science* (New York: University of Columbia Press, 1923–58), 3:58, 89–90 (hereafter *HMES*); Principe, *Aspiring Adept*, 153–55. Wil-liam Newman argues for the origins of the "mercury alone" approach in the thirteenth-century *Summa perfectionis* of pseudo-Geber; William R. Newman, ed., *The Summa perfectionis of Pseudo-Geber: A Critical Edition, Translation, and Study* (Leiden: Brill, 1991), 204–8; hereafter *Summa perfectionis*.

and stones. Despite the formulaic protestations of writers who insisted on metallic kinds, a diversity of practices in fact seems to have been the standard rather than the exception in late medieval England. Even mercurialist authorities admitted that minerals like vitriol and salt were necessary as "helpers" in the work, to prepare metals for further operations. Nor could one doubt the impressive chemical effects wrought by salts, spirits, and organic products—effects that were already in common use among artisans engaged in metalworking, winemaking, painting, and dyeing, among other crafts. From the dissolution of gold in *aqua regia* to the strange transformation of lead into a white, sweet-tasting gum using vinegar, metals repeatedly succumbed to the power of materials that differed from them fundamentally in nature.

Mercury's double life, as both metallic quicksilver and material principle, thus marks only the start of its identity crisis, as its nature was subjected to continual reinterpretation and debate. Like another ubiquitous term, *lapis* (stone), "mercury" came to signify either the starting matter of the alchemical work, or any liquid substance employed in its manufacture: encompassing a host of animal, vegetable, and mineral substances that ranged from metallic quicksilver and mineral acids to distilled alcohol and human blood. This diversity is reflected in the notion (inherited from Arabic alchemy) that more than one kind of stone existed: each stone made using different materials, and targeted toward different ends. By 1390, the latter view was sufficiently well known in England for the poet John Gower (ca. 1330–1408) to include it in the alchemical section of his Middle English poem, the *Confessio amantis*. In one passage, Gower describes a "vegetable stone" used in medicine and an "animal stone" for sharpening human senses, in addition to the more familiar mineral stone that transforms "the metalls of every mine."[33]

This diversity raises interpretative questions: not just what "mercury" means in a given text, but also what it means to a given reader, or community of readers, at distinct points in time. In this book, I focus on identifying, mapping, and analyzing one of the most distinctive and influential strands of English practice, which I term "sericonian" alchemy after its elusive prime matter—an inexpensive "mercury" drawn out of base metals, which Ripley and his followers called sericon.[34] This approach was initially formulated in the fifteenth century on the basis of fourteenth-century continental authorities, and continued to prosper in early modern England, particularly in the context of patronage suits. It also rested on uncontested philosophical

---

33. John Gower, *Confessio Amantis*, vol. 2, ed. Russell A. Peck, trans. Andrew Galloway, 2nd ed. (Kalamzoo, MI: Medieval Institute Publications, 2013), bk. 4, ll. 2553–54.

34. I discuss this aspect of alchemical terminology in Rampling, "Transmuting Sericon."

authority, as a practice apparently grounded in the largest and most influential of all alchemical corpora: the huge body of writings pseudonymously attributed to the Catalan philosopher Ramon Llull—or "Raymond," as he became known in England.[35]

Unlike another major strand of European practice, based on writings pseudonymously attributed to Jābir ibn Ḥayyān (the Latin Geber), the periconian approach offered a wide range of applications: not just transmuting metals, but also healing human bodies, prolonging life, and restoring youth.[36] On the other hand, it also differed from the primarily medical concerns of Paracelsus (1493–1541) and his followers, in offering an affordable route to gold-making.[37] As such, sericonian alchemy offered a versatile palette of products that proved attractive to practitioners from a range of backgrounds and with diverse practical and philosophical commitments. It also offered a tempting investment opportunity, adopted by generations of English alchemists who sought to attract prospective patrons with the promise of both health and wealth.

RECOVERING ENGLISH PRACTITIONERS

The meaning of "sericon" was not static. Like other alchemical cover names, or *Decknamen*, it changed form over the centuries as practitioners adapted

35. On pseudo-Lullian alchemy, see Michela Pereira, *The Alchemical Corpus Attributed to Raymond Lull* (London: Warburg Institute, 1989); Pereira, *L'oro dei filosofi: Saggio sulle idee di un alchimista del Trecento* (Spoleto: Centro Italiano di Studi sull'Alto Medioevo, 1992); Pereira, "*Medicina* in the Alchemical Writings Attributed to Raymond Lull (14th–17th Centuries)," in *Alchemy and Chemistry in the Sixteenth and Seventeenth Centuries*, ed. Piyo Rattansi and Antonio Clericuzio (Dordrecht: Kluwer, 1994), 1–15; Pereira, "Mater Medicinarum: English Physicians and the Alchemical Elixir in the Fifteenth Century," in *Medicine from the Black Death to the French Disease*, ed. Roger French, Jon Arrizabalaga, Andrew Cunningham, and Luis Garcia-Ballester (Aldershot: Ashgate, 1998), 26–52; William R. Newman, *Gehennical Fire: The Lives of George Starkey, an American Alchemist in the Scientific Revolution* (Cambridge, MA: Harvard University Press, 1994), 98–103. The key text of the corpus, the *Testamentum*, has been edited by Pereira; Michela Pereira and Barbara Spaggiari, eds., *Il Testamentum alchemico attribuito a Raimondo Lullo: Edizione del testo latino e catalano dal manoscritto Oxford, Corpus Christi College, 255* (Florence: SISMEL, 1999); hereafter *Testamentum*.

36. On the content and influence of pseudo-Geberian alchemy, see Newman, *Summa perfectionis*; Newman, *Atoms and Alchemy*.

37. On Paracelsian medicine, see Wilhelm Kühlmann and Joachim Telle, eds., *Corpus Paracelsisticum: Dokumente frühneuzeitlicher Naturphilosophie in Deutschland* (Tübingen: Max Niemeyer, 2001–); Didier Kahn, *Alchimie et Paracelsime en France à la fin de la Renaissance (1567–1625)* (Geneva: Librairie Droz, 2007); Allen G. Debus, *The Chemical Philosophy: Paracelsian Science and Medicine in the Sixteenth and Seventeenth Centuries*, 2 vols. (New York: Science History Publications, 1977). Debus's pioneering studies, while instrumental in developing the field, have to a large extent been superseded by more recent scholarship.

the medieval practice to accommodate new substances and techniques. Mapping these changes requires us to work primarily with manuscripts rather than print—an exercise in which we are aided by early modern readers, whose annotations and transcriptions (and occasional spillages) reveal the intensity with which they studied and discussed their medieval sources.

By tracing how these books circulated, we encounter previously unidentified networks of readers and practitioners, whose existence defies the stereotype of the solitary adept. While medicinal remedies might be quietly distilled at home, the labor and cost of chrysopoeia, not to mention its problematic legal status, meant that the quest for the mineral stone was often a corporate affair. The enterprise of alchemy saw monks and canons collaborating with secular priests, merchants, and artisans: exchanging books, debating ingredients, sharing space, and setting down their experience in treatises, poems, and recipe collections. Practitioners were no more "alone" than the mercury they professed to uphold, and their backgrounds were as diverse as their materials.

Within this mixed economy of alchemical collaboration, which often bridged crafts and communities, alchemical knowledge was mediated via Middle English as well as Latin. From the end of the fourteenth century, practitioners increasingly recorded their practices of reading and experiment in Middle English—although we should note that Latin texts still vastly outnumbered those available in English throughout the fifteenth century. Alchemy is the largest genre of Middle English scientific writing; the name of George Ripley alone is attached to more Middle English scientific and medical texts than that of any other author, outweighing Chaucer, Roger Bacon, Galen, and Hippocrates.[38] These writings were not produced only by clerics. English craftsmen and merchants also wrote vernacular commentaries that passed judgment on the learned Latin treatises of previous centuries, often imitating their style and philosophical framing, even as they stripped away conceptual material to privilege practical, replicable content.

Despite the attrition of the Reformation, large numbers of these texts survive in manuscript, few of which have received detailed scholarly attention.[39]

---

38. Linda Ehrsam Voigts, "Multitudes of Middle English Medical Manuscripts, or the Englishing of Science and Medicine," in *Manuscript Sources of Medieval Medicine: A Book of Essays*, ed. Margaret R. Schleissner (New York: Garland, 1995), 183–95. Voigts's findings are detailed in Linda Ehrsam Voigts and Patricia Deery Kurtz, comps., *Scientific and Medical Writings in Old and Middle English: An Electronic Reference* (Ann Arbor: University of Michigan Press, 2000), CD-ROM.

39. Although much remains unpublished, recent years have seen an encouraging increase in the number of critical editions, including the important corpus of interlinked alchemical verses now

Even in the case of well-known figures like Dee and Ripley, there is, there-fore, still much to learn, either from the books they owned and compiled or, in cases where the originals have not survived, from later sixteenth- and seventeenth-century copies. For instance, Elizabethan transcriptions—themselves evidence of intense interest among late sixteenth-century readers—allow us to reconstruct one of English alchemy's most import-ant "antiquities": Ripley's *Bosome Book*, a manuscript compendium of his writings on practical and philosophical subjects. While the recovery of this long-lost book caused a minor sensation among Elizabethan readers, soon communicated to the imperial court in Prague, its existence is now almost entirely forgotten. Yet, as Samuel Norton recognized in the 1570s, this manu-script offers a key to understanding Ripley's better-known works, including the *Compound of Alchemy*, one of the keystones of English alchemy. As Nor-ton knew, even the most puzzling "philosophical" works can reveal much when read alongside one another.

Tracing the reception of these materials offers other clues to the lives and habits of English alchemists, revealed through their annotations, addi-tions, and alterations to texts. In a science where success and credibility were viewed as contingent on sophisticated reading techniques, reader-practitioners approached their books with particular earnestness, pen in hand. This attitude will come as no surprise to historians of scholarship and of the book, who have long charted the efforts of humanist scholars both to dissect their reading matter using established readerly techniques, and to apply the bookish learning thus acquired to real-world situations and events—in our case, to chemical and medical practices.[40] In artisanal

edited by Anke Timmermann, *Verse and Transmutation: A Corpus of Middle English Alchemical Poetry* (Leiden: Brill, 2013). See also Robert M. Schuler, *Alchemical Poetry 1575–1700, from Previ-ously Unpublished Manuscripts* (New York: Garland, 1995); and Peter J. Grund, *"Misticall Wordes and Names Infinite": An Edition and Study of Humfrey Lock's Treatise on Alchemy* (Tempe: Ari-zona Center for Medieval and Renaissance Studies, 2011). Other important manuscript materials are discussed in Charles Webster, "Alchemical and Paracelsian Medicine," in *Health, Medicine, and Mortality in the Sixteenth Century*, ed. Charles Webster (Cambridge: Cambridge University Press, 1979), 301–34; Deborah E. Harkness, *The Jewel House: Elizabethan London and the Scientific Revo-lution* (New Haven: Yale University Press, 2007).

40. See particularly Lisa Jardine and Anthony Grafton, "'Studied for Action: How Gabriel Har-vey Read His Livy,'" *Past and Present* 129 (1990): 30–78. On the practices used by other learned readers of scientific texts, see William H. Sherman, *John Dee: The Politics of Reading and Writ-ing in the English Renaissance* (Amherst: University of Massachusetts Press, 1995); Renee Raphael, *Reading Galileo: Scribal Technology and the "Two New Sciences"* (Baltimore: Johns Hopkins Uni-versity Press, 2017). On "reading alchemically" in the context of a wider early modern library, see Richard Calis et al., "Passing the Book: Cultures of Reading in the Winthrop Family, 1580–1730,"

and household contexts, too, handbooks and recipe collections (particularly those kept within a single shop or family) may preserve modifications added over long periods of time, as each new generation adds its tweaks and changes to the page—a process that often preserves the contributions of women practitioners in ways seldom encountered in "philosophical" tracts.[41] Yet although alchemical treatises intersect with recipe literature, the rhetoric of the former clearly distinguishes philosophical writings from mere conglomerations of receipts, which (they claim) strip alchemical secrets of complexity and nuance.

Throughout the book, I draw on such readerly interactions as evidence for my own reconstruction of the relationship between reading and experiment. While I have attempted to do so in some detail, this has also required compromises in terms of what can reasonably be included in a book of this length. It has not been possible to discuss every English alchemist, and many interesting and important figures—from medieval religious like John Dastin (ca. 1295–ca. 1383) and John Sawtrey to such sixteenth- and seventeenth-century practitioners as the mathematician Thomas Harriot (ca. 1560–1621), physician Francis Anthony (1550–1623), and Margaret Clifford, Countess of Cumberland (1560–1616)—consequently receive short shrift here. For similar reasons I do not discuss alchemical imagery in detail, reserving this analysis for a future study.[42]

In place of these familiar names and themes, I have chosen to concentrate on material that is, for the most part, new. Many of the sources I discuss have not been previously associated with named practitioners, yet these connections reveal hitherto unknown circles of readers, correspondents, and

Past and Present 241 (2018): 69–141; on humanist reading more generally, Anthony Grafton, *Commerce with the Classics: Ancient Books and Renaissance Readers* (Ann Arbor: University of Michigan Press, 1997).

41. On cultures of English recipe books, see Elaine Leong, *Recipes and Everyday Knowledge: Medicine, Science, and the Household in Early Modern England* (Chicago: University of Chicago Press, 2019); Melissa Reynolds, "'Here Is a Good Boke to Lerne': Practical Books, the Coming of the Press, and the Search for Knowledge, ca. 1400–1560," *Journal of British Studies* 58 (2019): 259–88; Elizabeth Spiller, "Recipes for Knowledge: Maker's Knowledge Traditions, Paracelsian Recipes, and the Invention of the Cookbook," in *Renaissance Food from Rabelais to Shakespeare*, ed. Joan Fitzpatrick (Aldershot: Ashgate, 2009), 55–72. On distilling practices among elite women, see Alisha Rankin, *Panaceia's Daughters: Noblewomen as Healers in Early Modern Germany* (Chicago: University of Chicago Press, 2013).

42. Jennifer M. Rampling, *The Hidden Stone: Alchemy, Art, and the Ripley Scrolls* (Oxford: Oxford University Press, forthcoming). For excellent introductions to alchemical imagery, see Barbara Obrist, "Visualization in Medieval Alchemy," *HYLE—International Journal for Philosophy of Chemistry* 9 (2003): 131–70, www.hyle.org/journal/issues/9-2/obrist.htm; Principe, *Secrets of Alchemy*, chap. 6.

patronage-seekers, whose engagement with the writings of past adepts also sheds light on their own careers and practical commitments. For instance, newly identified texts allow us to revisit the trajectories of William Blomfild and Edward Kelley, two prominent alchemists who used their expertise as leverage while petitioning for release from prison, urgently penning treatises to King Henry VIII and Emperor Rudolf II, respectively. The marginal notes of another famous practitioner, Thomas Charnock, previously known only from seventeenth-century transcriptions, serve a different function on the page of his own fifteenth-century manuscripts. And our intuitions about the alchemical proclivities of the Tudor cosmographer Richard Eden, formerly reconstructed from court records and correspondence, can at last be tested against one of his own, previously unidentified manuscripts. To these well-known names we must add the contributions of English practitioners whose work has been, to a greater or lesser extent, overlooked—some anonymous, others whose names are still preserved, like Thomas Peter, who petitioned Henry VIII, and Richard Walton, who petitioned Elizabeth I.

The excavation of this alchemical tradition divulges something else as well: the role of personal, experimental practice in the broader context of national history. During periods of political, religious, and technological change, English men and women held onto alchemy as a source of knowledge and advancement. Their experience of alchemical reading altered their sense of what could be accomplished in nature, and what had been achieved in England's past—a sense cemented by their own autopsy of chemical transformations. To follow these alchemists as they acquired, applied, and marketed natural knowledge is, therefore, to build a bridge between the intellectual history of chemistry and the wider worlds of early modern patronage, medicine, and science.

PART ONE

The Medieval Origins of English Alchemy

CHAPTER ONE

Philosophers and Kings

Therfor take the stone animal, vegetable, and mynerall, the which is no stone,
neither hath the nature of a stone.[1]

According to an early modern legend, King Edward III of England (1312–
1377) once received a visiting alchemist from abroad. One version of the story,
translated from a French exemplar, introduces the alchemist simply as Raymond, a master of arts and doctor of divinity who "after long and paynfull
studdy" obtained the knowledge of alchemy. Seeking a virtuous prince who
would aid in the defense of Christendom, Raymond went to Edward and
offered to transmute enough gold and silver to finance a Crusade against the
Turks. But the young king, faithlessly reneging on his promise, instead used
his alchemical gold to fund self-aggrandizing wars against the French:

> The King so allway kept him as a prisoner, secretly in his contry, not suf-
> fering him to depart, and when his Army was reddye, the Kinge sent them
> into Fraunce instead of goeing against the Sarasones, whervpon great hurte
> ensued to Fraunce, vnder pretence of that title whiche Englishe yet say they
> haue to Fraunce.[2]

The Raymond in this story was no lowly clerk, but the Majorcan philosopher, logician, and theologian Ramon Llull (1232–1316), his name anglicized

---

1. Ashmole 396, in *Secretum Secretorum*, ed. Manzalaoui, 67.

2. J[ean] S[aulnier], "A doctrine Concerning the transmutation of Mettalls written by the most
reuerend Man Jo. S. & dedicated to his sonne," Sloane 363, fols. 19v-20r (seventeenth century). This
is an English translation of Jean Saulnier's French treatise, written in 1432. On the original text, see
Pereira, *Alchemical Corpus*, 44n42; J. A. Corbett, *Catalogue des manuscrits alchimiques latins* (Paris:
Office International de Labraire, 1939, 1951), 2:153.

as Raymond Lull. Depending on the version of the legend, the hapless philosopher was either imprisoned by Edward, or escaped back to the Continent.[3] Either way, Lull's misfortune resulted not from any lack of piety or skill, but from the revelation of his expertise to an unscrupulous prince. The moral of the story is clear: alchemical philosophers should take care not to allow their expertise to fall into the wrong hands, even the hands of an anointed king. A secondary moral, which would surely have been obvious to those who read this account in its original French, is that the English were not to be trusted, particularly when prosecuting their claim to the crown of France.

Unfortunately for this account, the historical Lull never visited England, and died before Edward III ascended the throne in 1327. Not only was Lull no alchemist, but his authentic works dismiss the possibility of transmutation.[4] Nonetheless, there is a sense in which Lull's work did shape the course of English alchemical practice. Between the fourteenth and seventeenth centuries, over 100 alchemical treatises were pseudonymously ascribed to the Majorcan philosopher, including some of the most influential works of Latin alchemy.[5] These Lullian pseudepigrapha were closely studied by English alchemists seeking to attain both metallic transmutation and medicinal elixirs, often with a view to securing royal patronage. In relocating Raymond physically to England, the legend of Lull embodies a relationship that in reality existed not between the king and the philosopher, but between a profusion of books and their readers.

The legend also emblematizes a political and economic truth: the English Crown's urgent need of bullion. Edward III never met the historical Raymond Lull, but he did seek out men with a reputation for alchemical expertise, and he did pursue transmutation as a potential remedy for England's pressing financial concerns. The evils of famine, pestilence, and war were accentuated by the hemorrhaging of silver coin overseas, and the perennial struggle against counterfeiting and clipping of the coinage, which continued into the fifteenth century despite strenuous and unpopular measures taken throughout the 1300s.[6] Against that background, alchemy constituted

3. On the formation of the Lull legend, see Pereira, *Alchemical Corpus*, esp. 39–40.

4. Ibid., 1–2n6.

5. Our knowledge of pseudo-Lullian alchemy stems largely from the very extensive scholarship of Michela Pereira: see particularly the works cited in the introduction, note 35.

6. On the various economic difficulties afflicting England and northern Europe in the fourteenth century, see David L. Farmer, "Prices and Wages, 1350–1500," in *The Agrarian History of England and Wales*, vol. 3, *1348–1500*, ed. Edward Miller (Cambridge: Cambridge University Press, 1991), 431–525; John Hatcher, "Plague, Population, and the English Economy, 1348–1530," in *British Population History: From the Black Death to the Present Day*, ed. Michael Anderson (Cambridge:

both an opportunity and a threat, a conflict that alchemists were themselves keenly aware of as they sought to position themselves and their art as a source of revenue rather than as a danger to the stability of the currency. The legend of Lull has its real-world analogues in a spate of cases throughout the fourteenth century that forced alchemists to redefine both their identities as practitioners, and the philosophical status of their science.

THE LEGAL STATUS OF ALCHEMY

Edward III was the first English king to patronize alchemical transmutation, as we learn from the unfortunate career of John of Walden. During the early 1340s, John received 500 gold crowns and twenty pounds of silver from the royal treasury, "to work upon for the benefit of the king by the art of Alkemie."[7] Presumably the king consented to the arrangement, since John received these funds from Philip Weston, the steward of Edward's chamber, and formerly his almoner and confessor. Yet John failed to persuade this vast sum to multiply. He subsequently languished in the Tower of London for seven and a half years, until he was discovered together with several other forgotten prisoners in the course of a 1350 audit, and his testimony recorded. Beyond these few details, however, we know nothing of the methods John employed in his practice, or the philosophical doctrines he endorsed.

If legal records are unhelpful in mapping the contours of individual practice, they nonetheless tell us a great deal about the state's response to transmutational alchemy: a reaction that would in turn shape how English alchemists chose to present their work in the fifteenth century and beyond. This response affected even how the art was named. The term most often encountered in fourteenth-century state papers and legal records is not the Latin *alchemia*, or even the English *alchemy*, but a word borrowed from the

---

Cambridge University Press, 1996), 9–94; William Chester Jordan, *The Great Famine: Northern Europe in the Early Fourteenth Century* (Princeton: Princeton University Press, 1997). On economic measures and their impact on the coinage, see Martin Allen, *Mints and Money in Medieval England* (Cambridge: Cambridge University Press, 2012).

    7. National Archives, Coram Rege Roll, 362, 25 Edward III, Hilary Term, Rex m.4d; cited in Dorothea Waley Singer and Annie Anderson, *Catalogue of Latin and Vernacular Alchemical Manuscripts in Great Britain and Ireland Dating from before the XVI Century* (Brussels: Maurice Lamertin, 1928, 1930, 1931), 3:777–80 (hereafter Singer): "Pro eo quod ipse recepit de Thesauro Domini Regis per manus Philippi de Weston quingenta scuta auri et viginti libras argenti ad comodum Regis inde faciendum per artem Alkemie." On Weston, see T. F. Tout, *Chapters in the Administrative History of Mediaeval England: The Wardrobe, the Chamber, and the Small Seals* (Manchester: Manchester University Press, 1928), 4:268.

French: *alconomie*. Possibly derived from an attempt to rank the new art alongside astronomy, *alconomie* seems to have served as the official term for both the practice and the products of alchemy before 1400.[8] A rare exception is the record of John of Walden's plight, which retains what is, surely, his own choice of word—"Alkemie"—and with it, a flicker of his philosophical aspirations.

Perhaps by John's time *alconomie* had already acquired distasteful associations with counterfeiting and fraud. By the end of the century the term had come to designate false metal as well as the art that made it, a fact that can hardly have recommended its use to the alchemists themselves. Although the word survives in a handful of later treatises, such as a fifteenth-century translation of the *Semita recta* pseudonymously attributed to Albertus Magnus, by 1400 it was rare for alchemists to use it on their own account.[9] Its use continued a little longer among nonpractitioners, migrating from official usage into vernacular literature, where it enjoyed a varied reception: criticized in Langland's *Piers Plowman* as a source of deceitful experiments, but praised in Gower's *Confessio amantis* as an art founded on nature.[10]

The records suggest that, by the start of the fourteenth century, *alconomie* was widely practiced but not yet formally regulated. Mints needed bullion, merchants needed ready cash, and alchemy offered a potential source of both these scarce commodities. Edward III was only the first in a succession of English monarchs to view alchemical transmutation as a potential prop for the kingdom's finances. The earliest reference in the English state papers dates from 1329, when John le Rous and Master William de Dalby acquired a reputation for making good silver "by the art of alconomie." Edward III ordered that both men should be brought before him with their instruments, willingly or otherwise.[11] Others besides the king were keen to secure expertise in this area, as appears from the kidnapping of the alchemist Thomas de Euerwyke (Thomas of York) by a London spicer, Thomas Crop, in 1336. Thomas had also claimed to be able to make silver plate "par la sience de Alconemie," a skill that his abductor hoped to extract by confiscating his

8. *OED*, s.v. "alconomie."

9. Sloane 513, fol. 155r (fifteenth century): "[th]is craft [th]at me clepud alkonomyȝe."

10. William Langland, *The Vision of Piers Plowman: A Critical Edition of the B-Text Based on Trinity College Cambridge MS B.15.17*, ed. A. V. C. Schmidt, 2nd ed. (London: J. M. Dent, 1995), Passus 10, l. 215: "Experiments of alkenamye the peple to deceyve"; Gower, *Confessio Amantis*, vol. 2, bk. 4, l. 2625: "Which grounded is upon nature."

11. National Archives, Patent Roll, 3 Edward III, pt. 1, m.21; cited in Singer, 3:777–78: "Sciatis quod cum datum sit nobis intelligi quod Johannes le Rous et Magister Willelmus de Dalby per artem alkemoniae sciant metallum argenti conficere . . . fuerint infra libertates, sive extra."

apparatus and the elixir-in-progress, and forcing Thomas to instruct him in their use.[12]

In their own writings, medieval alchemists seldom explicitly state whether they are responding to such local economic imperatives, preferring to treat the attainment of the stone as an end in itself, or else claiming morally unimpeachable goals such as helping the poor. On the other hand, English legal records do suggest that many practitioners had a clear, material sense of what they hoped to achieve—namely, precious metal of a quality suitable for coining. Until 1343 when Edward first struck gold in his own name, the country relied almost entirely on silver coinage.[13] Yet England, in common with most of Europe, suffered from a shortage of silver bullion throughout the fourteenth and fifteenth centuries. Without adequate supplies of legal money, trade was uncertain and economic growth curtailed. Finding a cheap source of silver was thus not only desirable, but an urgent matter of state.

When alchemists are mentioned in fourteenth-century legal records it is most often in relation to silver-making (argyropoeia). Probably we can detect a pragmatic aspect to this interest, since it is generally easier to whiten metals than to tint them yellow, a metallurgical bias that also appears in late medieval alchemical recipe collections. Yet even after the advent of gold specie, short supplies of "white money"—the lower-value coins made from silver alloyed with copper or tin, essential for day-to-day transactions—made silver-making an attractive option for both patrons and practitioners.[14]

The bullion shortage offered not only economic opportunities for alchemists, but also considerable danger. England was unusual in that its coinage was always issued by the king, unlike other parts of Europe where that right was often exercised by cities and local rulers.[15] Criminal activities that threatened to debase the coinage, including counterfeiting and clipping, were thus taken seriously by the Crown, particularly when metal was in short supply, raising anxiety about good, English groats being exchanged for cheap foreign coins, known as "Lushbournes," which had a lower silver content. While counterfeiting was a capital offense, even clipping merited stiff penalties and

12. National Archives, Patent Roll, 11 Edward III, pt 1. m.20d.; Singer, 3:778–79.

13. The exception is Henry III's unpopular and undervalued gold penny of 1257, which circulated for, at best, a couple of years: John Evans, "The First Gold Coins of England," *Numismatic Chronicle and Journal of the Numismatic Society* 20 (1900): 218–51; David Carpenter, "Gold and Gold Coins in England in the Mid-Thirteenth Century," *Numismatic Chronicle* 147 (1987): 106–13.

14. On the metallurgical content of forged coins, see M. B. Mitchiner and A. Skinner, "Contemporary Forgeries of English Silver Coins and Their Chemical Compositions: Henry III to William III," *Numismatic Chronicle* 145 (1985): 209–36.

15. Allen, *Mints and Money*, 381.

could result in capital sentences. A brutal and spectacular sting operation in 1278, which led to the arrest of around 600 Jews and numerous mint officials and goldsmiths, was aimed at curtailing clipping practices. Philip de Cambio, the moneyer of the London Mint, was among the hundreds sentenced to death, although the warden, Bartholomew de Castello, escaped by claiming benefit of clergy.[16] Yet, despite the terrific penalties for failure, the rewards for successful counterfeiting and clipping actually increased with the advent of the more valuable gold coinage. While counterfeiting the king's coin had always been treasonable, this status was formalized in the Treason Act of 1351/2.[17]

Illicit activity was not confined to laymen. Although Jews bore the brunt of the 1278 crackdown, plenty of English Catholics were engaged in the same practice, including the heads of religious houses. Guy de Mereant, prior of Montacute, a Cluniac house in Somerset, was fined for clipping coins in 1279 and again (with the additional offense of passing counterfeit money) in 1284.[18] No one order or geographical region held a monopoly on currency crime. William de Stoke, an Augustinian canon from the Essex priory of Little Dunmow, was charged, although probably not convicted, of counterfeiting gold and silver coins in 1369;[19] while in 1414 the abbot of Combermere, an impoverished Cistercian house in Cheshire, was accused of clipping gold coins.[20] Their religious profession did not prevent monks and canons from dabbling in dubious metallurgical practices, although it did partly insulate them from the penal consequences—a fact worth keeping in mind given the large number of English monks, canons, and friars, including heads of houses, who

16. Ibid., 68. The crackdown on Jews, accompanied by increased taxation of this community, was a precursor to the Edict of Expulsion of 1290; concerning which, see Robin R. Mundill, *England's Jewish Solution: Experiment and Expulsion, 1262–1290* (New York: Cambridge University Press, 1998).

17. "Declaration what Offences shall be adjudged Treason," 25 Edw. 3 Stat. 5 c.2. See J. G. Bellamy, *The Law of Treason in England in the Later Middle Ages* (Cambridge: Cambridge University Press, 1970), 85–86.

18. "House of Cluniac Monks: The Priory of Montacute," in *A History of the County of Somerset*, vol. 2, ed. William Page (London: Victoria County History, 1911), 111–15; *British History Online*, https://www.british-history.ac.uk/vch/som/vol2/pp111–115 (accessed 28 December 2014).

19. "Houses of Austin Canons: Priory of Little Dunmow," in *A History of the County of Essex*, vol. 2, ed. William Page and J. Horace Round (London: Victoria County History, 1907), 150–54; *British History Online*, https://www.british-history.ac.uk/vch/essex/vol2/pp150–154 (accessed 28 December 2014).

20. A. P. Baggs et al., "Houses of Cistercian Monks: The Abbey of Combermere," in *A History of the County of Chester*, vol. 3, ed. C. R. Elrington and B. E. Harris (London: Victoria County History, 1980), 150–56; *British History Online*, https://www.british-history.ac.uk/vch/ches/vol3/pp150–156 (accessed 28 December 2014).

practiced and wrote about alchemical transmutation both before and after the "multiplication" of metal was branded a felony by Henry IV in 1403/4.[21]

English alchemy in the late Middle Ages thus developed in a fraught context of state concern over bullion shortage, inadequate currency, and rampant counterfeiting. The criminal associations must have pained scholarly defenders of alchemy for whom transmutation, which entailed the transformation of metals in substance as well as outward appearance, was entirely distinct from the superficial artifice of coining. Philosophical treatises spell out the radical nature of such transmutations, which typically require raw metals to be rendered down into a more primitive material state before being reconstituted as a higher and subtler form of matter, through a series of procedures that far outstripped coining practices in complexity and cost. For instance, while a counterfeiter might use a mercury amalgamation technique to plate cheap metal discs with silver, "philosophers" viewed such primitive amalgamations as merely preparatory to more radical change. Chemical knowledge was not even required for many counterfeiting activities, which typically involved a mechanical process of hammering thin leaves of gold or silver over a core made of some base metal, rather than radical alteration of the metal itself.[22]

This difference, so apparent to any alchemical philosopher, was less obvious from the standpoint of magistrates and mint officials charged with policing the currency. Alchemical techniques could be implicated in coining activities through the production of "multiplied" metal: alloys that physically resembled gold and silver, but were considerably cheaper to produce. In 1393, a monk of Tewkesbury named John Pygas was hauled before the Bristol magistrates after he and his confederates "treasonably made sixty groats from the false metal called 'Alconamye,' made in the likeness of good coins," which they used to pay for local goods.[23] Since counterfeiting charges could no longer be ameliorated by benefit of clergy, at this point the forger's best hope was for a royal reprieve. The king, Richard II, actually did intervene in Pygas's case, asking the court to stay execution of any sentence pending fur-

---

21. See note 27 below.

22. Mitchiner and Skinner, "Contemporary Forgeries."

23. National Archives, Close Roll, C.54, No. 235, 17 Richard II; cited in Singer, 3:781–82: "Cum Frater Johannes Pygas, monachus Abbatie de Teukesbury, per nomen Johannis Pygas monachi Prioratus Sancti Jacobi de Bristollia; qui quidem Prioratus cella Abbatie predicte existit; de eo quod ipse, una cum aliis, die Veneris in septimana Pasche anno regni nostri Anglie sextadecimo, in villa Bristollie, in alto vico, sexaginta grossos de falso metallo vocato Alconamye, ad similitudinem bone monete regni nostri predicti fabricato, proditorie fecit."

ther instructions.[24] Probably Pygas had sued for a pardon, and the king may have been sufficiently intrigued by the composition of his alchemical metal that he was moved to "deal graciously" with an otherwise egregious case of illicit coining.

Yet, although the law perceived a connection between alchemy and currency crime, alchemical procedures technically applied only to the creation of the metal, not to the far more serious crime of forging coins. A less obviously treasonable alternative to coining alchemically produced metal oneself was to sell it directly to the mint—the course taken by the chaplain Willelmus de Brumleye in 1374. Using a process learned from William Shilchurch, a canon of the king's chapel of Windsor, Brumleye succeeded in selling a batch of metal made "by the art of alconomie" to the Tower mint.[25] The result was convincing enough that the keeper bought it for eighteen shillings. William was later arrested in possession of four pieces of counterfeit gold, which he had also tried to sell, although it is not clear whether this metal was of the same kind that had earlier impressed the keeper.

These cases reveal a certain official ambivalence with regard to alchemy. Successive governments sought to maintain the gold and silver content of English coin, the basic quality of which underwrote confidence in the currency. Counterfeiting coin was the most heinous offense, but men like William of Brumleye, who were not guilty of coining but who threatened to flood the mint with dubious, multiplied metal, still endangered the quality and reputation of the currency.[26] The problems associated with such practices undoubtedly underpin Henry IV's statute of 1403/4, which instructed that henceforth no one should "multiply Gold or Silver, nor use the Craft of Multiplication."[27]

24. Singer, 3:782: "Nos volentes cum prefato Johanne agere gratiose, vobis mandamus quod si contingat ipsum Johannem de prodicione predicta coram vobis per veredictum seu alio modo conuinci seu morti adiudicari, tunc execucioni iudicii in hac parte reddendi, quousque aliud a nobis habueritis in mandatis, supersedeatis."

25. National Archives, Coram Rege Roll, No. 448, 47 Edward III, Hilary Term, Rex m.15.d; also cited in Singer, 3:781: "cum arte Alconomie." For an English summary of the source, see H. G. Richardson, "Year Books and Plea Rolls as Sources of Historical Information," *Transactions of the Royal Historical Society*, 4th ser., 5 (1922): 28–70, on 39. On Brumleye, see Carolyn P. Collette and Vincent DiMarco, "The Canon's Yeoman's Tale," in *Sources and Analogues of the Canterbury Tales*, vol. 2, ed. Robert M. Correale and Mary Hamel (Cambridge: D. S. Brewer, 2005), 715–47, on 720–21.

26. Although the illegality of false metal was uncontroversial, some continental canon lawyers did speculate that genuinely transmuted gold could be sold legally: Nummedal, *Alchemy and Authority*, 151.

27. National Archives, Statutes of the Realm, 5 Henry IV, cap. IV; cited in Singer, 3:782: "Ordeignez est et establiz qu nully desorenavant use de multiplier or ou argent, ne use le art de

The wording of the statute acknowledged that multiplying was not identical to counterfeiting, since it affected the matter of the metal rather than its form. Yet, by forbidding the manufacture of unusual metal alloys, this measure closed a possible legal loophole that might have been exploited by practitioners who hoped to sell false metal to the Mint. Even if not intended for coinage, such metal could still be fraudulent if it resulted in overvaluation of goods, as John Herward of Rochester discovered in October 1414. Convicted of making false gold and silver bands for mazer cups, he was sentenced to stand in the pillory with the deceiving bands hung around his neck.[28]

The image of alchemy recorded in such cases is very different from the portraits of pious and learned adepts elaborated in philosophical treatises. By the end of the fourteenth century, a remarkable situation existed in which alchemy was simultaneously hailed as an elevated form of philosophy and damned as a fraudulent practice that threatened the integrity of English coin. The rhetorical distinction between philosophers and forgers would only widen during the fifteenth century, as the government struggled to both manage and exploit burgeoning interest in an art that was technically illegal, but still promised astonishing rewards.

THE PHILOSOPHICAL STATUS OF ALCHEMY

The history of English alchemy revealed by official records is one dominated by concern over transmutation and the multiplying of metals. As a history it is inevitably one-sided, since it records only practices that were illegal, or feared to be. Yet anxiety over alchemy's less salubrious associations also colored the vision of alchemy's past revealed in the writings of self-styled philosophers. Such writers typically emphasize the pious and philosophical over the pragmatic and profitable—a strategy that bolstered the reputation of the science as a branch of natural philosophy, while distancing its practitioners from associations with fraud. It is in the writings of devout reader-practitioners, many of whom were monks or friars, that we also find the medical and religious dimensions of alchemy most fully explored.

---

multiplication: Et si null le face et de ceo soit atteint qil encourage la peyne de felonie en ce cas." Translation in D. Geoghegan, "A Licence of Henry VI to Practise Alchemy," *Ambix* 6 (1957): 10–17, on 10n1.

28. "Folios cxxxi–cxlii: Feb 1413–14," in *Calendar of Letter-Books of the City of London: I, 1400–1422*, ed. Reginald R Sharpe (London: His Majesty's Stationery Office, 1909), 122–30; *British History Online*, https://www.british-history.ac.uk/london-letter-books/voli/pp122–130 (accessed 28 December 2014).

Alchemical knowledge was initially, if briefly, confined within a community of scholars. The first treatises arrived in the Latin West in the twelfth century, a product of the Arabic-to-Latin translation movement that galvanized scholars like Adelard of Bath and Gerard of Cremona.[29] A few recipe collections derived from Byzantine alchemical sources were present in Europe earlier than this, and European craftsmen were already familiar with a diverse range of chemical techniques, including methods for coloring the surface of metals.[30] It was not until the translation of the first Arabic treatises, however, that Latin readers truly encountered *alchemia* as a science: a field of knowledge distinguished by an authoritative philosophical provenance, an abstruse technical vocabulary, and an insistence on the importance of secrecy.

All of these characteristics are present in the pseudo-Aristotelian *Secretum secretorum* (Secret of Secrets), an early translation from Arabic that became one of the most widely read works of the Latin Middle Ages, and certainly one of the most influential contributions to the medieval vision of alchemical history.[31] The *Secretum* is presented as a letter from the Stagyrite philosopher to his pupil, Alexander the Great, during the latter's campaign

29. On the translation of Arabic alchemical writings into Latin, see Sébastien Moureau, "*Min al-Kīmiyāʾ ad Alchimiam*: The Transmission of Alchemy from the Arab-Muslim World to the Latin West in the Middle Ages," in *The Diffusion of the Islamic Sciences in the Western World*, ed. Agostino Paravicini Bagliani, Micrologus' Library 28 (Florence: SISMEL, 2020), 87–142; Robert Halleux, "The Reception of Arabic Alchemy in the West," in *Encyclopedia of the History of Arabic Science*, ed. Roshdi Rashed (London: Routledge, 1996), 3:886–902; Principe, *Secrets of Alchemy*, chap. 3. For an overview of the impact of the translation movement on scientific knowledge more generally, see Charles Burnett, "Translation and Transmission of Greek and Islamic Science to Latin Christendom," in *The Cambridge History of Science*, vol. 3, *Medieval Science*, ed. David C. Lindberg and Michael H. Shank (Cambridge: Cambridge University Press, 2013), 341–64.

30. One of the most important collections of late antique Greek recipes, pseudonymously attributed to the philosopher Democritus, is preserved in a series of Byzantine epitomes: Matteo Martelli, ed. and trans., *The Four Books of Pseudo-Democritus*, Sources of Alchemy and Chemistry 1 (Leeds: Maney, 2013). Other important collections include the *Mappae clavicula*, ca. AD 600, a compilation of terse craft recipes that survived through repeated recopying throughout the Middle Ages, and *De diversis artibus*, ca. 1125, a craft manual produced by the monk Theophilus that draws on earlier material as well as up-to-date techniques for pigment- and glass-making. See Cyril Stanley Smith and John G. Hawthorne, ed. and trans., *Mappae clavicula: A Little Key to the World of Medieval Techniques* (Philadelphia: AMS, 1974); Theophilus, *On Divers Arts: The Foremost Medieval Treatise on Painting, Glassmaking, and Metalwork*, ed. and trans. John G. Hawthorne and Cyril Stanley Smith (New York: Dover, 1979). On the context for transmission of craft knowledge and its associations with secrecy, see Long, *Openness, Secrecy, Authority*, esp. chap. 3; Eamon, *Science and the Secrets of Nature*, chap. 1.

31. On the *Secretum*, see Mahmoud Manzalaoui, "The Pseudo-Aristotelian Kitab Sirr al-asrar: Facts and Problems," *Oriens* 23–24 (1974 [1970–71]): 148–257; Steven J. Williams, *The Secret of Secrets: The Scholarly Career of a Pseudo-Aristotelian Text in the Latin Middle Ages* (Ann Arbor:

in Persia. The letter communicates secret knowledge that Aristotle supposedly took care not to disclose in his "official" works, including such topics as physiognomy, magic, alchemy, astrology, and the art of kingship. The cryptic nature of the work is exemplified by a famous passage in which the philosopher alludes to a stone that is not a stone, nor has the nature of a stone. This paradoxical substance is "animal, vegetable, and mineral"—a lemma that generations of alchemists took to signify Aristotle's own opinion on the prime matter of the alchemical work. Like the obscure prophetic verses of Merlin in Geoffrey of Monmouth's *History of the Kings of Britain*, this lemma had to be interpreted; and like prophecy, its interpretation changed in light of different historical contingencies.[32] The "animal, vegetable, and mineral" stone thus came to denote different practical traditions at different times, from a single, universal elixir to a wide palette of chemical products made from different ingredients and fit for different ends. From enigmas like these, Latin readers soon learned that the acquisition of alchemical knowledge entailed a particular approach to reading, which called for the kind of exegetical skill more commonly employed in interpreting scripture or prophecy than scientific and medical works.

They also learned something else: that alchemical knowledge might fitly be communicated from philosophers to kings. This lesson is cemented in one of the first purely alchemical treatises to appear in Latin, *De compositione alchemiae* (On the Composition of Alchemy), translated by the Englishman Robert of Chester in or around 1144.[33] The text describes the education of a Muslim prince, Khālid ibn Yazīd, by the Christian sage Morienus, a pious

---

University of Michigan Press, 2003); Williams, "Esotericism, Marvels, and the Medieval Aristotle," in *Il segreto*, ed. Thalia Brero and Francesco Santi, Micrologus' Library 14 (Florence: SISMEL, 2006), 171–91.

32. On the difficulty of interpreting prophetic literature, particularly the information on dates, places, and persons given in English political prophecy, see Lesley A. Coote, *Prophecy and Public Affairs in Later Medieval England* (Woodbridge: York Medieval Press, 2000), esp. 31–37. On more explicit connections between alchemy and prophecy, see Leah DeVun, *Prophecy, Alchemy, and the End of Time: John of Rupescissa in Medieval Europe* (New York: Columbia University Press, 2009); Chiara Crisciani, "Opus and sermo: The Relationship between Alchemy and Prophecy (12th–14th Centuries)," *Early Science and Medicine* 13 (2008): 4–24.

33. This is the date given in the colophon, although the complex transmission history of the text means this cannot be taken as conclusive. Morienus, *De compositione alchemiae*, in *BCC*, 1:509–19. One version of the text is available in an English translation: *A Testament of Alchemy: Being the Revelations of Morienus to Khālid ibn Yazid*, ed. and trans. Lee Stavenhagen (Hanover, NH: Brandeis University Press, 1974). For more recent work on Morienus, see Marion Dapsens, "De la Risālat Maryānus au *De Compositione alchemiae*: Quelques réflexions sur la tradition d'un traité d'alchimie," *Studia graeco-arabica* 6 (2016): 121–40.

recluse who is persuaded to come to court by the knowledge-hungry king. Like pseudo-Aristotle, Morienus cloaks his advice in philosophical speech, disguising the nature of his ingredients with a set of cover names, or *Decknamen*, such as the Green Lion, the White Fume, and the Stinking Water. More generous than Aristotle, he partly reveals their meaning: the Green Lion is glass, and the White Fume is mercury.[34]

For practitioners interested in seeking patronage on their own account, Aristotle's epistle to Alexander and Morienus's teaching of Khālid offered prestigious models for their own petitions. However, it is not until the fifteenth century that English writers are known to have sought alchemical patronage from royalty.[35] Prior to that, they were more likely to address senior ecclesiastical figures based abroad, reflecting the fact that most treatises from this period were still written by monks and friars, many of whom (particularly among the mendicant orders) enjoyed mobility and international connections. Among scholars of the first rank, the English Franciscan Roger Bacon, who taught at Paris, famously discussed alchemy in a series of treatises written at the request of his patron, Pope Clement IV. He hailed alchemy as a form of knowledge gained from experience (*scientia experimentalis*), reflecting on its prospects both for prolonging human life and for preparing for the coming apocalypse.[36] He wrote a commentary on the *Secretum secretorum*, and alludes to the pseudo-Aristotelian epistle throughout his writings, including in the *Opus maius* and *Opus tertium* addressed to

34. *Testament of Alchemy*, 38: "Leo viridis est vitrum et almagra est laton, quamvis in precedentibus terra rubea nominetur. Et sanguis est auripigmentum, et terra fetida est sulfur fetidum . . . Hic est modus fumi albi et leonis viridis et aque fetide." The use of *vitrum* suggests a possible misreading for *vitriolum* (vitriol).

35. I exclude from "alchemical patronage" presentation copies of the *Secretum secretorum*. As an exemplar of "mirrors of princes" literature, the *Secretum* was presented to royal patrons for many reasons besides interest in alchemical practice. On the text's popularity in medieval England, see Richard Firth Green, *Poets and Princepleasers: Literature and the English Court in the Late Middle Ages* (Toronto: Toronto University Press, 1980), 140–43.

36. Alchemy appears in the context of *scientia experimentalis* in Roger Bacon, *Opus majus*, ed. John Henry Bridges (Oxford: Clarendon Press, 1897), 2:214–15. On Bacon's apocalyptic concerns, see Amanda Power, *Roger Bacon and the Defence of Christendom* (New York: Cambridge University Press, 2013); Zachary Matus, "Reconsidering Roger Bacon's Apocalypticism in Light of His Alchemical and Scientific Thought," *Harvard Theological Review* 105 (2012): 189–222. On prolongation of life, Agostino Paravicini Bagliani, "Ruggero Bacone e l'alchimia di lunga vita: Riflessioni sui testi," in *Alchimia e medicina nel Medioevo*, ed. Chiara Crisciani and Agostino Paravicini Bagliani (Florence: SISMEL, 2003), 33–54; Faye M. Getz, "To Prolong Life and Promote Health: Baconian Alchemy and Pharmacy in the English Learned Tradition," in *Health, Disease, and Healing in Medieval Culture*, ed. Sheila Campbell, Bert Hall, and David Klausner (New York: Palgrave Macmillan, 1992), 141–50.

Clement.[37] Another English monk of the early fourteenth century, John Dastin, penned alchemical epistles to Cardinal Napoleon Orsini and may have spent time at the papal court at Avignon, although little is known of his life or whether he acquired his knowledge in England or abroad.[38] The Englishman John Dombelay seems to have prepared both of his attested works, the *Stella complexionis* (1384) and *Practica vera alkimica* (1386), at the request of the Archbishop Elector of Trier, Kuno II von Falkenstein (1320–1388).[39]

These international connections also kept England supplied with alchemical material, including new treatises and practical innovations from abroad. Sometimes continental manuscripts came to England with returning students. Sophie Page has reconstructed the book collections of English monks, including Michael de Northgate and John of London, who studied at Paris before settling at St. Augustine's abbey in Canterbury in the early 1320s, bringing with them manuscripts of astronomy, medicine, magic, and

37. Roger Bacon, *Secretum secretorum cum glossi et notulis, tractatus brevis et utilis ad declarandum quedam obscure dicta Fratris Rogeri*, in *Opera hactenus inedita Rogeri Baconis*, fasc. 5, ed. Robert Steele (Oxford: Clarendon Press, 1920), 1–175, on 117–18. For the influence of the *Secretum* on Bacon's alchemy, see Stewart C. Easton, *Roger Bacon and His Search for a Universal Science: A Reconsideration of the Life and Work of Roger Bacon in the Light of His Own Stated Purposes* (Oxford: Blackwell, 1952), on 30–31, 73–73, 77–86, 103–104; Pereira, *L'oro dei filosofi*; Eamon, *Science and the Secrets of Nature*; William R. Newman, "The Philosophers' Egg: Theory and Practice in the Alchemy of Roger Bacon," *Micrologus* 3 (1995): 75–101; Newman, "Alchemy of Roger Bacon"; Obrist, "Alchemy and Secret"; Moureau, "Elixir Atque Fermentum."

38. On the dating of Dastin's writings, see *HMES*, 3:85–102; W. R. Theisen, "John Dastin's Letter on the Philosopher's Stone," *Ambix* 33 (1986): 78–87. The alchemist may have been the same "Magister John Dastin" granted a canonry in Southwell in 1317 by Edward III at Orsini's request: José Rodríguez-Guerrero, "Un repaso a la alquimia del Midi Francés en al siglo XIV (parte I)," *Azogue: Revista electrónica dedicada al estudio histórico crítico de la alquimia* 7 (2010–13): 75–141, on 92–101.

39. Dombelay's name is not known exactly, as it is variously attested in manuscript copies— for instance, as Dumbaley, Dumbeler, Dumblerius, Bumbelem, and Bumbelam. He appears as "DUMBELEIUS [JOHANNES] de Anglia" in Thomas Tanner, *Bibliotheca Britannico-Hibernica: sive, de scriptoribus, qui in Anglia, Scotia, et Hibernia ad saeculi XVII initium floruerunt, literarum ordine juxta familiarum nomina dispositis commentarius* (London, 1748), 237. The *Practica vera alkimica* is explicitly dedicated to Kuno II: *TC*, 4:912. The *Stella complexionis* is dated to 1384 by colophon—e.g., Ashmole 1450, pt. 4, fol. 131v: "Explicit libellus vocatus Stella Alkimie compositus .A. Johanne Bumbulem de Anglia Anno domini 1384." Although the dedicatee is not mentioned by name in any of the manuscript copies I have examined, Dombelay addresses his patron as "Reverend Prince" (fol. 131v: "O Reuerende Princeps"), the appropriate form of address for the ruler of the ecclesiastical principality of Trier. Ashmole notes an alternative dedication at the end of his transcription of the text (itself copied from Christopher Taylour's 1584 transcription) in Ashmole 1493, fol. 97: "dedicated to K. Richard the 2d: King of England." However, he does not provide a source for this, and the internal and contextual evidence suggests that this dedication was proposed by a later reader on the basis of Dombelay's nationality and the date of his work, which happens to coincide with Richard II's reign.

alchemy, as well as theological texts.[40] The alchemist Philippe Éléphant, or Oliphant (fl. 1350s), who taught at Toulouse, seems to have been from the British Isles originally, possibly from Scotland.[41] John Dombelay was apparently in France when he composed the *Practica*, in which he incorporated material from another treatise (itself a commentary on an earlier work) written in Paris three decades earlier.[42] A century later, Richard Dove studied at both Orléans and Oxford before joining the Cistercian community at Buckfastleigh Abbey in Devon, where he compiled a manuscript, now Sloane 513, in which alchemical treatises sit alongside works on geometry, astronomy, and French verbs.[43]

Through such scholarly peregrinations, it did not take long for shifts in emphasis in alchemical theory to reach England from abroad. The same was true of interpretative methods. As new sources of authority arose, earlier treatises were reread in light of new information—an exercise that sometimes resulted in dramatic reinterpretations. The most consequential of these reconfigurations accompanied the rise of the "mercury alone" theory from the late thirteenth century, a doctrine whose impact can be seen by tracing readers' reception of the famous lemma from the *Secretum*, of the animal, vegetable, and mineral stone.

READING ALCHEMICALLY

By 1300 one of the most influential authorities for the use of organic ingredients in alchemy was *De anima* (On the Soul), compiled and translated from three lost Arabic treatises and pseudonymously attributed to Avicenna.[44]

40. Page, *Magic in the Cloister*, 11–12, 16, 18.

41. Guy Beaujouan and Paul Cattin, "Philippe Éléphant (mathématique, alchimie, éthique)," in *Histoire littéraire de la France*, vol. 41, *Suite du quatorzième siècle* (Paris: Imprimerie nationale, 1981), 285–363.

42. On Dombelay and the *Practica*, see *HMES*, 4:188–90.

43. Page, *Magic in the Cloister*, 127, citing David N. Bell, "A Cistercian at Oxford: Richard Dove of Buckfast and London," *Studia monastica* 31 (1989): 67–87.

44. The Arabic original of *De anima* (often referred to in scholarship as *De anima in arte alchimiae*) seems to have been composed in Spain in the twelfth century; Sébastien Moureau, *La "De anima" alchimique du pseudo-Avicenne* (Florence: SISMEL, 2016), 1:41–57. The combined text is important as evidence both for the reception of Jābirian alchemy in the Islamic West, and for the influence it exerted on Latin alchemical writing following its translation during the thirteenth century. For the early modern edition, see *De anima in arte alchimiae*, in *Artis Chemicae Principes, Avicenna atque Geber*, ed. Mino Celsi (Basel: Pietro Perna, 1572); for a modern critical edition, accompanied by an authoritative study of the text, see Moureau, *La "De anima" alchimique*. See also Moureau, "Questions of Methodology about Pseudo-Avicenna's *De anima in arte alchemiae*: Identification of a Latin Translation and Method of Edition," in *Chymia: Science and Nature in*

The writer justified his use of animal products as "stones" by employing the famous lemma from the *Secretum*, arguing that the true nature of the "stone that is no stone" is human blood, which, together with hair and eggs, provides one of the three essential ingredients in the alchemical work.[45] These natural products should be separated into their constituent "elements" through a process that we would now recognize as fractional distillation.[46]

It is not clear when *De anima* first became known in England. The earliest witness of the text is Vincent of Beauvais, who used it as one of his major sources on metals, and alchemy in particular, in his immense *Speculum maius*, finished around 1259.[47] The wide diffusion of Vincent's encyclopedia, in England as in the rest of Europe, doubtless contributed to the success of pseudo-Avicenna's model of organic alchemy.[48] For instance, Roger Bacon drew on pseudo-Avicenna in his own gloss on the *Secretum secretorum*, arguing that Aristotle used "stone" simply to refer to the starting matter of an alchemical operation. This *prima materia* might in practice derive from a wide range of animal, vegetable, or mineral products, including blood.[49] Elsewhere Bacon supported this position with reference to *De anima*, noting its use of different stones as *Decknamen* for organic products: thus "herbal stones" denoted hairs, "natural stones" eggs, and "animal stones" blood.[50]

A similar approach was taken up by Walter Odington, also known as

---

Medieval and Early Modern Europe, ed. Miguel López Pérez, Didier Kahn, and Mar Rey Bueno (Newcastle-upon-Tyne: Cambridge Scholars, 2010), 1–18; Moureau, "Some Considerations Concerning the Alchemy of the *De anima in arte alchemiae* of Pseudo-Avicenna," *Ambix* 56 (2009): 49–56; Paola Carusi, "*Animalis herbalis naturalis*: Considerazioni parallele sul 'De anima in arte alchimiae' attribuito ad Avicenna e sul '*Miftāh al-hikma*' (Opera di un allievo di Apollonia di Tiana)," *Micrologus* 3 (1995): 45–74; Newman, "Philosophers' Egg."

45. Pseudo-Avicenna goes so far as to work the saying into one of his own procedures. *De anima*, ed. Moureau, 2:361: "Accipe lapidem qui non est lapis, et non est de naturis lapidum, et diuide, et fac de eo spiritum, et animam & corpus" ("Take the stone that is no stone, and nor has the nature of a stone, and divide it, and make from it a spirit, and a soul, and a body").

46. This process is discussed in detail in Newman, "Philosophers' Egg"; *De anima*, ed. Moureau, vol. 1.

47. Vincent cites "Avicenna" in the *Speculum naturale*, books 7–8, and the *Speculum doctrinale*, book 11. On Vincent's sources, see Sébastien Moureau, "Les sources alchimiques de Vincent de Beauvais," *Spicæ: Cahiers de l'Atelier Vincent de Beauvais*, n.s., 2 (2012): 5–118.

48. The earliest surviving manuscripts of *De anima* are later than the composition of Vincent de Beauvais's *Speculum maius*, probably dating from the late thirteenth to early fourteenth century. None of these manuscripts can be convincingly linked to England, although Glasgow University Library, MS Hunter 253 may conceivably have been produced there, or, more plausibly, in northern France. I am grateful to Sébastien Moureau for confirming this point.

49. Bacon, *Secretum secretorum*, 117–18.

50. Bacon, *Opus tertium*, 85: "Et lapides herbals sunt capilli. Lapides naturales sunt ova. Lapides animales sunt sanguis, sicut Avicenna dicit primo libro de Anima."

Walter of Evesham (fl. ca. 1280–1301), a monk at the Benedictine abbey of Evesham, near Worcester.[51] As with Bacon, something is known of Walter's broader interests, which ran the full gamut of the mathematical arts (attested by a treatise on musical theory, the *Summa de speculatione musicae*); works on optics and arithmetic; and an almanac for his abbey that starts in the year 1301.[52] Walter's range of interests is also reflected in the multiplicity of ingredients that he considers in his alchemical treatise, the *Ysocedron*—a title reflecting the work's division into twenty chapters.[53] Drawing heavily on *De anima*, Walter selects his starting materials from all the kingdoms of nature:

> The matter of the medicine is drawn from three things: that is, from animals, vegetables, and minerals. From animals we take human blood, hairs, and the eggs of chickens, and these are called "stones" by philosophers.[54]

The *Ysocedron* also discusses medicinal applications. For Walter, an affinity exists between minerals and human bodies. On the one hand, minerals make excellent medicines, as shown by the value of gold as a treatment for leprosy. On the other, human blood is excellent for treating metals.[55] Nor does Walter's concern with proportion detract from the practical bent of the treatise. Among his many recipes, for instance, he describes a red oil made by distilling egg yolks, noting that this operation works for eggs, but not for hair.[56]

Not all readers were convinced by *De anima*'s plurality of stones, however. For a medieval natural philosopher, the notion that metals could be

---

51. Walter Odington is not to be confused with the Walter Evesham who was attached to Merton College, Oxford, in the second quarter of the century: Frederick Hammond, "Odington, Walter (*fl. c.*1280–1301)," *ODNB* (accessed 11 May 2012).

52. Hammond, "Odington, Walter." Odington also composed a work on the age of the earth, *De aetate mundi*; see J. D. North, "Chronology and the Age of the World," in *Stars, Minds, and Fate: Essays in Ancient and Medieval Cosmology* (London: Hambledon, 1989), 91–115; Carl Philipp Emanuel Nothaft, "Walter Odddington's *De etate mundi* and the Pursuit of a Scientific Chronology in Medieval England," *Journal of the History of Ideas* 77 (2016): 183–201.

53. This treatise survives in five manuscripts, including an almost complete copy transcribed in 1474 by a Welshman, David Ragor: British Library, MS Add. 15549, fols. 4r–20v; edited by Phillip D. Thomas, *David Ragor's Transcription of Walter of Odington's "Icocedron"* (Wichita: Wichita State University, 1968), 3–24; hereafter *Icocedron*. See also *HMES*, 4:127–32.

54. *Icocedron*, 5: "Materia medicine a tribus elicitur, videlicet, ab animalibus, vegetabilibus, et mineralibus. Ab animalibus accipimus sanguinem hominis, capillos, et ova gallinarum, et ista lapides vocantur a philosophis."

55. Ibid., 7: "Et sunt affinia corpus hominis et mineralia quia mineralia sunt meliores medicine pro corpore hominis ut aurum pro leproso propter suam temperanciam, ita sanguis hominis pro metallo."

56. Ibid., 15: "De ovis sequirtur [*sic*] . . . Vitellos pone in descensorio, et descendet oleum totaliter rubeum . . . De capillis vero non curo in hoc opere."

generated from entirely separate species—from blood, eggs, or even mineral salts—was a breach of philosophical decorum. The sulphur-mercury theory inherited from Arabic alchemy held that metals were generated within the earth through the mixtion and slow decoction of two primordial substances, a natural process that could, in principle, be replicated (but in a shorter time) by an alchemist working above ground.[57] While it followed from the theory that pure "mercuries" and "sulphurs" could be obtained by reducing metals to their more primitive constituents, this approach left little place for ingredients whose nature was foreign to that of the metals. From the thirteenth century on, Latin authors expressed increasing skepticism over the use of nonmineral (and even nonmetallic) ingredients in alchemical practice. While this skepticism did not entirely displace organic substances—blood, for instance, remained a popular object of medicinal distillation well into the fifteenth century—it did have a signal effect on the rhetoric of philosophical treatises, particularly those concerned with transmutation rather than healing.[58]

This shift emphasizes philosophical coherence, and hence the status of alchemy as a science, but at a price. By focusing attention on quicksilver and other metallic bodies, and on methods for their analysis (for instance, through dissolution in mineral acids), "mercurialist" alchemists sacrificed a richer choice of ingredients, and with it the potential for more varied chemical knowledge. The trend is exemplified by one of the most influential medieval alchemical texts, the Summa perfectionis magisterii (Sum of the Perfection of the Magistery). This treatise, probably composed toward the end of the thirteenth century, is pseudonymously attributed to the eighth-century Arabic authority Jābir ibn Ḥayyān (the Latin Geber), but, as William Newman has argued, was more likely written by a Franciscan friar named Paul of Taranto.[59] Geber is primarily interested in gold-making, and his concern

57. On the sulphur-mercury theory, see the introduction, note 31.

58. On the use of distilled blood in alchemical medicine, see especially Peter M. Jones, "Alchemical Remedies in Late Medieval England," in Alchemy and Medicine from Antiquity to the Eighteenth Century, ed. Jennifer M. Rampling and Peter M. Jones (London: Routledge, forthcoming). A treatise on the topic was pseudonymously attributed to Arnald of Villanova, the Epistola ad Jacobum de Toleto de distillatione sanguis humani; see Michela Pereira, "Arnaldo da Vilanova e l'alchimia: Un'indagine preliminare," in Actes de la I Trobada internacional d'estudis sobre Arnau de Vilanova, vol. 2, ed. Josep Perarnau (Barcelona: Institut d'Estudis Catalans, 1995), 95–174, on 165–71; Antoine Calvet, Les oeuvres alchimiques attribuées à Arnaud de Villeneuve: Grand œuvre, médecine et prophétie au Moyen-Âge (Paris: S.É.H.A., Archè, 2011), 42, 572–79.

59. See Newman, Summa perfectionis; Newman, "New Light on the Identity of Geber," Sudhoffs Archiv für die Geschichte der Medizin und der Naturwissenschaften 69 (1985): 76–90; Newman, "The Genesis of the Summa perfectionis," Archives internationales d'histoire des sciences 35 (1985): 240–302. The term "magistery" encompasses more than mere mastery, and is not easily translated;

with metallurgical rather than medicinal effects is apparent in his preference for metallic ingredients. However, even Geber has to admit the use of certain nonmetalline substances, particularly those volatile spirits like mercury, sulphur, and arsenic that seem to unite most readily with metallic bodies.[60] Since these spirits are required to bring about certain changes in the metals, practitioners "cannot therefore be freed from their use," for they are "the true medicine of the bodies' alteration."[61]

Thus, although Geber does not absolutely exclude other ingredients from alchemical practice, he does regard their use—and their users—with skepticism, noting that many practitioners "start from diverse principles." In consequence, "some affirm this science and magistery to be found in spirits, others in bodies [i.e., metals], others in salts and alums, niters and boraxes, and others in all vegetable matter."[62] Geber asserts that he has personally tested these claims, exposing their errors only by dint of "long, tedious experience, and with the deposition of much money."[63] His own work is intended to correct such mistakes, "and teach the truth in this science."[64] In this way, he contrasts his method, grounded in coherent natural principles and rigorous practice, with that of alchemists who lack either clear principles or practical skill. A true philosopher cannot conduct the work entirely from his study: personal experience of materials and processes, gained from personal testing, is essential for the proper conduct of the science.

Geber's privileging of metals, particularly quicksilver, marks a shift in the tenor of alchemical writing. Many subsequent works, including some of the most influential treatises of the fourteenth century, are characterized by hostility toward the use of vegetable and animal products of the kind presented in earlier writings like *De anima*, or the work of Roger Bacon and Walter Odington. Thus, while Geber concedes that it is possible to bring about change in these substances, he notes that this is "extremely hard." Even nonmetalline minerals, such as alums and salts, are increasingly singled out for mockery. For instance, the use of alums and salts is robustly criticized in the

---

on its various early modern connotations, see Martin Ruland Jr., *Lexicon Alchemiae sive Dictionarivm Alchemisticvm, Cum obscuriorum Verborum, & Rerum. Hermeticarum, tum Theophrast-Paracelsicarum Phrasium* . . . (Frankfurt, 1612), 310–13.

60. On Geber's theory of metallogenesis and transmutation, see Newman, *Summa perfectionis*, chap. 4.

61. Newman, *Summa perfectionis*, 682, 738.

62. Ibid., 651.

63. Ibid., 652.

64. Ibid.

pseudepigraphic writings attributed to the Montpellier physician Arnald of Villanova (ca. 1238–1311). Pseudo-Arnald remarks that only a fool would seek in nature that which is not in nature: "Therefore because neither gold nor silver is in alums or salts, we should not seek it in them."[65]

In their place, writers stress the importance of mercury in the alchemical work, and the necessity of generating "kind from kind"—metal from metal alone, achieved by extracting essences (also called "mercuries") out of metalline bodies, including from quicksilver. Yet this mercurialist philosophy still had to accommodate the long-standing and authoritative tradition of employing vegetable and animal ingredients in the alchemical work, exemplified by De anima's multiple stones. One solution was to read these authorities "alchemically," asserting that although their enigmatic words appeared to describe organic matter, they in fact concealed references to mineral substances. Such metaphorical revisionism maintained the consensus of the philosophers, while simultaneously denying authority to frauds, rivals, and other fools who lacked the perspicacity to unlock the riddles of their forebears.

As it happens, pseudo-Avicenna had previously discussed exactly this kind of reading in De anima, but in the service of the opposite practical agenda. In what would become a familiar move in alchemical writing, he warns practitioners not to be deceived by the terminology employed by philosophers. For instance, they should not assume that the naming of the four mineral "spirits"—auripigmentum, sulphur, sal ammoniac, and "quick gold" (i.e., mercury)—provides a straightforward description of the work's ingredients. These terms should instead be read as cover names, denoting the elements of the stone: its earth, water, air, and fire. Yet even these readings may vary with context:

> Where you come upon "auripigmentum," use in its place the fire of the stone; and when you come upon "sulphur," understand air, and sometimes fire; and by "sal ammoniac not dissolved," understand earth; by "quick gold," water; and sometimes just quick gold by itself.[66]

65. Arnald of Villanova [pseud.], De secretis naturae, ed. Antoine Calvet, in Calvet, Les oeuvres alchimiques, 496: "Ratio autem quare non fit in salibus et aluminibus est quia fatuus est qui querit a natura quod in ea non est. Igitur quod in aluminibus et salibus non est aurum vel argentum, hoc non queramus in eis."

66. Avicenna [pseud], De anima, 107: "[V]bi inueneris auripigmentum, mitte in loco suo ignem lapidis, et quando inueneris sulphur, intellige aerem, et quandoque ignem, et sal armoniacum non solutum intellige terram, aurum viu. aquam, et aurum viuum quandoque per se."

Here, pseudo-Avicenna preserves the double identity of mercury, as both common quicksilver and the "water" of the stone—a fluid substance that clearly does not correspond to water as usually encountered in nature.

De anima thus offers a way of reading that is distinctively alchemical in nature, in which an earlier text (like the *Secretum*) may be interpreted in line with an anticipated practical outcome. But problems arise if we imagine applying this approach more generally to alchemical writing. Using pseudo-Avicenna's lexical advice as a guide to interpreting ingredients would rapidly transmute a straightforward recipe into a completely different text, one predicated on a different kind of chemistry and hence likely to produce different practical results.

This is exactly what happened when mercurialist alchemists began to interpret earlier texts, including *De anima* itself, by assuming that seeming references to organic ingredients in fact signify mercury or the elements. The strategy is particularly evident in the body of pseudepigrapha attributed to Arnald of Villanova. In influential treatises such as the *Rosarius philosophorum* (Rosary of the Philosophers) and *De secretis naturae* (On the Secrets of Nature), pseudo-Arnald sets out the case for mercury as the sole prime matter of alchemy.[67] He supports this position by arguing that ancient adepts spoke metaphorically when they characterized the stone as animal, herbal, or natural, or claimed that it was found in blood, hair, or eggs. For instance, the stone may be called "animal" because it has a spirit, and therefore also a soul. It may also be called "blood," since blood is red, like the stone.[68] Yet these words are misunderstood by ignorant practitioners who assume that the philosophers literally mean eggs and blood. These fools, "understanding only the letter," then attempt to work with eggs, blood, alums, salts, and the rest, yet find nothing. Reading their sources out of context, they circulate the bowdlerized fragments in the form of recipes—"and with these receipts," says Arnald, "they deceive the whole world."[69]

Failure to succeed in alchemy is thus presented as an *exegetical* failure as much as a lack of practical skill. Since the ability to read texts on multiple levels, including allegorically, is the remit of the scholar, this approach effectively excludes artisans and other unlearned practitioners from the science. Pseudo-Arnald claims that "no one should therefore come to this art unless

---

67. On pseudo-Arnald's mercurialist approach to alchemy, see Calvet, *Les oeuvres alchimiques.*

68. Arnald of Villanova [pseud.], *De secretis naturae*, 500: "Ratio est quia quod habet spiritum habet animam, lapis igitur noster habet spiritum, ergo habet animam. . . . Unde dictus est sanguis propter rubidinem quia sanguis est rubeus, similiter lapis noster est rubeus."

69. Ibid., 512: "[E]t decipiunt totum mundum cum ipsis receptis."

he has first learned logic, and afterwards philosophy, and knows the causes and natures of things. Otherwise he wearies his soul and body to no purpose."[70] Such a program implies that the successful alchemist must be educated at least to the level of a master of arts—a distinction that buttressed alchemy's own natural philosophical pretensions while distancing it from artisanal practice. Reading alchemically meant reading *philosophically*, and reading like a philosopher meant reading metaphorically. Through their skill in reading texts in this way, alchemical philosophers thereby distinguished themselves from literalist craftsmen. Importantly, they also distinguished their work from the taint of currency crime, a matter of increasing concern to ecclesiastical and secular authorities throughout the fourteenth century.

The outcome of this rhetoric was to emphasize the importance of alchemical *reading*. Only the wise could determine the abstruse words of the philosophers, who typically concealed the truth of their processes beneath metaphors and parables. Accordingly, English philosophers sought to ally themselves reputationally with authorities whose treatises invited sophisticated reading techniques. Yet the process of reading alchemically could have unexpected consequences for both the original text and the practical outcome. As alchemists learned to read their sources on many levels, the range of possible interpretations of a given text increased, creating space for innovative readings based on personal experience, including new experimental observations. In the case of the pseudo-Lullian corpus, such endeavors eventually fused into a new approach to alchemy, in which a traditional emphasis on gold-making was productively blended with developing interest in the medicinal applications of distilled wine. English alchemy would eventually achieve its greatest heights through the melding of these two desiderata—metal and medicine—under the definitive authorship of Raymond Lull.

THE COMING OF RAYMOND

Chronologically speaking, the story of Lullian alchemy begins not with Raymond's mythical imprisonment by Edward III, but with a text. The earliest work in the corpus is the *Testamentum*, probably composed in Latin during the 1330s and translated into Catalan and French by the end of the fourteenth

---

70. Ibid., 490: "Nullus ergo ad hanc scientiam veniat nisi primo audiverit logicam, postea philosophiam, et sciat causas et naturas rerum. Aliter frustra fatigaret animam suam et corpus suum."

century.[71] Although its structure varies from copy to copy, the original version seems to have consisted of three books, a "Theorica," a "Practica," and the "Book of Mercuries," accompanied by a song in Catalan, the *Cantilena*. From the work's Catalan associations, and its use of circular figures to express the principles of the art, it appears that the author—whom Michela Pereira has dubbed the "Magister Testamenti"—regarded himself as in some sense a follower of the historical Lull. Over time, readers came to assume that Lull was himself the author of the treatise, an ascription cemented by later works in the corpus that are deliberately framed as works of the Majorcan philosopher. Of these, the most influential is the *Liber de secretis naturae, seu quinta essentia* (Book Concerning the Secrets of Nature, or Quintessence; hereafter *De secretis naturae*), whose author not only claims to be Lull, but also asserts his authorship of the *Testamentum* and other Lullian works.

While scholars now recognize that these treatises were written at different times and in different places, and consequently recount diverse and even contradictory methodologies, the pseudepigraphic component meant that late medieval readers tended to view all as authentic works of Lull. This view also demanded a particular approach to reading, in order to demonstrate coherence between texts that actually said quite different things. Incongruities were readily explained by the conventions of alchemical reading and writing: for instance, the technique of *dispersio*, already present in Arabic alchemical writings, whereby necessary information is provided across a series of writings rather than being concentrated in a single book.[72] To assemble the writer's true meaning therefore requires wide reading and careful study. This strategy can also transform the sense of a text, in ways that, while distorting the author's original intention, nonetheless resulted in some highly innovative and fruitful experimental outcomes.

The grounds for flexibility of future interpretation are already present in the *Testamentum*, particularly in the Magister's wide-ranging definition of alchemy. Although the contents of the book are primarily concerned with transmutation, this definition explicitly creates space for other applications, including medicine and the making of precious stones:

71. At some point, perhaps during the fifteenth century, the Catalan version was then translated back into Latin. For the origins of the *Testamentum* and its linguistic peculiarities, see *Testamentum*; Pereira, "Alchemy and the Use of Vernacular Languages in the Late Middle Ages," *Speculum* 74 (1999): 336–56, on 354–55.

72. The technique of dispersion of knowledge is employed in the corpus of works attributed to Jābir ibn Ḥayyān; Paul Kraus, *Jābir b. Ḥayyān, contribution à l'histoire des idées scientifiques dans l'Islam* (Cairo: Institut français d'archéologie orientale, 1943), 1:XXVII–XXX. On its use in the European alchemical tradition, see Newman and Principe, *Alchemy Tried in the Fire*, 186–87.

Alchemy is a secret and most necessary part of natural philosophy, from which one art is made which is not open to everyone: the which teaches how to change all precious stones and to reduce them to true temperament [i.e., an equal balance of qualities]; and to bring every human body to the noblest health; and to transmute all bodies of metals into true Sun and true Moon [i.e., gold and silver] by means of one body, the universal medicine, to the which all particular medicines are reduced.[73]

Here the writer of the *Testamentum* touches on many of the problems and paradoxes that we have already encountered. He confirms alchemy as a major component of natural philosophy, and hence as part of learned rather than artisanal knowledge. It is, furthermore, one fundamental art rather than a diffusion of practices: a science aimed at producing a single, universal medicine that may be used to achieve many particular ends. Lastly, alchemy is a secret art: its privileged knowledge is "not open to everyone."

In addition to offering a slate of significant and desirable outcomes, from transmutation to medicine, the definition may also be read as a defense of alchemy. The Magister seems to allude to fraudulent practice when he emphasizes that his elixir will not produce merely superficial change, but *true* gold and *true* silver—ends that are, however, available only to learned men. To decipher and reproduce the chemistry of the *Testamentum* will, in fact, require all of the skills of alchemical reading outlined above, for it is not just one of the longest works of philosophical alchemy, but also one of the densest.

Although the Magister speaks of only one, universal stone in his definition of alchemy, he conceives of more than one kind of mercury. The density and complexity of the text, coupled with the Magister's continual switching between different sets of cover names, make it difficult to clearly distinguish between these multiple identities. However, two particular mercuries, or "argent vives," may be separated out from the general *massa confusa*. One is a "mineral mercury" drawn from metals; the other a "vegetable mercury" whose origin is less clear. The identification of these two mercuries—mineral and vegetable—provides the spine of the pseudo-Lullian corpus. Later writings would build on this apparent distinction, bifurcating practice

73. *Testamentum*, 2:306: "Alchimia est una pars celata philosophie naturalis magis necessaria, de qua constituitur una ars, que non apparet omnibus, que docet mutare omnes lapides preciosos et ipsos reducere ad verum temperamentum; et omne corpus humanum ponere in multum nobilem sanitatem; et transmutare omnia corpora metallorum in verum solem et in veram lunam per unum corpus medicinale universale, ad quod omnes particulares medicine reducuntur."

into a "metallurgical" strand directed toward making metals and gems, and a "medicinal" strand concerned with healing human bodies. In what follows, I shall tease out a single practice from this complicated work, to illustrate both the ambivalent nature of the Magister's double mercury, and the difficulties it posed for late medieval readers.

MINERAL MERCURIES

To make sense of the *Testamentum*'s abstruse philosophy requires mastery of all the techniques of alchemical reading. As well as traditional strategies of alchemical composition, such as the use of cover names and *dispersio*, the Magister employs novel forms of presentation influenced by his familiarity with other fields, including scholastic medicine and the authentic philosophy of Raymond Lull.[74] He employs extended biological metaphors to describe the stone, detailing its conception and "medical" regimen, as well as the nuances of its complexion. He enhances the authority of chemical procedures by reducing them to principles, each assigned a letter of the alphabet, which he then plots onto diagrammatic figures similar to those employed in authentic Lullian works.[75] A host of *Decknamen* describe the various ingredients used in the work—in particular, sufficient permutations of "sulphur" and "mercury" to baffle even a practiced reader.

Beneath these techniques, however, we can discern a familiar approach grounded in the triumvirate of mercury, gold, and silver. For instance, in the first part of the work, the *Theorica*, we learn that the Magister's mineral mercury is not common quicksilver at all: it is a substance that "in the first place is extracted from the body"—that is to say, from a metal.[76] This mercury is an essential substance drawn out of gold (Sol) and silver (Luna)—an extraction

74. On the philosophical background of the *Testamentum*, see Pereira, *L'oro dei filosofi*. By scholastic medicine, I refer to the incorporation of medicine into European universities as a discipline grounded in natural philosophical principles; see pp. 103–104, below.

75. On the use of Lullian figures in the pseudo-Lullian alchemical corpus, see Pereira, *Testamentum*, cxxxvii–clxiv; Pereira, "Le figure alchemiche pseudolulliane: Un indice oltre il testo?," in *Fabula in tabula: Una storia degli indici dal manoscritto al testo elettronico*, ed. Claudio Leonardi, Marcello Morelli, and Francesco Santi (Spoleto: Centro italiano di studi sull'alto Medioevo, 1994), 111–18; Marlis Ann Hinckley, "Diagrams and Visual Reasoning in Pseudo-Lullian Alchemy, 1350–1500 (MSt thesis, King's College, University of Cambridge, 2017). On George Ripley's use of a related figure in the *Compound of Alchemy*, see also Jennifer M. Rampling, "Depicting the Medieval Alchemical Cosmos: George Ripley's Wheel of Inferior Astronomy," *Early Science and Medicine* 18 (2013): 45–86.

76. *Testamentum*, 1:196: "Quapropter extrema nostri lapidis in sua prima creacione est argentum vivum, quod est extractum a corpore in primo latere; et in secundo est elixir completum."

that is accomplished using a mysterious but powerful solvent called the "Water of the Green Lion."[77] Such procedures, cloaked in the language of multiple mercuries and especially of the Green Lion (a favorite cover name, already encountered in De compositione alchemiae), are not so far removed from those outlined in the works of pseudo-Geber and pseudo-Arnald, as we see when attempting to translate them into practice.

The Magister reveals the method for extracting his gold and silver mercuries in the Practica, although readers should not expect straightforward clarification of the process. He describes chemical combinations with reference to a diagram: an alphabetical wheel that sets out the various principles of matter. This is not an authentic Lullian combinatorial wheel but a simpler figure, in which letters of the alphabet designate the various material components of the work (fig. 1). At the beginning, A denotes God, as the first cause of all things and hence the appropriate first step in the process. B is argent vive, here defined as "the common substance existing in all corruptible bodies"—a reference to mercury's role as a primordial constituent of metals. C is saltpeter, and D "vitriol azoqueus" (a substance that, in the Theorica, is linked to the Green Lion[78]). By combining B, C, and D, one obtains E: the solvent, or menstruum, which "encapsulates the nature of the aforesaid three things in one."[79]

How might a practicing alchemist have read this process? For anyone used to working with mercury and its compounds, the figure must have looked, at first glance, like a rather laborious way of expressing the process for corrosive sublimate (mercury chloride in modern parlance), using an aqua fortis made from vitriol and saltpeter to dissolve and sublime quicksilver.[80] Lullian gravitas is bestowed on the prosaic recipe format through alphabetical substitutions: "In the virtue of A, first take one part of D and a half of C"—a formula that translates as "In the name of God, take one part of vitriol and

77. Ibid., 1:196–98: "Postquam diximus extrema nostri lapidis . . . ad primum latus est aqua leonis viridis cum metallo coniuncta. Et in secundo est lapis, qui creatus est. Et medium illorum est sol et luna, unde exit nostrum argentum vivum, quod est corpus liquefactum, fusum et putrefactum, de quo creatur lapis, quando purgatum est a sua macula originali."

78. Ibid., 1:198: "Fili, leo viridis azoqueus, qui dicitur 'vitriolum', per naturam fit de substancia argenti vivi communis, quod est radix nature, unde creatur metallum in sua propria minera."

79. Ibid., 2:310: "B significat argentum vivum, quod est substancia communis, extans in omni corpore corruptibili . . . C significat salem petre . . . D significat vitreolum azoqueum . . . Et postea menstruale significatur per E, quod continet naturas dictorum trium totaliter in unum."

80. When made from vitriol and saltpeter, aqua fortis broadly equates to modern nitric acid. However, in medieval Europe fewer distinctions were made between the different mineral acids, which in practice may have differed considerably as the result of material impurities and variation in manufacture.

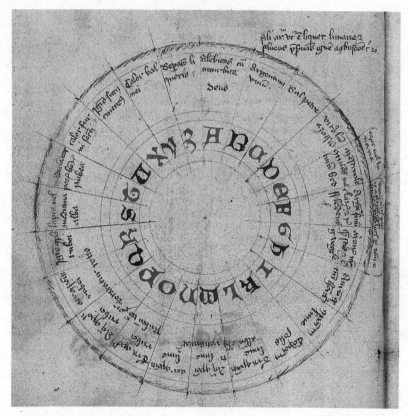

FIGURE 1. Pseudo-Lullian wheel, *Practica Testamenti*. The wheel begins with A ("Deus"), signifying God. The practice starts with B ("Argentum viuum"). Yale University, Beinecke Rare Books & Manuscripts Library, MS Mellon 12, fol. 97v. By permission of the Beinecke Rare Books & Manuscripts Library.

half a part of saltpeter."[81] This nod to philosophical principles is succeeded by minutely detailed instructions on the manufacture of *aqua fortis*, starting with the subtle grinding and mixing of the vitriol and saltpeter on a marble stone, and continuing to the appropriate choice of lute for sealing and protecting the glass vessel, and advice on the regulation of the fire. Having produced the solvent E, the Magister warns his disciple to store it safely: "For now you can say that you have the Stinking Menstruum at your command . . . by which all bodies are quickly reduced into their first matter."[82]

81. *Testamentum*, 2:316: "Cum virtute de A primo accipies unam partem de D et mediam de C; et totum sit pistatum super marmore et subtiliter pulverizatum et subtiliter mixtum pones in una cucurbita vitri."

82. Ibid., 2:318: "Quia nunc potes dicere quod menstruale fetens habes in tuo precepto . . . per quam omnia corpora in suam primam materiam reducentur breviter."

The sublimation of mercury is one of the most widespread processes used in late medieval alchemy, occurring in hundreds of recipes and scores of philosophical treatises, including major authorities like the *Summa perfectionis* of pseudo-Geber and *Rosarius philosophorum* of pseudo-Arnald, as well as earlier works of Arabic provenance.[83] That the Magister would employ corrosive sublimate as the starting point for his own mercurialist alchemy is therefore entirely in character with the general trend of fourteenth-century alchemical theory. Yet, complicating matters, mercury also plays an analogical role in the text, indicating not only liquids but also the properties of volatile spirits. At one point, for instance, the Magister describes all three ingredients of the Stinking Menstruum as "mercuries," perhaps alluding to their volatility when heated, which requires the practitioner to seal his flask securely:

> And you shall lute the joints [of the alembic] with bands of cloth covered with a paste made from flour blended with egg white, so that the properties of the three united mercuries, that is to say, of salinity, glassiness, and wateriness conjoined into one, are not lost.[84]

According to this passage, the solvent E inherits the properties of salinity from saltpeter, glassiness from vitriol, and a moist or watery quality from quicksilver. We might also read this as a commentary on the appearance of corrosive sublimate, which sublimes in the form of delicate white crystals that have to be scraped from the head of the vessel.

Yet the real significance of the passage arises not so much from the novelty of the procedure as from the vagueness of the language, which still allows room for doubt—and, in consequence, space for different interpretations. In practical terms, the identity of the mineral menstruum BCD, or E, depends on the nature of B. For instance, if the Magister intends B to signify common quicksilver, C saltpeter, and D vitriol, then the resulting menstruum might indeed produce corrosive sublimate. But what if B were not intended to be interpreted literally, but instead used to denote the essential "mercury" drawn out of a base metal, such as lead? In that case, E would

---

83. The *Liber de aluminibus et salibus* pseudonymously attributed to al-Rāzī, an influential twelfth-century treatise written in Arabic and later translated into Latin, includes several processes for subliming mercury that would probably result in (or were intended to result in) corrosive sublimate. See Jennifer M. Rampling, "How to Sublime Mercury: Reading Like a Philosopher in Medieval Europe," *History of Knowledge*, 24 May 2018, https://wp.me/p8bNN8–23p.

84. *Testamentum*, 2:318: "Lutabisque iuncturam cum benda panni liniti cum pasta facta de flore farine, distemperata cum albumine ovi, ut proprietates unite trium mercuriorum, videlicet salsuginei, vitrei et aquatici, coniunctorum in unum, non perdantur."

be something else entirely: the product of dissolving "mercury of lead" in an *aqua fortis* made from saltpeter and vitriol (C and D). In this example, the fluidity of mercury as a technical term would permit diverse readings of the passage, potentially validating alternative approaches to the underlying chemistry. In fact, just such a reading would later be adopted by Raymond's fifteenth-century commentator George Ripley, who, as we will see, argued for the use of base metals like lead in his own alchemical work.

So far we have traced this process only in relation to mineral ingredients. Yet what are we to make of the seventh substance in the practical wheel—the substance denoted as G? Having come this far, it is infuriating to discover that the Magister defines this vital ingredient simply as "the mercury that you know."[85]

Unlike the other substances, the Magister does not disclose the identity of G in this section of the *Practica*. His wording, "that you know," indicates the opposite: that we are not intended to know, or not easily. Yet the use of this expression, a typical formula for an encoded alchemical substance, indicates that G is an ingredient of great importance to the work—important enough that its identity should not be readily disclosed. In fact, G signifies the second of the two major "argent vives" that the Magister alluded to earlier in the *Theorica*. This substance possesses a vegetative character, defined as the capacity to grow, or to enable other metals to grow. In the *Practica* it is called "G. vegetable," or simply "our mercury" (*mercurius noster*). Once combined with the Stinking Menstruum, E, this vegetable mercury helps to draw another kind of "mercury" out of gold and silver. These metallic mercuries constitute the Magister's sought-after "mineral" argent vive: a substance that thereby comes into being through the action of the "vegetable" argent vive, its elusive cousin.

As readers of the *Testamentum* quickly grasped, the identity of G must be deciphered in order to obtain the precious mercuries of gold and silver. While the Magister is generally vague on this point, he does drop a hint in his recipe for a "corruptible water" (*aqua corruptibilis*), supposedly able to dissolve gold:

> Take 2. oz. of G. and draw out its humor with an alembic with two ounces of common nature, which is water of wine (*aqua vini*); and put in one ounce of the gold which you wish to dissolve . . . After, congeal this matter, separating the water by alembic, and afterwards put into it again of the juice of "larien,"

85. Ibid., 2:310: "Et per G significatur mercurius, quem scis."

otherwise called lunaria, as much as you will; and you shall see the gold dissolved into a vegetable water the color of the sun. And thus of three things we have formed the third circular figure, signified by K.[86]

Here, G seems to be equated with yet another cover name: the juice of "larien," or "lunaria." Although this term is never expounded, it does occur earlier in the *Theorica*, where the Magister instructs that "You may take of the juice of lunaria and draw forth its sweat with a light fire, and you shall have in your power one of our argent vives in liquor."[87] In this setting, lunaria might be regarded as a cover name for quicksilver; in which case, its "juice" ought to correspond to corrosive sublimate, which actually is soluble in "water of wine" (ethanol), as the recipe suggests. Yet the context of its appearance in the *Practica* leaves its identity uncertain. For instance, lunaria could just as easily be envisaged as a product derived from wine, such as tartar, or perhaps a metal compound capable of solution in alcohol. Either way, the Magister's enigmatic account leaves the door wide open for an interpretation based on the use of wine-based solvents. Since the appearance of the *Testamentum* coincided with increasing interest in distilled wine as a medical remedy, later readers were swift to seize this opportunity—and, in the process, to realize alchemy's promise as an art that healed both metals and human bodies.

VEGETABLE MERCURIES

The notion that mercury possessed a "vegetable" quality was not new in European alchemy: mercurialist treatises frequently allude to the vegetable nature of quicksilver or other base metals, with reference to their raw, undigested character. Quicksilver, for instance, might be regarded as the crudest or least digested metal, since it remains closest in character to the primordial "mercury" principle from which all metals arise. Yet over the course of the fourteenth century, developments in medicine suggested a far more exciting explanation for the vegetable character of the philosophers' prime matter: the inclusion of a literally vegetable ingredient, made by distilling wine.

86. Ibid., 2:324–26: "Recipe duas uncias de G et extrahe suum humorem per alembicum cum duabus uncis nature communis, que est aqua vini; et in una uncia proice solem, quem vis dissolvere.... Post congela istam materiam, separando aquam per alembicum. Post proice intus de succo de larien, alias lunarie, quantum volueris; et videbis aurum dissolutum in aquam vegetalem in colorem solis. Et sic de tribus rebus formamus terciam figuram circulatem, per K significatam."

87. Ibid., 1:38: "Accipias de succo lunarie et trahe suum sudorem cum parvo igne, et habebis in tua potestate unum de nostris argentis vivis in liquore per formam aque albe."

By the 1330s, the use of wine for therapeutic purposes was already well established in Western medical practice, as was the distillation of wine to extract ethanol—"water of life" (*aqua vitae*) or "burning water" (*aqua ardens*).[88] From the second half of the thirteenth century, however, medical practitioners began to produce much higher-proof distillates than previously, partly as the outcome of new techniques and apparatus that enabled distillers to make almost absolute alcohol. Prominent among them was Taddeo Alderotti (d. 1295), the celebrated professor of medicine in Bologna, who wrote seven *consilia* in praise of highly rectified spirit of wine. New products required new equipment, and Alderotti also described a method for producing high-proof *aqua vitae* using an apparatus of his own devising: a flask with external pipes (*canale serpentinum*) that allowed faster cooling of the distillate.[89] The method produced a clear, inflammable water with the ability to preserve organic matter from corruption and to draw out the beneficial "essences" of herbs and spices. Alderotti hailed this *aqua vitae* as "of inestimable glory, the mother and mistress of all medicine"—a line that the Magister would later quote in the *Testamentum*.[90]

The significance of the distillate as an alchemical (rather than merely medical) product rests on the claims of the Franciscan tertiary John of Rupescissa, who was impressed by the unusual qualities of spirit of wine, which seemed to protect bodies against age and sickness. John composed his famous *Liber de consideratione quintae essentiae* (Book Concerning the Quintessence; hereafter *De consideratione*) in the early 1350s, only a few years after the Black Death commenced its depredations on the European population.[91] For many people, the arrival of the pestilence on the heels of

---

88. Linda E. Voigts, "The Master of the King's Stillatories," in *The Lancastrian Court: Proceedings of the 2001 Harlaxton Symposium*, ed. Jenny Stratford (Donington: Shaun Tyas, 2003), 233–52; Lu Gwei-Djen, Joseph Needham, and Dorothy Needham, "The Coming of Ardent Water," *Ambix* 19 (1972): 69–112; R. J. Forbes, *A Short History of the Art of Distillation* (Leiden: Brill, 1970). On these developments and their relationship to alchemical practice, particularly in Italy, see Chiara Crisciani and Michela Pereira, "Black Death and Golden Remedies: Some Remarks on Alchemy and the Plague," in *The Regulation of Evil: Social and Cultural Attitudes to Epidemics in the Late Middle Ages*, ed. Agostino Paravicini Bagliani and Francesco Santi (Florence: SISMEL, 1998), 7–39.

89. Edmund O. von Lippmann, "Thaddäus Florentinus [Taddeo Alderotti] über den Weingeist," *Archiv für Geschichte der Medizin* 7 (1913–14): 379–89; Gwei-Djen, Needham, and Needham, "Coming of Ardent Water," 70–71; Nancy G. Siraisi, *Taddeo Alderotti and His Pupils: Two Generations of Italian Medical Learning* (Princeton: Princeton University Press, 1981), 300–301.

90. Taddeo Alderotti, *I consilia*; cited in Siraisi, *Taddeo Alderotti*, 301: "[Aqua vitae] est igitur eius gloria inextimabilis, omnium medicinarum mater et domina."

91. On John of Rupescissa (Jean de Roquetaillade) and the quintessence, see F. Sherwood Taylor, "The Idea of the Quintessence," in *Science, Medicine and History: Essays on the Evolution of Sci-*

Europe-wide famine and schism in the church suggested the imminence of the end of days: an anxiety that also colored thinking about the transformative power of alchemy.

The Magister Testamenti had previously reflected on the ability of the philosophers' stone to withstand the purging fires of Judgment Day.[92] Rupescissa, a millenarian prophet influenced by the apocalyptic prophecies of Joachim da Fiore, viewed alchemical techniques as offering more immediate assistance. For John, the repeatedly rectified spirit of wine provided an effective yet inexpensive medicine that might enable his spiritual brethren to fortify themselves against the anticipated coming of Antichrist. By referring to this substance as "our heaven," he implied that it provided a terrestrial analogue for the immutable ether, or "fifth essence" (*quinta essentia*) of the celestial bodies. Just as the heavens resisted change in Aristotle's cosmology, so John's own earthly quintessence seemed to preserve organic matter from corruption. It was also a powerful solvent, more penetrating than normal *aqua ardens*, and capable of drawing out the essence not only from animal and vegetable matter, but also from metals, including antimony and gold.

John viewed the quintessence as belonging to the sphere of medicine rather than traditional alchemy. Throughout the book he refers to his practice as "medicine," contrasting "alchemical gold," a toxic substance made using corrosives, unfavorably with the quintessence, which is not only safe to ingest, but capable of healing the most intractable diseases, including leprosy and the pestilence.[93] This distinction, coming at a time when physicians were already paying increased attention to alcohol-based remedies, helped to establish the quintessence as the basis of a new school of alchemical pharmacology.

It would also become a staple of pseudo-Lullian alchemy. John of Rupescissa may have disdained transmutation (at least in this text), but his quintessence offered a promising line of inquiry for readers seeking to decipher the riddle of the *Testamentum*'s chrysopoetic vegetable mercury, the myste-

---

entific *Thought and Medical Practice Written in Honour of Charles Singer*, ed. Edgar A. Underwood (Oxford: Oxford University Press, 1953), 1:247–65; Robert P. Multhauf, "John of Rupescissa and the Origin of Medical Chemistry," *Isis* 45 (1954): 359–67; Robert Halleux, "Les ouvrages alchimiques de Jean de Rupescissa," *Histoire littéraire de la France* 41 (1981): 241–77; DeVun, *Prophecy, Alchemy*.

92. *Testamentum*, 1:14–16; discussed on pp. 178–80, below.

93. John of Rupescissa, *De consideratione Quintae essentiae rerum omnium, opus sanè egregium* (Basel: Conrad Waldkirch, 1597), 22: "Et aurum alchimicum, quod est ex corrosiuis compositum, destruit naturam."

rious G.[94] This line was taken up by the fourteenth-century pseudepigrapher writing as Raymond Lull, whose *Liber de secretis naturae, seu quinta essentia*— a title that reveals the influence of Rupescissa's tract—provides one of the most influential examples of synthetic alchemical writing.[95] The identity of the author is unknown, but he was evidently deeply familiar with both authentic Lullian writings and earlier alchemical works written in Lullian style, such as the *Testamentum*. Thus, although the first two books of *De secretis naturae* are largely derived from Rupescissa's medical approach in *De consideratione*, the third book, on transmutation, draws from the alchemy of the *Testamentum*. *De secretis naturae* is therefore the outcome of Raymond's deliberate "splicing" of two distinct bodies of work: the mineral alchemy of the Magister Testamenti, and the wine-based medicine of John of Rupescissa.

This splicing of separate textual traditions to produce a working practice relies on a remarkable example of reading meaning *into* a text: here, the presumption that John of Rupescissa's "quintessence of wine" and the *Testamentum*'s "vegetable mercury" refer to essentially the same material product. This being the case, the quintessence may be safely substituted into the interpretative space provided by the Magister's enigmatic use of cover names like G and "lunaria," without affecting the chemical outcome of the process. Besides providing a form of commentary on the text, this substitution also results in a new kind of practical alchemy. Nor is this reading necessarily fraudulent, since the identity of the vegetable mercury is never explicitly stated in the *Testamentum*, leaving open the possibility that spirit of wine was, in fact, the intended ingredient. The fact that the Magister was also familiar with distilled alcohol of the kind promoted by Alderotti may have provided further support for this interpretation.

To strengthen the connection between the two traditions, the Raymond of *De secretis naturae* adopts the Magister's expressions "vegetable mercury"

94. John of Rupescissa wrote another work on transmutation, the *Liber lucis*, which appears to describe the manufacture of corrosive sublimate from mercury, vitriol, and saltpeter: *BCC*, 2:84–87. This text is discussed in Principe, *Secrets of Alchemy*, 64–67.

95. The prologue has been edited by Michela Pereira: "Filosofia naturale lulliana e alchimia: Con l'inedito epilogo del *Liber de secretis naturae seu de quinta essentia*," *Rivista di storia della filosofia* 41 (1986): 747–80. On the complicated history of the text, see Pereira, *Alchemical Corpus*, 11–20; Pereira, "Sulla tradizione testuale del *Liber de secretis naturae* seu de quinta essentia attribuito a Raimondo Lullo: Le due redazioni della *Tertia distinctio*," *Archives internationales des sciences* 36 (1986): 1–16. Core doctrines are discussed in Pereira, "'Vegetare seu transmutare': The Vegetable Soul and Pseudo-Lullian Alchemy," in *Arbor Scientiae: Der Baum des Wissens von Ramon Lull. Akten des Internationalen Kongresses aus Anlaß des 40-jährigen Jubiläums des Raimundus-Lullus-Instituts der Universität Freiburg i. Br.*, ed. Fernando Domínguez Reboiras, Pere Villalba Varneda, and Peter Walter (Turnhout: Brepols, 2002), 93–119.

and "lunaria" to describe his own quintessential waters. But although this use of a shared terminology helps paper over the cracks between the two strands of the corpus, Raymond's substitution of the quintessence of wine for the Magister's vegetable (yet probably minerally derived) mercury fundamentally alters the underlying alchemy of the *Testamentum*, which is still primarily concerned with mineral substances such as metals and salts. While his adoption of vegetable mercury implies a shared substrate, the amalgamation of two practical approaches in fact subjects the term to considerable hermeneutical stress. Indeed, to borrow an alchemical analogy, this term would simultaneously stabilize the corpus, and make it volatile.

MANY MERCURIES?

Raymond's substitution brought the Magister's mercurialist alchemy up to date by blending a more traditional approach to transmutation with a recent and popular pharmacological innovation. Yet the variegated origins of the pseudo-Lullian corpus presented subsequent generations of alchemists with an exegetical dilemma. On the one hand, the *Testamentum* and *Codicillus* preached a metallic-based alchemy that focused on mineral solvents and referred to vegetable products only tangentially. On the other, *De secretis naturae* freely advocated the use of plants and herbs, both as the basis for making quintessence of wine and as a means of "sharpening" a variety of medicinal and transmuting elixirs. In fact, Raymond goes farther than Rupescissa in strengthening the medical credentials of his work, by claiming that the quintessence was formerly known to foundational authorities like Hippocrates and Galen, thereby seeking to reconcile John's radical alchemical pharmacology with orthodox Galenic medicine in a way that the iconoclastic Franciscan would surely never have condoned.[96]

One outcome of this disjunction was the evolution of a new tradition based on two "waters": a toxic mineral solvent used to make alchemical gold and silver; and the heavenly quintessence used in medicine, which in some cases could also be adapted for use in transmutation. This material and functional distinction, apparently supported by the *Testamentum*'s use of separate mineral and vegetable mercuries, became one of the most characteristic features of the pseudo-Lullian corpus. Rather than envisaging a single, multi-

---

96. On Raymond's subversion of medical authorities, see Jennifer M. Rampling, "Analogy and the Role of the Physician in Medieval and Early Modern Alchemy," in Rampling and Jones, *Alchemy and Medicine*.

purpose stone, commentators increasingly described a *multiplicity* of stones, each grounded in different principles and prepared for different ends. This metaphorical turn came full circle when, in addition to mineral and vegetable stones, the corpus also grew to assimilate the notion of an animal stone made by distilling blood or urine.[97]

This multiplicity is made explicit in the *Epistola accurtationis*, a short, practically focused work, probably composed late in the fourteenth century, which discusses abbreviated procedures, or "accurtations," for the animal, vegetable, and mineral stones. The *Epistola* circulated widely in England, and even earned the distinction, rare among pseudo-Lullian writings, of being translated into English before the end of the fifteenth century.[98] Its popularity may relate to the fact that, for the first time in the pseudo-Lullian canon, it assumes the existence of three separate stones, made from different "waters" and having different applications. Nature, composition, and function must agree: thus the mineral stone, based on a corrosive water made from vitriol and vermilion, is suitable only for transmuting metals, while the medicinal vegetable stone is drawn from a vegetable water, and enables the "restytucyon of growth and conseruacyon off mannys body frome all corrupcyon accydentall."[99] The animal stone, made from blood, is described in mysterious and vague terms. It serves to transmute all things, but is also a perfect medicine for man's body. Although it contains more science than any other stone, Raymond spends little time on it, and, despite contemporary interest in the distillation of blood, we may suspect that his animal elixir was included primarily to round out the set of stones alluded to by Aristotle, Bacon, Arnald, and others.[100]

Such a bundling together of diverse practices, goals, and material prin-

97. The pseudo-Lullian *Liber de investigatione secreti occulti* (Book Concerning the Investigation of the Hidden Secret) identifies human urine rather than mercury as the first material principle, rationalizing this choice in terms of the stone's "vegetable" and "animal" nature: it can both grow and reproduce itself, and its first principle should therefore be drawn from living things—particularly from that noblest of all creatures, man. See Michela Pereira, ed., "Un lapidario alchemico: Il *Liber de investigatione secreti occulti* attribuito a Raimondo Lullo; Studio introduttivo ed edizione," *Documenti e studi sulla tradizione filosofica medievale* 1 (1990): 549–603, on 578–79. Although the colophon of this treatise dates it to 1309, this is clearly spurious; the work is more probably placed around the end of the fourteenth century.

98. For the Middle English translation, see Sloane 1091 (fifteenth century), fols. 97r-101r; a later copy with some revision of spelling and syntax is in Ashmole 1508, fols. 266r-68v (transcribed by Elias Ashmole). On the dating of the *Epistola*, see Pereira, *Alchemical Corpus*, 9–10.

99. Sloane 1091, fols. 97r-101r.

100. Ibid., fol. 97r. On distilling blood, see note 58 above.

ciples beneath the umbrella-like authority of Lull allowed these writings to achieve a natural philosophical respectability denied to humbler recipe collections. Although some pseudo-Lullian treatises, such as the *Repertorium*, retained the Magister's primarily metallurgical focus, others laid out a veritable banquet of alchemical ingredients before their readers. One outcome was to impute a new sense to pseudo-Aristotle's elusive "animal, vegetable, and mineral stone": no longer just an analogy for a single, universal elixir, but a manifesto for diversity in ingredients and alchemical pursuits, which defied facile comparison with the counterfeiting of metals. Yet the varied material principles of the stone were still veiled in the language of mercury, to the extent that they were sometimes defended in what may seem, at first glance, to be aggressively "metallurgical" readings. No surprise, then, if commentators right up to the present day have struggled to identify precisely what alchemists intended by their prime matter—or their mercury.

TELLING ALCHEMICAL HISTORY

While *De secretis naturae* is now recognized as a Lullian forgery, this was obviously not the perception of Raymond's late medieval readers. For those who accepted both the *Testamentum* and *De secretis naturae* as authentic productions of Lull, the books provided mutual support, allowing for easier interpretation and hence a better chance of successful recovery of their practical contents. *De secretis naturae* shed light on the *Testamentum* by revealing the identity of lunaria as quintessence of wine, proving that "vegetable mercury" was more than just a metaphor. Together, the components of the Lullian corpus seemed to embody the alchemical technique of *dispersio*, in which separate parts of a process were distributed across diverse texts in order to conceal the nature of the whole—a strategy illustrated by the aphorism "One book opens another" (*liber librum aperit*).

While cross-referencing between texts improved readers' prospects for reconstructing past practices, it also allowed them to reconstruct the history of alchemy itself. In addition to technical content, manuscript copies of alchemical works frequently offered snippets of bio-bibliographical detail, or hints that might be interpreted as such. For instance, one of the earliest surviving copies of the *Testamentum*, Oxford, Corpus Christi College, MS 244, includes a colophon that would provide grist for several later legends. It states that the book was written in London in 1332, at the Hospital of St. Katherine by the Tower, and dedicated to "King Edward of Woodstock . . .

into whose hands we send the present *Testament* for safekeeping."[101] The king of England in 1332 was, of course, Edward III.

For fifteenth-century readers the connection between Lull and Edward need not have seemed fantastical, since it was attested by apparently reliable documentary evidence. Conceivably the *Testamentum* was the work of a Catalan immigrant to England, who optimistically dedicated it to Edward in the hope of securing royal investment—a possibility supported by the fact that Edward was demonstrably interested in transmutation. However, since the earliest extant manuscript copy was produced in 1455, more than a century after the alleged date of composition, we cannot assume that either the colophon or the reference to Edward of Woodstock is original to the work.[102]

Nonetheless, for later readers the reference served as evidence for a relationship between England's warrior king and one of Europe's most renowned adepts. The relationship readily evoked analogies with Alexander the Great and his tutor, Aristotle, just as Raymond's treatises offered a plausible, practical interpretation of pseudo-Aristotle's animal, vegetable, and mineral stone. The connection with Lull also supplied a pious underpinning to the pursuit of alchemy, which set it apart from the reckless pursuit of gain associated with fraud and currency crime. The historical Lull devoted most of his life and scholarship to the task of converting Muslims to Christianity, a theme echoed in the *Codicillus*, a fourteenth-century treatise that may also have been written by the Magister Testamenti.[103] As the title hints, this work is framed as a codicil to the writer's supposed will (the *Testamentum*), which he claims to have composed at the behest of that "renowned King Edward, in whose aforesaid divine safe-keeping and protection the great memory or knowledge of philosophy will not benefit the wicked." Thanks to this royal support, the writer trusts that his knowledge will be employed not for personal gain, but "for the conversion of pagans and the preservation of the faith, on which the salvation of the faithful depends . . . that it may

101. *Testamentum*, 3:513–14: "Fecimus nostrum 'Testamentum' per voluntatem de A in insula Anglie in Ecclesia Sancte Katerine prope Londonum versus partem castri ante Tamisiam regnante iam Rege Eduuardo de [Woodstoc] per graciam Dei. In cuius manus mittimus in custodiam per voluntatem de A presens 'Testamentum' anno post Incarnacionem domini 1332 cum omnibus suis voluminibus, que nominata fuerunt in presenti 'Testamentum,' cum 'Cantilena,' que sequitur."

102. On the authenticity of the colophon, see Pereira, *Alchemical Corpus*, 3–4; Pereira, "English Physicians," 34.

103. Pereira, *Alchemical Corpus*, 10–11.

redound not only for the good of the body, but for the everlasting good of mind and soul."[104]

What present-day scholarship might simply interpret as the Magister's attempt to emulate authentically Lullian piety was taken by later readers as evidence that Raymond had, at some point, consented to make gold for Edward in order to further his Christian mission. For English readers who took this king to be their own Edward III, this evidence was highly suggestive. Everyone knew that Edward had financed not a Crusade but the invasion of France, an enterprise that would occupy him and his descendants for the next century.[105] All the elements were thus in place for an alchemical view of history in which Lullian transmutation allowed Edward to finance both the new golden coinage and the opening salvos of the Hundred Years' War. In the Lull legend, the economic importance of alchemy and the philosophical credentials of Raymond blended into a single narrative, in which alchemy and its learned practitioners contributed, in the most material way possible, to the making of England.

Neither alchemical practice nor alchemical history arose from a vacuum. In each case readers conflated textual clues with evidence from other sources, including their own experience of testing chemical substances. In the case of alchemical history, readers of the *Codicillus* who "knew" the work was written by Lull, but also knew that Edward did not fight a Crusade, finagled history in order to supply a plausible explanation for Raymond's relationship with the English king. This history, although fictive, is nonetheless compatible with Edward's known treatment of John of Walden—although, as befits a historical rather than a legendary personage, John was not detained because of his success at transmutation, but more likely on account of his failure. John's misadventure, imperfectly recalled, may even have contributed to the later story of Lull's English enterprise: a kernel of historical fact elaborated and improved in its mythic retelling.

The kind of substitution and splicing that generated alchemical histories also resulted in new visions of practice. As we will see in subsequent chap-

104. *Codicillus*, 5: "Ideo instante inclyto Rege Eduardo, in cuius custodia ac protectione diuina praeposita, ne tanta Philosophiae memoria vel cognitio impiis largiantur, hoc opus stricto ligamine commendamus, in conuersionem paganorum, in conseruationem fidei, a qua salus fidelium dependet, cui praesse dignoscitur, non solum vt corpori, sed vt menti et animae redundet in commodum sempiternum."

105. For a contextual study of the Hundred Years' War that includes both English and French perspectives, see Christopher Allmand, *The Hundred Years War: England and France at War c. 1300–c. 1450*, rev. ed. (1988; Cambridge: Cambridge University Press, 2001).

ters, the practical content of pseudo-Lullian alchemy, although revered, was not read at face value, but tested and reinterpreted in light of alchemists' own experience of chemical substances and operations, as well as their own methods of alchemical reading. These practitioners found ways of reading their own practical findings back into their sources, reconfirming the authority of past adepts even as they generated new insights about the nature of matter and its interactions. This way of doing philosophy served both the past and the present of alchemy. Bravura displays of alchemical reading demonstrated the subtlety of authorities and commentators: the first for having encoded their terms, and the second for having ingeniously revealed their true sense. This technique shows what was at stake in using one book to open another, although readers must have hoped that their books would also open practice. Correct decipherment satisfied more than the demands of logical consistency; it also resulted in a replicable procedure. From the fifteenth century onward, the exposition of obscure texts and the reconstruction of difficult practices would provide two sides of the same golden coin, as English practitioners embarked on the quest for alchemical patronage.

CHAPTER TWO

Medicine and Transmutation

Playne unto your Hyghnes it shall declared be.[1]

In 1415, the practice of artificially "multiplying" precious metals was illegal in England and Wales. The prohibition did not prevent one alchemist from setting up his furnace at the priory of Hatfield, near the town of Chelmsford in Essex, a subordinate priory of the great Benedictine abbey of St. Albans. The alchemist, William Morton, was not a member of the religious community, but a "wooleman" from Newcastle-upon-Tyne, who had established a collaborative relationship with the former prior. Morton and his business partners, who included both religious and laymen, used the priory not only as a site for alchemical practice, but also as a platform for more ambitious bids for patronage. Their goal was to make two alchemical powders, or elixirs: one for transmuting "red" metals, such as copper and brass, into gold; the other for turning "white" metals, including lead and tin, into silver.

The enterprise at Hatfield is very different from the image of alchemy encountered in the learned treatises discussed in the previous chapter. Such writings typically present alchemy as part of philosophy, requiring a sound grasp of theoretical principles as well as operative skill. Its masters are portrayed as pious men, drawn from the ranks of the learned rather than the craft guilds. Beyond this, philosophical treatises stress the need for secrecy in communicating the science. Alchemical wisdom, encapsulated in layers of "philosophical" obfuscation, is to be passed down from master to disciple, rather than hawked abroad to the ignorant or undeserving.

Morton and his collaborators do not fit this idealized philosophical tem-

1. George Ripley, *Epistle to Edward IV*, in *TCB*, 110.

plate. Despite the recent statute against multiplication, they did not keep their practice secret, asserting "to various of the king's people" that their powder could produce real gold and silver—that is to say, bullion of a quality suitable for coinage.[2] Probably they were angling for serious patronage, since they presented their work to one of the greatest landowners of Essex: Joan de Bohun, Countess of Hereford, maternal grandmother to King Henry V. In the end, Morton's most successful suit was a plea addressed to Henry himself after the collaborators ended up before the King's Bench in 1418, indicted under the recent statute. The king, then campaigning in France, wrote from the recently captured city of Bayeux to grant him a pardon, while the prior, John Bepsete, also escaped after the court deemed his participation insufficient to merit punishment.

As cases like Morton's reveal, the practice of alchemy seldom lived up to philosophical hyperbole. Yet in other ways the Hatfield enterprise was entirely typical of pre-Reformation English alchemical practice. Alchemy was not the province of either the secular or the religious sphere, but spilled across these permeable boundaries to create a mixed economy of practice featuring a variety of sites and actors. Although its practitioners saw it as the means to many ends, including medicine, alchemy was also perceived as a royal art whose aspirations were particularly suited to the princely business of striking coin and financing wars. As the century progressed, alchemy appeared ever more frequently in patronage suits and legal records connected with English monarchs and their councillors. Morton was neither the first nor the last English alchemist to seek royal support for his practice; his error lay in failing to secure royal permission first.

With the need for bullion heightened by war, political instability, and economic hardship, the fifteenth century emerges as an extraordinary period of development and consolidation in English alchemical practice. Yet only one alchemical treatise survives from this century that is explicitly framed as a petition to an English patron—and it addresses not the king, but one of his senior bishops. The text is the *Medulla alchimiae*, written in 1476 by George Ripley, an Augustine canon regular who, like Morton, hailed from northeast England.[3] A century after Ripley's death, the *Medulla* still reigned

---

2. National Archives, Coram Rege Roll, 6 Henry V, Trinity Term, rot. 18d (KB 27/629). The case is summarized by R. C. Fowler, "Alchemy in Essex," in *The Essex Review: An Illustrated Quarterly Record of Everything of Permanent Interest in the County*, vol. 16, ed. Edward A. Fitch and C. Fell Smith (Colchester: Behnam & Co., 1907), 158–59.

3. On Ripley, see Jennifer M. Rampling, "Establishing the Canon: George Ripley and His Alchemical Sources," *Ambix* 55 (2008): 189–208; Rampling, "The Catalogue of the Ripley Corpus:

as one of England's most influential and widely cited treatises, particularly in its popular English translation, the *Marrow of Alchemy*. Cited in later treatises, recipes, and hundreds of marginal notes, the *Medulla* provided the interpretative lens for a generation of Elizabethan alchemists.

Yet it is not Ripley's best-known work. That honor goes to his celebrated Middle English poem, the *Compound of Alchemy* (1471), which may have been presented to Edward IV, together with a separate poem, the *Epistle to Edward IV*.[4] Unfortunately no accompanying petition survives. Thomas Norton's *Ordinal of Alchemy* (1477), the only English work of the fifteenth century to rival Ripley's *Compound* in posthumous celebrity, was probably also intended for presentation, but once again the vital documentary context is lacking. These four works—one Latin treatise and three English poems—constitute the entirety of surviving "philosophical" material that we can say, with some confidence, was written to secure patronage in fifteenth-century England. All four were produced within the space of seven years, and all three of the Ripleian works double as commentaries on the alchemy of pseudo-Raymond Lull. Although later viewed as singular and canonical, these works thus capture a particular moment in English alchemical history, when high-ranking patrons, both temporal and spiritual, were presumed to be receptive to alchemical lore.

Two of the major themes of this study—alchemical patronage and pseudo-Lullian alchemy—are united in the corpus of works by and attributed to Ripley. Ripley's own early reputation rested largely on his success as an expositor of "Raymond," whose reading of the pseudo-Lullian animal, vegetable, and mineral stones would became a touchstone of English alchemical practice throughout the next century. But when we delve into his conclusions, a new picture emerges. Ripley's alchemy is less a faithful commentary on Raymond than it is a synthesis of authorities, achieved through exegetical manipulations and resulting in coherent, practical outcomes, tested against his own experience. Through Ripley's attempt to make sense of the many-stranded Lullian corpus, Lullian alchemy was itself transformed, in a process of mutually reinforcing textual and practical innovation that I term "practi-

Alchemical Writings Attributed to George Ripley (d. *ca.* 1490)," *Ambix* 57 (2010): 125–201; hereafter CRC.

4. First published in English as George Ripley, *The Compound of Alchymy . . . Divided into Twelue Gates*, ed. Raph Rabbards (London, 1591); reproduced with some spelling emendations in *George Ripley's Compound of Alchemy (1591)*, ed. Stanton J. Linden (Aldershot: Ashgate, 2001). Ripley's collected works (several of which are spurious) were published in his *Opera omnia chemica*, ed. Ludwig Combach (Kassel, 1649); hereafter *OOC*.

cal exegesis." Reconstructing this process allows us to trace the journey by which Ripley, an insular commentator on a continental European tradition, was eventually reinvented as the exemplar of a distinctively "English" way of doing alchemy.

A LICENSE TO PRACTICE

The success of pseudo-Lullian alchemy, particularly as represented in influential commentaries like Ripley's *Medulla*, reflects changing priorities among readers, practitioners, and patrons alike during the fifteenth century. Philosophical treatises continue to dwell upon the language of mercury and "one thing," and to critique the use of diverse animal, vegetable, and mineral ingredients. Yet they also reveal widespread interest in the medicinal aspects of alchemical practice, as exemplified by the "quintessential" writings of two Franciscans, John of Rupescissa and Raymond, which seem to have been circulating in England early in the fifteenth century.[5] As Peter Jones has shown, friars provided an important early vector for disseminating these texts by incorporating medical recipes gleaned from *De consideratione* and *De secretis naturae* into their own medical compendia.[6] These texts complemented existing medical trends: for instance, the surgeon John of Arderne (1307–1392) was already recommending distilled oils and waters in the 1370s.[7] In the following century, John Argentein (ca. 1443–1508), royal physician and pro-

5. Although the historical Lull was a Franciscan tertiary, the identity of the pseudo-Lull is of course unclear. On Franciscan interest in alchemy, see Chiara Crisciani, "Alchimia e potere: Presenze francescane (secoli XIII–XIV)," in *I Francescani e la politica: Atti del convegno internazionale di studio, Palermo 3–7 dicembre 2002*, ed. Alessandro Musco (Palermo: Biblioteca Francescana—Officina di Studi Medievali, 2007), 223–35; Michela Pereira, "I francescani e l'alchimia," *Convivium Assisiense* 10 (2008): 117–57; DeVun, *Prophecy, Alchemy*; Zachary A. Matus, *Franciscans and the Elixir of Life: Religion and Science in the Later Middle Ages* (Philadelphia: University of Pennsylvania Press, 2017). On medical distillation practices in friaries, particularly in Italy, see Angela Montford, *Health, Sickness, Medicine, and the Friars in the Thirteenth and Fourteenth Centuries* (Aldershot: Ashgate, 2004).

6. Peter Murray Jones, "Mediating Collective Experience: The *Tabula Medicine* (1416–1425) as a Handbook for Medical Practice," in *Between Text and Patient: The Medical Enterprise in Medieval and Early Modern Europe*, ed. Florence Eliza Glaze and Brian K. Nance, Micrologus' Library 39 (Florence: SISMEL, 2011), 279–307; Jones, "The Survival of the *Frater Medicus*? English Friars and Alchemy, *ca.* 1370–*ca.* 1425," *Ambix* 65 (2018): 232–49; Jones, "Alchemical Remedies."

7. See, for instance, John of Arderne, *Liber medicinarum sive receptorum liber medicinalium*, in Glasgow University Library, MS Hunter 251, fol. 16r; Peter Murray Jones, "Four Middle English Translations of John of Arderne," in *Latin and Vernacular: Studies in Late-Medieval Texts and Manuscripts*, ed. A. J. Minnis (Cambridge: D. S. Brewer, 1989), 61–89.

vost of King's College, Cambridge, included distilled remedies (one involving the distillation of blood) in his own medical compendium.[8]

Distillation and distilled products, including liqueurs and medicinal
waters, also came to occupy a shared space between alchemy and medicine
as traditionally defined. The evolving character of distillation as a practice
distinct from either appears from Henry VI's appointment of one Robert
Broke as "master of the kyngis styllatorys and maker of hys excellent wateris"
between 1432 and 1455, as identified by Linda Voigts.[9] Broke may have been
a vintner by training, and there is no evidence to suggest that he was interested in alchemy as distinct from his own distillation practice. Here, developments in the technology of distillation provided a mean of sorts between
medical and alchemical practice, creating space for a new kind of practitioner. A proliferation of recipes for distilled waters in English manuscript
collections that are not otherwise obviously "alchemical" testifies to the
spread of these pharmacological practices, which, since they did not concern the multiplication of precious metals, did not necessarily fall within the
compass of Henry IV's statute.[10]

While distillers pursued their quasi-medical activities outside these legal
bounds, economic concerns prompted kings to formally examine the potential of transmutation. Henry IV's grandson Henry VI issued licenses to practice alchemy and instituted commissions to investigate it.[11] On 18 August
1452, the king instructed that "multipliers" should be apprehended, and
their materials and instruments examined—an indication that chrysopoetic alchemy was still being illegally practiced, but also evidence for the
king's own interest in the outcomes of this activity.[12] Such interest could be
exploited, and the danger of felony averted, if practitioners obtained a royal
license to practice. The licenses issued to merchants, physicians, and gentlemen during the reigns of Henry VI and his Yorkist successor, Edward IV,
testify to widespread interest in alchemy in England from the 1450s onward,
unconfined by region or social class.

8. On Argentein's interest in alchemical distillation, of blood in particular, see Jones, "Alchemical Remedies." On Argentein, see also L. D. Riehl, "John Argentein and Learning in Medieval Cambridge," *Humanistica Lovaniensa* 33 (1984): 71–85.

9. Voigts, "Master of the King's Stillatories," 235.

10. See Voigts, "Master of the King's Stillatories," 250–52, for a representative selection.

11. On the composition of these commissions, see Singer, 3:788–91; Wendy J. Turner, "The
Legal Regulation and Licensing of Alchemy in Late Medieval England," in *Law and Magic: A Collection of Essays*, ed. Christine A. Corcos (Durham, NC: Carolina Academic Press, 2010), 209–25,
on 218–24.

12. National Archives, Patent Roll, 30 Henry VI, pt. 2, m.9d; cited in Singer, 3:787–88.

While the interest is evident, the alchemists' proposed methods are not. Under Henry VI licenses took a standard form, which granted the petitioners protection from legal harassment in order to focus on the "transubstantiation" of metals—but unfortunately without providing details of the intended practice. If petitions for royal licenses were once associated with treatises that outlined the intended approach, these have long since become detached.

Some clues survive in legal records of a slightly earlier period, which usually refer to vitriol and a variety of salts. In 1374, Willelmus de Brumleye worked on "gold and silver and other medicines, to wit sal armoniak, vitriol and solermonik."[13] The records of William Morton's case also preserve a list of his materials, revealing a practice that was certainly not restricted to "one thing" alone. As well as mercury and charcoal powder, Morton employed vermillion, realgar, verdigris, sal niter, sal alkali (a salt produced from ashes, used in glassmaking), "sawundiuer" (sandiver, a salt skimmed from the surface of molten glass), vitriol (metal sulphates), arsenic, a substance called "sakeon," and "various other things and powders unknown to the jurors."[14] Many of the ingredients on this list were staples of late medieval craft practice, including the vitriol and sal niter used to make mineral acids and mercury sublimate. While Morton proposed to use these ingredients "to make the said powder, called Elixir," his procedures actually created "a black matter burnt and congealed in a round glass."

For the most part, the licenses tell us nothing about the substance of practice, although they do suggest that petitioners were concerned to present themselves as philosophers. Royal licenses abandon the language of "alconomie," multiplication, and in many cases even alchemy, in favor of "philosophy." The language of debasement and diminution associated with illicit multiplication is replaced by evocative expressions with theological overtones: "transubstantiation" and "translation," terms that imply profound and fundamental change.

Thus the earliest license recorded in the Patent Rolls, awarded to John

13. Coram Rege Roll, No. 448, 47 Edward III, Hilary Term, Rex m.15.d: "cum arte Alconomie de auro et argento et aliis medicinis, videlicet sal armoniak, vitrio, et solermonik." The latter term sounds like a second reference to sal ammoniac; it could also mean "bol ammoniac." The clerk may have been confused by the profusion of substances. The term "medicines" is frequently used in works of transmutational alchemy to refer to substances used in preparing the stone.

14. KB. 27/629: "Ibidem laborauit & operabatur tam in arte ignis qm^co alijs diuersis rebus & pulueribus videlicet Mercurio puluere carbonum sakeon vermelion Resalger vertegrees sal niter sal alkale sawundiner vitriall arsenyk & alijs diuersis rebus & pulueribus dictis Iur' ignotum ad predictum puluerem vocatur Elixer conficiendi."

Cobbe on 6 July 1444, allows him "by the art of philosophy, to work upon certain metals to translate imperfect metals from their own kind, and then to transubstantiate them by the said art into perfect gold or silver."[15] Alchemy is further defined as the "art *or science* of philosophy" in the otherwise almost identical license granted to two Lancashire knights, Sir Edmund de Trafford and Sir Thomas Ashton, on 7 April 1446.[16] This document offers no clue as to how Trafford and Ashton—two friends whose families subsequently intermarried—became interested in alchemy. Since the license is extended to include the petitioner's "servants," it is possible that these gentlemen had forged a collaboration with practitioners whose names are now lost, and that their own petition thus represented only the upper rung of a ladder of patronage whose feet rested lower down the social hierarchy.[17]

As the formula makes clear, royal interest focused on the fitness of the transmuted metals for minting coin, and hence on their ability to withstand any trial to which "natural" gold and silver might be subjected. All of these licentiates are held to the same standard:

[to] translate imperfect metals from their own kind, and then to transubstantiate them by the said art or science, as they say, into perfect gold or silver, unto all manner of proofs and trials, to be expected and endured, as any gold or silver growing in any mine.[18]

This focus of course reflects the purpose of the original statute, as a means of regulating the multiplication of metal for coinage. Yet the standard phraseology also smooths away any further details concerning the aims of individ-

15. National Archives, Patent Roll, 22 Henry VI, pt. 2, m.9 (C66/458). On Cobbe and other licentiates under Henry VI, see Turner, "Legal Regulation and Licensing of Alchemy," 214–15.

16. National Archives, Patent Roll, 24 Henry VI, pt. 2, m.14 (C66/462); my emphasis.

17. Ibid. English translation in Thomas Fuller, *The History of the Worthies of England: A New Edition*, ed. P. Austin Nuttall (London: Thomas Tegg, 1840), 2:216: "We, considering the premises, willing to know the conclusion of the said working or science, of our special grace have granted and given leave to the same Edmund and Thomas, and to their servants, that they may work and try the aforesaid art and science, lawfully and freely, without any hindrance of ours, or of our officers whatsoever."

18. Fuller, *History of the Worthies of England*, 2:216. Almost identical licenses were issued to William Hurteles, Alexander Worsley, Thomas Bolton, and George Horneby (occupations unspecified) on 4 July 1446; to "Robert Bolton of London Gentilman" on 15 September 1447; to John Mistelden and his three servants on 30 April 1452; and to William Sauvage, Hugh Hurdelston, and Henry Hyne, with their three servants, on 3 September 1460. See, respectively, National Archives, Patent Roll, 24 Henry VI, m.5 (C66/475); Pell Records, 27 Henry VI; Patent Roll, 30 Henry VI, pt. 2, m.27 (C66/475); Patent Roll, 39 Henry VI, m.23 (C66/490); all cited in Singer, 3:784–92.

ual petitioners or their particular attitudes toward practice and *prima materia*, leaving just the bare fact of their interest and involvement.

The grants and licenses issued by Edward IV reveal the same interest in transmutation, although during his reign the standard formula was dropped, allowing glimpses of specific projects. In some cases, the wording—probably supplied by the petitioners themselves—suggests a concern with the status of alchemy as a branch of natural knowledge, or *scientia philosophiae*. Thus in 1463, Sir Henry de Grey, Baron Grey of Codnor (ca. 1435–1496), who later served a short and ill-fated tour as Edward IV's lord deputy of Ireland, was granted the authority either to practice alchemy himself or to oversee its practice, a development that would have enabled Grey to establish himself as a patron in his own right. The grant allowed Grey to pursue transmutation according to the "knowledge of philosophy," provided that he did so at his own expense and reported any positive progress to the king.[19]

Probably the wording reflects the social status of the petitioner, since Richard Carter's "full license," granted in 1468, presents alchemy as merely an "art or occupation," although it does permit him to practice on all species of metals and minerals.[20] Unusually, Carter's practice was to be conducted at the king's own manor of Woodstock, possibly as a precaution intended to facilitate supervision of the practice, although Carter may himself have requested this provision in order to secure adequate working space. Premises suitable for elaborate chemical practices, particularly those involving multiple furnaces and complex distillation equipment, were not easily secured; indeed, Carter's contemporary Thomas Norton counted "a perfite worchynge place" among the conditions necessary for unhindered practice.[21] The same need may have driven Morton to seek out space at Hatfield Priory, or William of Brumleye to secure lodging with the prior of Harmondsworth in the previous century. Working space was still an issue by 1565, when Thomas Charnock petitioned Elizabeth I for permission to set up his own practice in the Tower of London, partly to allow for uninterrupted practice, but also that "the Queenes maiestie and her honourable counsell showlde have [th]e more assuerance for the accomplyshinge off my promes."[22]

Whatever the outcome of Carter's enterprise at Woodstock, it seems not to have dampened Edward's interest in transmutation as a possible remedy

---

19. National Archives, Patent Roll, 3 Edward IV, pt. 2, m.17 (C 66/506).
20. National Archives, Patent Roll, 8 Edward IV, pt. 2, m.14 (C66/522).
21. Norton, *Ordinal*, 84 (ll. 2701–2).
22. Lansdowne 703, fol. 11r.

for his economic problems, nor his willingness to endorse the philosophical pretensions of practicing alchemists. In 1476 he licensed the first explicitly mercurialist project, permitting David Beaupre and John Marchaunt to generate gold and silver from mercury, through the "artificial, natural knowledge of philosophy."[23] And in 1477/8 he extended his protection to John French, an alchemist of Coventry, who was allowed to "practise a true and a profitable conclusion in the Cunnying of transmutacion of metals," for the king's "profyte and pleasure."[24] These records testify to the Crown's ongoing need to secure bullion—still a serious concern given the steady erosion of English possessions in France under Henry VI, and ongoing political instability at home. With one exception, Lancastrian and Yorkist licenses refer exclusively to the transmutation of metals rather than medicine.

The exception is the outcome of a famous petition made to Henry VI on 31 May 1456 by a surprisingly mixed group of twelve men, comprising physicians, clerics, and London guildsmen.[25] Among them were John Kirkeby, the king's chaplain, and three doctors who had recently attended the king during his debilitating "lethargy": Gilbert Kymer, John Faceby (or Fauceby), and William Hatclyff.[26] The petitioners supplied the king with a Latin draft of their own suggested wording, most of which was retained in Henry's letters patent. Rather than identifying the art as alchemy, the petitioners record their hopes of extracting "glorious and noteworthy medicines" from various animal, vegetable, and mineral ingredients, including "wine, precious stones, oils, vegetables, animals, metals, and certain minerals."[27]

The license is unique both in alluding to nonmetallic materials and in subordinating transmutation to medicinal outcomes: healing sickness, prolonging life, restoring health and vigor, preserving memory and intellect, curing wounds, and protecting against poison. Indeed, the goal of transmutation (which made a license necessary in the first place) is mentioned almost in passing: "also many other benefits, most useful to us and the well-being of our kingdom . . . such as the transmutation of metals into true gold and very fine silver."[28] While this diversity of outcomes and ingredients was, by this

23. National Archives, Patent Roll, 16 Edward IV, pt. 1, m.20 (C66/538).

24. Corporation of Coventry, Leet Book, 6 January 1478; cited in Singer, 3:793–94.

25. On the petitioners, see Geoghegan, "Licence of Henry VI," 11–13; Pereira, "Mater medicinarum," 27, 41–42nn6–9; Voigts, "Master of the King's Stillatories."

26. On Kymer's treatment of Henry VI, see Faye Getz, "Kymer, Gilbert (d. 1463)," ODNB; Robert Ralley, "The Clerical Physician in Late Medieval England" (PhD diss., University of Cambridge, 2005), chap 2.

27. Translation in Geoghegan, "Licence of Henry VI," 15.

28. Geoghegan, "Licence of Henry VI," 16.

time, a standard feature of treatises and recipe collections, particularly those associated with pseudo-Lullian works, it is highly unusual to find medicinal goals privileged in a royal petition of this early date. In fact, their appearance here points to the growing credibility of alchemical medicine. Among the petitioning physicians, Kymer is known to have used distilled remedies in his own medical practice, even recommending "lunaria" and the Rupescissan quintessence of wine in his *Dietarium de sanitatis custodia* (Dietary on Guarding Health), a work on medical regimen written for Henry VI's uncle, Duke Humfrey of Gloucester, more than three decades earlier, in 1424.[29]

The license is also unusual in explicitly linking the practice of alchemy to the interpretation and extraction of alchemical knowledge from textual sources, which almost certainly included pseudo-Lullian works. The wording mimics the rhetoric of medieval alchemical treatises by stressing the difficulty of interpreting and implementing ancient wisdom, while rejecting false practices. The petitioners' stated aim is to recover—that is, to *replicate*—the marvelous alchemical medicines discovered by "the sages and most famous philosophers of ancient times." Such recovery is not straightforward, since the authorities recorded their secrets "in their writings and books under signs and symbols."[30] Overcoming the "arduous difficulties" of translating enigmatic words into successful practice therefore requires a special kind of practitioner: "talented men, sufficiently learned in natural sciences, and willing and disposed to practise the said medicines; men who fear God, seek truth, and hate deceitful work and the false tincturing of metals." This combination of devoutness, trustworthiness, book learning, and experience in applying medicines maps conveniently onto the character of the petitioners, six of whom were physicians. An obvious distinction is drawn between the philosophical credentials of the petitioners and the deceitfulness of multipliers who merely alter the outward appearance of metals.

The petitioners also employ another method characteristic of pseudo-Lullian alchemy to distance their philosophical alchemy from multiplying: a pious emphasis on health over wealth. Following John of Rupescissa, the Raymond of *De secretis naturae* had stressed the value of the "Greater Work"

---

29. Jones, "Alchemical Remedies." As Jones notes, Kymer quotes from *De retardacione senectutis*, which is attributed to Roger Bacon, suggesting that he was already interested in alchemical methods for prolonging life during the 1420s. The *Dietarium* may itself have been modeled on the pseudo-Aristotelian *Secretum secretorum*; Faye Getz, *Medicine in the English Middle Ages* (Princeton: Princeton University Press, 1998), 86. Kymer is the only contemporary alchemist named by Thomas Norton, who rates his medical knowledge over his alchemical expertise; Norton, *Ordinal*, 50 (ll. 1559–62).

30. Geoghegan, "Licence of Henry VI," 15.

of medicine over the "Lesser Work" of transmutation. This priority recurs in the petitioners' praise of one remedy: "a most precious medicine, which some have called the mother of philosophers and Empress of medicines (*philosophorum matrem et imperatricem medicinarum*)."[31] As Michela Pereira has noted, this wording evokes the "mother of medicines" (*mater medicinarum*) described in the *Testamentum*, suggesting that the petitioners were already familiar with this important work. Pereira has also convincingly argued that the petition is connected to a magnificent compendium of pseudo-Lullian works, MS 244, preserved in Corpus Christi College Library, Oxford. This substantial manuscript, which includes both Latin and Catalan versions of the *Testamentum*, was compiled by the royal chaplain John Kirkeby in 1455, the year before he and his eleven associates presented their petition to the king.[32] Although several other licenses refer to alchemy as a "philosophy," the 1456 petition is thus the only one plausibly linked to a specific authority or practical tradition, suggesting that the petitioners hoped to produce distilled remedies of the type popularized by John of Rupescissa and Raymond, and already employed by Kymer.

As Kirkeby and his colleagues were well aware, the possession of such authoritative literature was meaningless without the ability to both interpret and act upon its contents. In this respect, the diverse origins and practical underpinnings of the pseudo-Lullian corpus posed particular challenges for alchemical exegetes. If two conflicting texts are regarded as stemming from the pen of a single author, Raymond, then it falls to Lull's commentators to disentangle the seemingly contradictory doctrines of their *auctor*. More is at stake here than the philosophical coherence of the text: if the underlying process is incorrectly read, then the work itself will fail. For reader-practitioners keen to accurately reproduce the *experimenta* of their authorities, it is therefore essential to lay bare the intended sense of a text.[33] One must identify not only ingredients, but also their proportions, method of

31. Ibid.; my interpolations. Kymer quotes the same passage in the *Dietarium*; Jones, "Alchemical Remedies."

32. Pereira has speculated that Kirkeby's encounter with the *Testamentum* might itself have provided the catalyst for the petition, and that this handsome volume may have been planned as a presentation copy for the king; Pereira, "Mater Medicinarum," 35–36. On Kirkeby, see also Linda Ehrsam Voigts, "The 'Sloane Group': Related Scientific and Medical Manuscripts from the Fifteenth Century in the Sloane Collection," *British Library Journal* 16 (1990): 26–57, on 34–37.

33. The medieval term *experimenta*, often used to denote practical recipes and descriptions of procedures, is closer to the modern sense of "experience" than of "experiment." On *experimenta*, see Katherine Park, "Observation in the Margins, 500–1500," in *Histories of Scientific Observation*, ed. Lorraine Daston and Elizabeth Lunbeck (Chicago: University of Chicago Press, 2011), 15–44.

preparation, and a hundred other details that are all too often glossed over, omitted, or disguised in written accounts.

Operative success is thus framed as an extension of textual engagement. For practitioners hopeful of securing patronage, petitions and presentation volumes like CCC 244 offered an opportunity to demonstrate that technical information could indeed be extracted and mastered from authoritative texts. The 1456 petitioners worked both the exegetical and the experimental dimensions of their practice into the license itself, which directs them "to inquire, investigate, begin, pursue, complete and test absolutely, all and singular, the said medicines, according to their science and discretion and the doctrines and writings of the sages of old."[34]

We cannot know exactly how the petitioners—only three of whom, including Kirkeby, were actually granted licenses—intended to address the practical challenges of pseudo-Lullian alchemy. The difficulty of prising technical information from the core Lullian treatises helps to explain the sheer scale of the corpus, bulked out by dozens of commentaries in which successive "Raymonds" seek to provide an *Elucidatio* (Elucidation), *Apertorium* (Opening), or *Clavicula* (Little Key) to unlock the core texts and hidden processes of the canon. At one extreme, these provide little more than textual concordances between earlier writings. At the other, alchemists used pseudo-Lullian doctrines merely as a framework for practices and philosophies imported from other sources and traditions. While such adaptations are sometimes obvious, they are often harder to spot, particularly in cases where a commentator has elected to cloak his own innovations under the mantle of Lull's authority.

In England, no commentator recrafted his source material with the aplomb of George Ripley, Canon of Bridlington. The artfulness of Ripley's writings, in which Lullian theory is used to underpin a practical program, shows how the techniques of alchemical reading could be deployed to attract interest from potential sponsors. Although there is no record of Ripley ever applying for or receiving a license to practice alchemy, his best-known works, the *Compound of Alchemy* and *Medulla alchimiae*, were written with a view to attracting patronage. Both are thoroughly grounded in the authority of past adepts, particularly Raymond, as well as a lesser-known master, Guido de Montanor. Their content also transcends gold-making to include the confection of alchemical medicines and compound waters. Beyond this, they show that Ripley had a sharp eye for a logical contradiction or practi-

---

34. Geoghegan, "Licence of Henry VI," 16.

cal faux pas. Indeed, his talent for deciphering and reconciling intractable authorities would help establish his reputation as the greatest of all Lullian expositors, to the extent that his own commentator, Samuel Norton, would one day hail him as the "Raymond of the English." In order to map the relationship between pseudo-Lullian doctrines, patronage, and English practice, it is therefore fitting that we take Ripley as our guide.

THE MAKING OF AN ALCHEMIST

We know surprisingly little about the man who would later be hailed across Europe as the doyen of English alchemy. George Ripley lives on primarily in his own writings, of which the best known is the *Compound of Alchemy*, a Middle English poem of 1,976 lines, composed in rhyme royal. Owing to the exigencies of manuscript survival, this is also one of very few works attributed to Ripley that may be convincingly dated to his own lifetime, thanks to its survival in three late fifteenth-century manuscripts.[35] After the *Compound*, the work most reliably attributed to Ripley is the *Medulla alchimiae*, a Latin treatise dated by colophon to 1476.[36] This work was translated into English in 1552 by the clerk David Whitehead, and thereafter seems to have circulated more often in English than in the original Latin, as the *Marrow* or *Mary of Alchemy*.[37]

According to his own testimony in the *Compound*, Ripley was a canon regular of the Augustinian priory of St. Mary in Bridlington, a coastal town in east Yorkshire. Thanks to a papal letter of 1458/9, we know that one "George Ryphey" indeed served as a canon at the priory, and that he probably traveled in pursuit of learning, for the letter gave Ryphey the right to

---

35. CCC MS 172; Ashmole 1486, pt. 3; and Trinity O.5.31 (CRC 9.34, 9.30, and 9.5, respectively). References to the "Canon de brydlingtone" are also found in a fifteenth-century manuscript, associated with verses from the *Compound*: Sloane 3579, fols. 4r, 11r, 18r-v, 20r, 39v-40r (CRC 9.xx).

36. Trinity R.14.58, pt. 3, fol. 6r (hereafter *Medulla*): "Explicit tractatus Medulla Alkimie dictus per G.R. compilatus Anno domini 1476." The *Medulla* is CRC 16.

37. Sloane 3667 (after 1572/3), fol. 104v: "Here endyth the treates called the Marye of Alkamye compiled by gorge Rypley the yere of our Lord 1476 And turned into Englyse by Mr Davye Whithede clarke. anno 1552." Another Elizabethan "Marye" provides only initials: "translated into English D. W. anno 1552" (Ashmole 1480, pt. 3, fol. 15v). The translator is perhaps the "Mystar Whithed" described as an alchemist by John Stow, who may in turn correspond to the Protestant divine David Whitehead (1492–1571); "Introduction: Documents Illustrative of Stow's Life," *A Survey of London*, by John Stow, reprinted from the text of 1603 (1908), XLVIII-LXVII; *British History Online*, http://www.british-history.ac.uk/report.aspx?compid=60007&strquery=alchemy (accessed 13 May 2009).

leave Bridlington "and to dwell for seven years in a university, even without the realm of England, and study theology."[38] The license granted the canon *in commendam* benefices to support him during his absence from Bridlington, on the proviso that he return to his priory at the end of the period. This provision accords with Ripley's own claim to have learned his alchemy abroad. Yet it is unclear whether Ripley ever returned to Bridlington after his travels. The autobiographical details at the start of the *Compound* suggest that most of his knowledge was acquired in Italy rather than Yorkshire:

> Here foloweth the Compend of Alkymye
> made by a channon of Brydlyngton
> affter hys lernyng in Italye
> at Exnyng for tyme he there dyd wonn
> in whych ben declared openly
> both the secretys of mone and sonn
> how they her kynd had to multyply
> in one body together must wonn
> whych channon syr George Reppley hyght
> exempt from clausturall observance
> for whom ye pray both day and nyght
> syth he dyd labor yow to advaunce.[39]

The topos of travel in search of knowledge is a common one in alchemical writings, and Ripley's claim to have studied in Italy, the center of pseudo-Lullian practice, can only have enhanced his own authority. Yet the claim is also consistent with the known dispensation, and his observation later in the poem that "I cowde never fynde hym wythin *Englond*" who could teach the art of fermentation.[40] The preface to the *Medulla* is more explicit still, referring to the learning that Ripley acquired during his travels in Italy and the surrounding regions over the space of nine years.[41] It may have been during

38. *Calendar of Entries in the Papal Registers Relating to Great Britain and Ireland: Papal Letters*, vol. 11, *1455–1464*, prepared by J. A. Twenlow (London: Her Majesty's Stationery Office, 1893), 530–31.

39. "Titulus operis," *Compound of Alchemy*, in CCC 172, fol. 12v. While it has been suggested that these verses were added to the manuscript in the sixteenth century (Linden, *George Ripley's Compound of Alchymy*, 106n20), they are in fact written in the original, fifteenth-century hand.

40. Ripley, "Fermentation," in *TCB*, 177.

41. *Medulla*, fol. 1v: "Tractaturus de secretis Alkimice que progressu et indagatu annorum nouem in Italia circumvicinisque ipsius partibus nanciscebar medullam quodammodo nature ipsa grossiori feculentiorique substancia carnium resecata ex ipsius interioribus secretioribus."

these travels that he met his master, the unknown "Doctour" referred to in the *Compound*.[42]

The reference to a sojourn in "Exnyng" evidently perplexed later readers, who attempted to identify it with Oxford, or a location in Italy.[43] A more likely candidate is the fenland parish of Exning in Suffolk, a few miles northeast of Cambridge, and a flourishing village during Ripley's time. The exemption from claustral observance might conceivably have been obtained in Rome, but the extension of Ripley's tenure outside his priory is not, in any case, implausible given that regular canons were not bound to enclosed life within a convent, and were in consequence eligible to manage parishes in the absence of an incumbent priest.[44] One alchemical text attributed to Ripley even gives his occupation as that of farmer and curate, suggesting that he did indeed have management of a parish—possibly Flixborough in Lincolnshire, a living held by his kinsfolk, the Willoughby family, from the bishop of Durham.[45] While this source should be treated with circumspection, the occupation is not incompatible with Ripley's claimed exemption.[46] Priestly activity also explains the canon's elevation to "Sir George Ripley" by the later sixteenth century: the honorific "Sir" denoting the profession of priest.[47]

42. Ripley, "Calcination," in *TCB*, 131: "The same my *Doctour* to me did shew."

43. Tanner, *Bibliotheca Britannico-Hibernica*, 633: "Post reditum Oxoniae studuit, et scripsit *Compendium alchymiae*." CCC 172, fol. 12v: "Ixninge in Italy" (marginal note in later hand). The Oxford connection is dismissed in A. B. Emden, *A Biographical Register of the University of Oxford to A.D. 1500*, vol. 3 (Oxford: Clarendon Press, 1959), 1577. The town is given as "Ixninge" in Raph Rabbards's 1591 edition, and "Yxning" in *TCB*.

44. Canons from Bridlington had been serving vacant churches since the late twelfth century; one, for instance, served as vicar of Grinton in northwest England. J. C. Dickinson, *The Origins of the Austin Canons and Their Introduction into England* (London: S.P.C.K., 1950), 228–40. On the distinction between monks and canons—clerks living according to rule (*regulariter viventes*)—see Dickinson, 197–223.

45. "Breviation," CRC 4. Variant spellings of the parish are included in the CRC—for example, "farmour & Curate of Flixburch Churche" (Sloane 83, fol. 2r; CRC 4.4). On the Willoughby family's relationship to Flixborough (also spelled Flyxburgh and Flixburrow), see Sir William Dugdale, *The Baronage of England, or, An Historical Account of the Lives and Most Memorable Actions of our English Nobility in the Saxons time to the Norman Conquest* . . . (London: Thomas Newcomb, for Abel Roper, John Martin, and Henry Herringman, 1675–76), 83–84.

46. *Papal Letters*, 11:530–31: "To receive and retain in *commendam* during the said period a benefice with or without cure wont to be held by secular clerks, even if it be in a parish church or its perpetual vicarage or a chantry, and be of lay patronage . . . and to resign such *commenda* as often as he pleases, and receive and retain another benefice in *commendam*, and accept and hold a stipend from any regular place or monastery."

47. Even Fuller was uncertain on this point. Fuller, *History of the Worthies of England*, 363: "Sir George Ripley (whether Knight or Priest, not so soon decided)." Sandys, amending Fuller's entry, put the canon back in his place: "never more then Sir Priest, and Canon of Bridlington"; George Sandys, *Anglorum Speculum, or, The Worthies of England in Church and State alphabetically digested*

It is in the *Medulla alchimiae* that we find the most poignant clues to Ripley's social status and personal aspirations. One of just two fifteenth-century English treatises to have been clearly written for a specific patron (the other is the *Epistle to Edward IV*), it is addressed to a high-ranking prelate, Ripley's "honored father and lord, the Lord Bishop."[48] The patronage relationship is emphasized from the very first line of the Latin prefatory poem, here in a sixteenth-century translation by William Bolisse, or Bolles:

> Right noble lord, and prelate deare.
> vouchsafe of me these verses take
> which I present vnto you heere
> that mention of the stone doth make.[49]

Ripley continues to address the bishop directly throughout the treatise, never more purposefully than in the epilogue that provides the meat of his petition. Whitehead's English translation captures the emotional fervency of this appeal, in which Ripley approaches the bishop with fitting humility, as "a poore servant of Christ, an humble chanon, taken to the banquets of phi*losopher*s, not by deserving, but by the gift of god."[50] He offers an account of his family's misfortunes, and a movingly worded plea to be received into a religious house in the bishop's diocese for the remainder of his life, "notwithstanding that I haue a licence to liue without the cloister." This desire to retreat once more from the world has its roots in the brutal northern aftermath of the Wars of the Roses:

> [M]y carnall parents being dead, & also my kinsfolk being gentlemen of Yorkeshire & Lincolneshire, as Yeuersley, Ripley, Hedley, Welley, Willoughby, Burnham, Waterton, Fleming, Tailboy, by violence of the conquering sword, & the mighty hand of God so *per*mitting it, of *our* soueraigne lord king Edward, within this realme, w*h*ich were of the partie of Henry, by whose gifts in times past I was refreshed, w*h*ich now are lame*n*tably dead, with many other; what can now helpe my heavines, or what can swaye my

---

48. *Medulla*, fol. 1r: "Dedicatus honorando patri et d*o*mino d*o*mino Ep*iscopo*."

49. The original Latin, in *Medulla*, fol. 1r, reads: "Haec mea preclare, presul dictamina care / Suscipe quo puro, metra tibi dicere curo / En tibi de petra." Unless otherwise stated, I use Bolles's English translation of the verses, as recorded in Trinity O.2.33, fols. 2r-3r. On Bolles, see chap. 6, below.

50. Trinity O.2.33, fol. 16v.

secret wepings & sighings, day & night, though I resist them: what should I intangle my selfe with, worldly vanities, & noisom observations, or with the pleasures of the world, vaine, and transitory. Vanities of vanities, all is vanity, all things passe but the loue of god.[51]

The families Ripley names as his kinsfolk were all prominent in northeast England during the fifteenth century, and Ripley's account is consistent with the fate of these families following the Lincolnshire rising against Edward IV in 1469 and 1470.[52] Richard, Lord Welles and Willoughby, and his son, Lord Robert Welles, were the leaders of the rising of 1470; the latter losing to Edward at Losecoat Field.[53]

The loss of these connections may well have proved catastrophic for a canon living outside the bounds of a priory, who perhaps depended upon his family for support. This episode also suggests a pragmatic reason for Ripley's decision to return to the refuge of the cloister, although his appeal hints at an emotional exhaustion that is compatible with a sincere desire to withdraw from the world. It is clear, however, that Ripley did not foresee an end to alchemical activity following his withdrawal. He offers both his services and his discretion in preserving the bishop's health through alchemical means:

51. Ibid., fol. 17r. The Latin text of the epilogue in *Medulla*, fol. 6r, reads: "Et quidem non me fateor *philosophu*m aut inter *philosop*hos computandum, sed paup*erem* [christi] *ser*uulum humil*em* canonicu*m non* meritis, sed dono dei ad *p*hilosophorum assumptum epulas quas stilo incultissimo hoc in tractatu exiguo vobis ex intimis *ani*me mee votis com*m*unicare veraciter ia*m* sategi. M*i*hi igitur paup*er*culo religioso viuendi statu*m* temere, *h*abita foris claustra co*n*u*er*sandi licentia no*n* obstante sedulo disquisita a*n*imi affectanti, vostr*e* grat*ios*i*o*se supportacio*n*is asilu*m* animat*i*s quo sic viua*m*, q*u*o me v*l*tra secularib*us* negotijs no*n* immisceam, quo cu*m* optimu*m* sit grat*i*a stabilire cor intra claustra a seculo me iteru*m* vt opto clauda*m h*abeamque si dignemini secreti*us* secre*tu*m & me semp*er* secretissimu*m com*perietis. Vnde aliq*u*a premissor*um* vostr*a* pro com*m*oditate, conseruanda q*u*oq*ue* corporis sospitate p*er*ficia*m* secretorum. Eo *en*im me posse, puto affectas mei complem*entu*m perfecti*us* me expediti*us* obtinere q*u*o deo me disponam corpore, et o*mn*ia quietius adherere, parentib*us* carnalib*us* defu*n*ctis, co*n*sanguineoru*m*que meorum generosoru*m* Eboracen*tium* & lincoln*ensi*um com*itatu*s yeu*er*sley, Ripley, Medeley, Welley, Willeby, Burnham, Watirton, fflemmyng, Taylbus in ore gladij victrici et p*r*eualida manu deo p*er*mittente *d*omini *n*ostri Regis Edwardi infra regnum istud ex p*ar*te R[egis] H[enrici] existent*er* quorum olim largici*o*nib*us* hones*ti*us* exibebat [*sic*] flebiliter cum pluribu*s* ia*m p*eremptis quid mee mederetur m[a]estitie quid secre*tis fletib*us* inuitisq*ue* diurnis suspirijs et nocturnis s*ec*ularib*us* vanitatibus. nociuis oblectaci*o*nibus me ve mu*n*di hui*us* solatijs ta*m* vanis ta*m* transitorijs v*l*teri*us* *im*plicare. Vanitas vanitatum et o*m*nia vanitas. O*m*nia *pr*etereunt preter amare deu*m*." Ripley cites Ecclesiastes 1:2 (Douay-Rheims): "Van*ity of vanities, said Ecclesiastes vanity of vanities, and all is vanity."

52. John H. Tillotson, ed. and trans., *Monastery and Society in the Late Middle Ages: Selected Account Rolls from Selby Abbey, Yorkshire, 1398–1537* (Woodbridge: Boydell and Brewer, 1988); A. J. Pollard, *North-Eastern England during the Wars of the Roses* (Oxford: Clarendon Press, 1990).

53. Pollard, *North-Eastern England*, 307.

And seing it is best to stablish the hart with grace, [tha]t I may shut my selfe againe within the cloister, from the world according to my desire, & [tha]t I may be hid, if you thincke so good, being well assured, [tha]t you shall finde me always most secrett. And some of these secrets aforesaid I shall performe, both for this comodity, & also for to conserue the health of your body, for by [tha]t meane I thincke I shalbe the better able, more perfitly & most speedily to attaine to the accomplishment of my desire, wherby I may dispose my selfe the more quickly to sticke to god both body & soule.[54]

The early modern perception of Ripley's vita presents the canon as an adviser and éminence grise to kings and popes, or dwells on improbable legends, such as his supposed donation of £100,000 per year to the Knights of St. John on Rhodes, reported by Elias Ashmole.[55] The epilogue to the *Medulla* suggests that the reality was more modest. Yet, while Ripley never attained the political and economic influence that later biographers conferred on him, he did seek to use his alchemical knowledge to secure ecclesiastical and possibly royal patronage though works that would preserve his memory more powerfully than any elixir.

The *Compound* may have been his first attempt. The poem was composed in 1471, according to the short "Explicit Alchimicae" attached to the late fifteenth-century copy in Trinity O.5.31:

Here ends the treatise of alchemical philosophy of which George Ripley, Canon, was the author, which was composed and set in order in the year 1471. Reader, I beg, give aid to the author with prayer, that after life he may have gentle purging. Amen.[56]

The date is significant. If Ripley lost the support of his family connections after the 1470 rebellion, circumstances may have forced him to seek prefer-

---

54. Trinity O.2.33, fol. 17v.

55. The seeds of this vita are detectable in early modern accounts by the antiquarians John Leland, John Bale, and John Pits: see, for instance, John Bale, *Scriptorum illustrium maioris Brytanniae . . . Catalogus* (Basel: Johannes Oporinus, 1557), 622–23. It was developed by Ashmole, *TCB*, 444, 456–59, and had attained its mature form by the mid-eighteenth century, in Nicolas Lenglet-Dufresnoy, *Histoire de la Philosophie Hermétique* (Paris: Coustelier, 1742), 264–66, and Tanner, *Bibliotheca Britannico-Hibernica*, 633.

56. Trinity O.5.31 (late fifteenth century), fol. 37v: "Explycyt Alkimice tractatus philosophie. Cuius Rypla george canonicus quis auctor erat .M. quadrigentes septuaginta vnusque tenerat [sic]. Annis qui scriptis compositusque fuit. Auctori lector praebe praece queso iuuamen. Illi purgamen leue post vitam sit ut. Amen." This early version differs slightly from that printed in *TCB*, 193, where it is accompanied by Elias Ashmole's translation into English verse.

ment elsewhere. The *Compound* was almost certainly written for presentation to a patron, and its composition in 1471, the year of Edward IV's triumphant restoration, makes the king a likely recipient.

A relationship between Ripley and the king is also suggested by the *Compound*'s association with another work, the so-called *Epistle to Edward IV*. This anonymous Middle English poem was already being attributed to Ripley in manuscript copies by the second half of the sixteenth century, and was printed with the *Compound* under Ripley's name in both 1591 and 1652. It comprises thirty stanzas, of which the first nine clearly address the newly restored king. The first verse alludes to Edward's recent victory:

> O Honorable Lord, and most victoryous Knyght,
> With Grace and Fortune abundantly endewed,
> The savegard of *England*, & maynteyner of right;
> That God you loveth indeede he hath well shewed:
> Wherefore I trust this Lond shalbe renewed
> With Joy and Riches, with Charyty and Peace,
> So that old ranckors understrewed,
> Tempestuous troubles and wretchednes shall cease.[57]

There is some internal evidence to associate the poem with Ripley, starting with the interest in overlooking "olde ranckors," which might well reflect Ripley's dubious status as the poor relation of an attainted clan. The author also alludes to foreign travel, promising to disclose "Great secretts, which in farre countries I did learne." This includes a period spent at the University of Louvain, the *studium generale* founded in 1425 in the duchy of Brabant.[58] If the author is indeed Ripley, this suggests that he used his papal dispensation to study in the Burgundian Netherlands. Yet it also begs the question of why he failed to mention such studies in the autobiographical sections of the *Medulla* and *Compound*, where he drew attention rather to his "lernyng in Italye."

Indeed, the *Epistle* provides an excellent example of the hazards associated with attributing medieval alchemica. The poem accompanies no early manuscripts of the *Compound*, including the three surviving fifteenth-century copies.[59] And while a truncated version survives anonymously in three late fifteenth-century codices (all written in the same hand), this lacks the nine dedicatory stanzas that allow it to be pinpointed to a particular time

57. Ripley, *Epistle to Edward IV*, in *TCB*, 109.

58. Ibid., 110.

59. See note 35 above.

and patron. Thus, while we certainly cannot exclude Ripley's authorship, it seems that we cannot assume it either—particularly given the number of licenses issued during Edward's reign, which shows that Ripley was far from being the only English alchemist to pursue royal support.

THE *MARROW OF ALCHEMY*

Ripley may ultimately have failed in the patronage game. Although the *Compound* won him lasting fame, there is no evidence that it gained him any material benefits, or, indeed, that the king even received a copy of the poem. By 1476, apparently disillusioned, Ripley switched his poetic talents from English to Latin to provide prefatory verses for a Latin prose treatise, the *Medulla alchimiae*.

Unlike the petitions for royal licenses discussed above, Ripley's *Medulla* was written for an episcopal patron. Although Ripley does not address his lord, the bishop, by name, by the late sixteenth century this was generally assumed to have been Edward's disgraced former Chancellor and Archbishop of York, George Neville (1432–1476).[60] Neville lost the king's favor after supporting the defection of his brother Richard Neville, Earl of Warwick and "Kingmaker," in 1469, and, despite a pardon, it took only a hint of treason to prompt his arrest and imprisonment in Calais in 1472.[61] Ripley penned the *Medulla* a year after Neville's release in 1475. By evoking his own family's support for Henry VI and subsequent ill fortune at Edward's hands, Ripley may have hoped to receive a sympathetic hearing from the head of his former diocese.

Although Ripley does not refer directly to the *Secretum secretorum* when addressing the bishop, he may have viewed the pseudo-Aristotelian epistle as a model for his own letter. In his prefatory verses, he points to the stone's threefold nature: animal, vegetable, and mineral. Unlike pseudo-Aristotle,

---

60. See, for instance, Ashmole 1487, pt. 2 (ca. 1569), fol. 172r; Copenhagen, Royal Library, GKS 1746 (ca. 1570–1600), fol. 1r; The Hague, Royal Library of the Netherlands, Bibliotheca Philosophica Hermetica MS 46 (seventeenth century), fol. 1v. A more cautious attribution is suggested by William Bolles, the scribe of Glasgow University Library, MS Ferguson 102 (ca. 1525–75), fol. 68r: "This treatise was writen vnto a certen Bisshop of England by Sir george Riplay chanon." Two copies suggest alternative dedicatees: Neville's successor as Archbishop of York, Lawrence Booth, in Royal College of Physicians of Edinburgh, MS Anonyma 2, vol. 1 (seventeenth century), 187; and Thomas Ruthall, Bishop of Durham, in Glasgow University Library, MS Ferguson 91 (seventeenth century), fol. 37r. However, Ruthall was not appointed to the bishopric until 1509.

61. Michael Hicks, "Neville, George (1432–1476)," *ODNB*. Neville personally took Edward into custody after the king's defeat at Edgecote Moor.

however, he identifies three distinct stones rather than a single, tripartite substance. The mineral stone may be used for transmutation but is harmful to ingest, unlike the animal and vegetable stones, which offer medicinal benefits: "To cure all things their vertue is."[62] This is the aspect of the work that Ripley chooses to emphasize in the closing lines of the poem, as he offers his services as both alchemist and healer, using a neat chemical analogy:

If thou vnbroken long wouldst keepe
in perfect health thy vessell still
then for thy chanon looke thou seeke
remember him that hath good will.[63]

The preface thereby provides an apt introduction to both the multiplicity of stones and the structural conceit of the treatise, which is divided into three chapters that deal with each stone in turn. Ripley presents the stones in the same order in which they appear in the pseudo-Lullian *Epistola accurtationis*, which he cites in his introduction and which supplies the *Medulla* with some of its practical content.[64] However, this ordering also contributes to the usefulness of the *Medulla* as a patronage suit, starting with the Lesser Work of transmutation, and building toward the Greater Work of medicine, which, Ripley hints, will be of greatest value to his patron.[65]

THE MINERAL STONE

When Ripley wrote the *Medulla*, attempts were already afoot to reconcile some of the apparent discrepancies in the pseudo-Lullian corpus. One of the easiest solutions was to divide the Lullian corpus into separate mineral, vegetable, and animal stones. A special category for the "mixed stone" was also suggested, based on a practice in the *Epistola* that combines the mineral and vegetable stones to make a powerful transmuting agent that is nonetheless unsuitable for medicine.[66] These texts also suggested concordances between

62. *Medulla*, fol. 1r: "Omnibus in curis, hec subueniunt valeturis."
63. Ibid., fol. 1v: "Infractum vi vas, longo bene tempore viuas / Canonici memor esto tui sibi ferto iuvamen." Translation from Trinity O.2.33, fol. 3r.
64. Ibid., fol. 3r: "Sed quia de primo elixire in hoc primo capitulo agemus: ideo de hoc igne contra naturam que est aqua mineralis fortissima et mortalis . . . vt Raymundus dicit in epistola accurtacionis."
65. Discussed further on pp. 112–13, below.
66. A good example is the *Tractatus de duabus nobilissimis aquis* (Treatise on the Two Most Noble Waters), also called the *Secreta secretorum Raymundi* (Secrets of Secrets of Raymond),

otherwise confusing pseudo-Lullian *Decknamen*: the vegetable mercury, the natural fire, the resolutive menstruum, and, of course, the Green Lion. Ripley's own writings reveal his ability to deftly extract and collate practical information from such outwardly conflicting sources as the *Testamentum, De secretis naturae,* and the *Epistola accurtationis.* The ongoing quest to harmonize the pseudo-Lullian corpus provides necessary context for understanding Ripley's practical program.

The *Medulla* is characterized by a synthetic approach to the pseudo-Lullian corpus. More so than in the *Compound*, Ripley takes the opportunity to engage with the conflicts and problems of his source material, perhaps hoping to impress the bishop with his expertise in obtaining efficacious results through textual problem-solving. Ripley's technical innovations are nevertheless very subtly integrated with the Lullian material, requiring close comparison of his own works and those of his authorities. The overall impression is of a faithful commentary on Raymond's writings, an impression cemented by Ripley's assertion in the preface that the work is a compilation (*compilatus*) of authorities—a typical expression of authorial modesty that in fact serves to increase the authority of the overall text. Ripley's own contribution, he suggests, rests on his ability to extract the sense from these difficult sources: to draw out "the marrow of nature from its inner and more secret bones, the grosser and more feculent substance of the flesh being cut away."[67]

In practice, this meant engaging with the multiple waters of the Lullian corpus. Ripley does so by introducing the three stones, as set out in the *Epistola*: "mineral, vegetable, and animal, so called because they are made from the waters of mineral, vegetable, and animal things, serving unto them."[68] The mineral stone, for instance, is made from mineral substances that are unsuitable for medicine, such as vitriol. Ripley turns here to a passage in the *Epistola* in which Raymond warns his patron that the manufacture of the mineral stone is particularly dangerous, since it requires two waters with

---

which is preserved in several late fifteenth-century manuscripts: Trinity O.8.32, pt. 1, fols. 12r-39r; Trinity O.8.9, fols. 32v-36r; CCC MS 136, fols. 52r-54v; see also Paris, Bibliothèque Nationale, MS Lat. 14007, fols 82r-83v (late fifteenth to sixteenth century). At least one early modern reader even suspected that Ripley had written this tract himself, as noted in Bologna, Biblioteca Universitaria di Bologna, MS 457, vol. XXIII, pt. 3, fol. 55r: "Haec collectis aquarum tributa Ripleo."

67. *Medulla*, fol. 1v: "Nanciscebar medullam quodammodo nature ipsa grossiori feculentiorique substancia carnium resecata ex ipsius interioribus secretioribus quoque \ossibus/."

68. Ibid., fol. 3r: "Est enim triplex elixir, minerale, vegetale et animale. sic dicta: eo quia per aquas mineralium vegetalium & animalium rerum sibi deseruientes fiant."

contrary operations: one to volatilize and one to congeal the stone. While the first can be safely used, the second is a powerful corrosive:

> This water is drawn, as you know, from a stinking menstruum compounded (*immasati*) of four things, and it is the stronger water of the world and mortal; whose spirit wholly multiplies the tincture of the ferment.[69]

This passage provides the basis for Ripley's own exposition of the mineral stone. However, he significantly expands on Raymond's terse statement:

> Because in this first chapter we will treat about the first elixir, let us therefore disclose somewhat more of this fire against nature, which is a mineral water, most strong and mortal, which serves to that elixir. And this water is drawn by elemental fire from a certain stinking menstruum compounded of four things, as Raymond says in the *Epistola accurtationis*. And it is the strongest water in the world, whose spirit alone augments and multiplies the tincture of the ferment.[70]

The key difference between the two passages is that whereas the original refers only to a "water," Ripley identifies this substance as "fire against nature." Here he has silently glossed his source by borrowing a term from the metallurgical strand of the pseudo-Lullian corpus as represented by the *Testamentum* and *Codicillus*. In these works the "fire against nature," made from corrosive minerals such as vitriol, is contrasted with the "natural fire," used to denote the mercuries and sulphurs drawn out of precious metals. As Ripley recognizes, these liquid fires, or waters, are distinguished from true, elemental fire: "that which fixes, calcines and burns away, being nourished by combustible things."[71] Ripley in fact makes great play of the diversity of "fires," both in the *Medulla* and in the *Compound*:

---

69. *BCC*, 1:863: "Haec aqua extrahitur, ut nostri ex quodam menstruali foetenti immassati ex rebus quatuor, & fortior est aqua mundi & mortalis: cujus spiritus totam tincturam fermenti ampliat."

70. *Medulla*, fol. 3r: "Quia de primo elixire in hoc primo capitulo agemus: ideo de hoc igne contra naturam que est aqua mineralis fortissima et mortalis elixeri deseruiens illi, aliquid vlterius disseramus. Hec autem aqua extrahitur igne elementali a quodam menstruali fetenti immassata ex rebus .[quattu]or. vt Raymundus dicit in epistola accurtacionis et est fortior aqua mundi cuius solus spiritus tincturam fermenti ampliat et multiplicat."

71. Ibid., fol. 2v: "Elementalis est qui fixat calcinat et comburit ex combustibilibus enutritus."

Fower Fyers there be whych you must understond,
Naturall, Innaturall, against Nature, alsoe
Elementall whych doth bren the brond;
These foure Fyres use we and no mo.[72]

In his own writings, Ripley uses such language play to show his readers how a philosophical treatise must be read. For instance, he warns against facile interpretations: Raymond's fires are really waters, while the Green Lion—the true prime matter of the stone—should never be taken as vitriol, since this nonmetalline substance not only is useless in medicine, but also destroys the active virtues of metals.

Yet, although Ripley employs the same language as his pseudo-Lullian sources, he does not necessarily interpret it in the same way. In the *Testamentum*, the Green Lion refers to vitriol *azoqueus*, an ingredient of the "stinking menstruum." Ripley, however, rejects this identification, observing that since only ignorant practitioners would seek their prime matter in vitriol, it is more properly called the "Green Lion of Fools." Despite rejecting vitriol as prime matter, Ripley still acknowledges that the active fire within vitriol is useful as a helper to the work, specifically when preparing common quicksilver for use in the mineral stone:

From the Green Lion of Fools is drawn with strong fire that which we call *aqua fortis*. . . . And the thing from which this *aqua fortis* is drawn is vitriol, green and *azoqueus*, that is, not artificial but natural (namely, the droppings of copper). . . . How secret a virtue and power is this fire, is apparent enough in the constriction of the body of the volatile spirit, in the form of a snowy whiteness, when commonly sublimed by it.[73]

Ripley's hints allow us to identify the fire against nature as corrosive sublimate, made from "four things" (copper vitriol, saltpeter, quicksilver, and an unknown fourth ingredient, possibly sulphur). When distilled over strong

---

72. Ripley, "Separation," in *TCB*, 142. See also *Medulla*, fol. 2v: "Ignis autem pluripliciter differt: Quidam autem est naturalis, quidam innaturalis: quidam elementalis, et quidam contra naturam."
73. *Medulla*, fol. 3r: "De fatuorum tum leone viridi extrahitur ea quam diximus aqua fortis: igne forti. . . . Res autem de qua trahitur hec aqua fortis est vitriolum viride & azoqueum, hoc est non artificiale: sed naturale scilicet stillicidium cupri. . . . Quam secrete virtutis et potentie sit ignis iste: in constriccione corporis spiritus volatilis in speciem candoris niuei ab eo sublimati vulgariter, satis patet." In modern chemistry green vitriol usually corresponds to iron sulphate and blue to copper sulphate. Ripley's green vitriol, however, is clearly stated to be derived from copper, demonstrating that care must be taken when considering the materials available to early modern practitioners.

heat, vitriol and saltpeter immediately react to form a mineral acid, which in turn acts upon the "volatile spirit" (quicksilver) to constrict its fluidity and fix it in the form of white corrosive sublimate. Elixirs made from this fire are therefore prohibited for internal use—indeed, Ripley warns, quoting Raymond, "it would be safer for a man to eat the eyes of a basilisk than gold made with our fire against nature."[74] In this way, Ripley preserves the *Testamentum*'s process for corrosive sublimate (including Raymond's terms "*azoqueus*" to describe the vitriol and "stinking menstruum" for the sublimate) while subtly demoting it, from the "Green Lion" to a mere helper in the work.

But if the true Green Lion is not vitriol, what is it? For Ripley, the "Green Lion of Philosophers" is linked with generation rather than corruption; it therefore corresponds to Raymond's *natural* fire. It is a quintessential principle, generated through the influence of the stars, which encapsulates the "spirits of ardent waters, and the potential vapors of minerals, and natural virtues of living things."[75] Since it exists in all the kingdoms of nature, this natural fire may be employed not only for transmutation, but also for medicine. In this respect it differs from the fire against nature, so named because it destroys the specific form of bodies, and is "against all natural operations."[76] Unlike the fire of nature, however, Ripley is far more coy about showing how this elusive fire should be obtained. Only in the next chapter of the *Medulla*—on the vegetable stone—does he provide further clues to its identity.

Although this profusion of names and interpretations may seem confusing (and was certainly intended to be so), such wordplay allows us to trace changes in the way that reader-practitioners conceived of chemical substances, particularly when hunting for the true prime matter of alchemy. Ripley's views on the proper identification of the Green Lion show how attitudes toward vitriol had moved on since the 1330s, when the *Testamentum* was written. Then, corrosive sublimate was still promoted as an exciting avenue for alchemical research. By the end of the fourteenth century, however, a wide variety of wine-based, "vegetable" products had come into vogue,

74. *Medulla*, fol. 3r: "Tutius quidem homini esset comedere oculos basilisci quam aurum cum igne nostro contra naturam factum." Precursors of this saying are found in the *Testamentum* and *De secretis naturae*.

75. Ibid., fol. 2v: "Naturalis est qui corporibus influxus est a sole et luna et stellis: vnde et spiritus aquarum ardentium et vapores potentiales mineralium, ac naturales virtutes animalium generantur."

76. Ibid.: "Sic dictus contra naturam: eo quod eius operatio sit contra omnes operaciones naturales ... hoc totum quia natura componit: iste ignis discomponit. Et portat ad corrupcionem nisi ei superaddatur ignis nature."

whereas "alums and salts" had been consistently rejected, by alchemical authorities ranging from Geber, Dastin, and Arnald of Villanova to Raymond himself in the wine-based strand of the Lullian corpus. On the other hand, the "vegetable" alchemy of *De secretis naturae* presented a promising new identification for the Green Lion, and one that rejected corrosives entirely.

This is the route that Ripley would pursue, while relegating vitriol to a walk-on role in the making of a mineral water drawn from quicksilver. The Lullian doctrine of Lesser and Greater Works, itself the product of a schism in the pseudo-Lullian corpus, achieves clear expression in Ripley's identification of Raymond's natural and contra-natural fires with, respectively, the "true" and "false" Green Lions—vegetable elixir and deadly mineral corrosive. Yet Ripley's final process for the vegetable stone would still differ from that of Raymond—the outcome of a strategy that legitimated his method in terms of authoritative sources, while still obtaining practical results that he could square with his own experience.

THE VEGETABLE STONE

It is not until the second chapter of the *Medulla* that Ripley introduces his patron to the vegetable stone and the secret of its healing power. The root of this practice is Raymond's natural fire: "All the benefit of the vegetable stone is had by virtue of the fire of nature . . . therefore we intend to treat clearly of this fire, and the manner of working with it, solely for your Lordship's pleasure."[77] As hinted earlier in the treatise, Ripley anticipates the bishop's approval not only because of the stone's value for transmutation, but also because it provides the basis for the legendary *aurum potabile*:

> And then it has the power to turn all bodies into pure gold, and to heal all infirmities above all the potions of Hippocrates and Galen, for this is the true potable gold, and no other, which is made from "elementated" gold.[78]

As with the mineral stone, the preparation (or "elementation") of gold for the vegetable stone requires a philosophical solvent, or mercury. Although Ripley does not mention *De secretis naturae*, his discussion of the vegetable

---

77. Ibid., fol. 5r: "Totum beneficium lapidis vegetabilis fit virtute ignis naturae . . . ideo de composicione huius ignis et de modo operandi cum eo solo lucide ob vestre dominationis complacentiam intendimus pertractare."

78. Ibid., fol. 5v: "Et tunc habet potestatem conuertendi omnia corpora in aurum purum et sanandi omnes infirmitates supra omnes potaciones Ypocratis et Galieni quia illud est verum aurum potabile, et nullum aliud, quod de auro elementato."

mercury is clearly based on this core pseudo-Lullian work, although perhaps not in the way we might expect.

As we have seen, *De secretis* was composed of two books based on John of Rupescissa's quintessence of wine, as well as a third one, the *Tertia distinctio*, primarily concerned with gold-making. In this *Third Distinction*, Raymond describes the extraction of mercuries from a variety of metals and vegetable substances, including wine. The making of the elixir relies on two "menstrua," named for their complementary and contrary effects. Importantly, these menstrua do not correspond to the two contrary fires described in the metallurgical-based strand of the pseudo-Lullian corpus, although Ripley would later identify one of them with natural fire.

One problem for readers of the *Tertia distinctio* is that it describes the two menstrua only in vague terms. Raymond explains that one, which is called *resoluble*, first exists in a potential form within metallic bodies, including gold and silver. This form is actualized only when the metals are dissolved in the second menstruum:

> The resoluble menstruum is the quintessence of metals, proceeding into act. . . . It is also defined as follows. The menstruum is a potential vapor present in every metal, by whose odor alone quicksilver is converted into metal, which is not brought into act unless by the resolutive. And this menstruum springs from gold and silver.[79]

While the first menstruum is drawn from metals, the second is of vegetable origin. It is called *resolutive*, for it dissolves what has been dissolved previously:

> This menstruum is a burning water [*aqua ardens*] sprung from wine, perfectly rectified, by virtue of which . . . all those metals are dissolved, putrefy, and are purified; and the elements are divided from them, and the earth exalted into a foliate earth which is called sulphur of nature, by its attractive virtue.[80]

---

79. *De secretis naturae*, fol. 107r: "Et menstruum resolubile quinta essentia metallorum, deuenit in actum. . . . Item menstruum resolubile diffinitur sic. Menstruum est uapor potentialis existens in quocunque metallo, in cuius solo odore conuertitur argentum uiuum in metallum, cuius est qui non nisi resolutiue ducitur ad actum. Et menstruum resolubile oritur ex auro & argento."

80. Ibid.: "Quod aliud resolutiuum menstruum diffinitur sic. Menstruum est aqua ardens, perfecte rectificata, orta à uino cuius uirtute . . . illa omnia metalla dissoluuntur, putrefiunt & purificantur, & elementa ab eis diuiduntur, & terra exaltatur in terram foliatam, quae dicitur sulphur naturae sua uirtute actiua."

On the face of it, Raymond's resolutive menstruum appears to denote a simple quintessence of the kind described by John of Rupescissa, obtained by repeated distillation of wine and sharpened by additional ingredients such as tartar. Although vegetable in nature, this "sharpened *aqua ardens*" is capable of dissolving gold and silver in order to extract and once more dissolve the "quicksilver of the essence of metals." Together, the two menstrua may create any stone, "philosophical as well as precious"—a reference to their multiple applications for both transmutation and healing.[81]

Thus far, Raymond's terminology seems to offer nothing more than an elaborate disguise for the use of distilled alcohol to dissolve precious metals. Yet to anyone familiar with the impermeability of gold, such an account would be difficult to accept. Gold and silver simply cannot be dissolved in spirit of wine, unless the term has been used to denote something else— immediately raising an interpretative issue. Ripley describes the problem at the start of his chapter on the vegetable stone:

> Some assert that this fire is a water drawn from wine, according to the common way, and should be rectified, being distilled as many times as possible . . . yet, when water of this kind (which fools call the pure spirit), even if rectified a hundred times, is put upon the calx of whatever body, however well prepared, nevertheless we see it will be found weak and entirely insufficient for the act of dissolving our body with conservation of its form and species. Wherefore it seems there is an error in the choice of this principle, which is called the resolutive menstruum.[82]

To complicate matters further, in another work Raymond bans nonmetallic ingredients altogether, stating that the necessary solvent should be drawn from a metallic body. Ripley fastens onto the contradiction:

81. Ibid., fol. 62r: "Menstruum quod est in uegetabilibus, est liquor cum quo dissoluuntur metalla; & menstruum quod est in mineralibus dissoluitur à uegetabili. . . . Primum dicitur aqua ar[dens] acuata: secundum dicitur argentum uiuum de essentia metallorum. Et ex istis duobus tanquam natura componitur lapis quiuis, tam philosophicus, quàm preciosus."

82. *Medulla*, fol. 5r: "Quidam autumant ignem istum aquam esse a vino tractam vulgari modo rectificarique debere eam multotiens distillando vt possit ab ea eius aquosum flegma vires et potentias sue igneitatis impediens, penitus extirpari. Sed cum talis aqua centies rectificata quam dicunt fatui spiritum esse purum mittitur super calcem corporis optime preparatam: videmus quod ad actum dissoluendi corpus cum conseruacione sue forme et speciei impotens ac omnino insufficiens reperitur Quare videtur quod in electione huius principij quod menstruum resolutiuum dicitur error fit."

If, as Raymond says, the resolutive menstruum springs from wine or the tartar thereof, how is what the same philosopher says to be understood: "Our water is a metalline water, because it is produced only from a metalline kind"?[83]

Although Ripley once more neglects to name his source, this passage includes an almost direct quotation from the *Repertorium*, a short exposition based on the *Codicillus*.[84] The conflict therefore marks an interface between the "quintessential" and "metallurgical" strands of pseudo-Lullian alchemy, a fact of which Ripley, convinced of the unity of the corpus, was of course unaware. To Ripley, the contradictory statements suggest that Raymond has deliberately resorted to riddling language in order to catch out his less perspicacious readers. It therefore falls to an equally skilled reader and practitioner—Ripley himself—to resolve the contradiction and reveal Raymond's true intent.

Moments such as this capture alchemical reading in action, and with it, the power of commentary as a tool for dissecting arguments. The outward meaning of a term cannot be squared with experiential knowledge— therefore it *must* be wrong. Since the authority is unlikely to have erred, a new reading must be imposed. In this case, having pointed out the apparent conflict, Ripley settles down to the business of resolving it, continuing to draw upon a range of pseudo-Lullian texts to support his argument.

Ripley first notes that the "metalline water" mentioned in the *Repertorium* must refer to one of Raymond's menstrua: either the resolutive (that is, the burning water) or the resoluble (the potential vapor of metals). If the former, Raymond cannot mean "metalline" in the literal sense, since he has elsewhere clearly stated that the resolutive menstruum derives from wine. Yet, ingeniously, a vegetable menstruum might still be regarded as metalline "after a certain manner" (*secundum quid*), since, like metal, it is both sulphurous and mercurial: sulphurous because it burns like fire, and mercurial

---

83. Ibid.: "Sed si a vino oritur menstruum resolutiuum vt vult Raymundus vel a tartaro eius: quomodo intelligitur quod idem philosophus dicit. Aqua nostra est aqua metallina, quia ex solo genere metallico generatur."

84. "Conclusio summaria ad intelligentiam Testamenti seu Codicilli Raymundi Lullij . . . quae aliter Repertorium Raymundi appellatur," *TC*, 3:731: "Patet per aquam nostram philosophicam, quae dicitur metallina, eo quod ex solo genere metallico generetur." That Ripley knew this text appears from his reference to the "Reportory" in the *Compound*; Ripley, "Calcination," in *TCB*, 131.

because tartar of wine, when dried in the sun, shines like mercury.[85] The resolutive menstruum may therefore be regarded as metalline by analogy.

Ripley's use of *secundum quid* suggests that this solution should not be taken too seriously. The expression, a standard item in the medieval logician's vocabulary, denotes the logical fallacy of moving from the general to the particular without accommodating special circumstances.[86] Its appearance here underlines the playful aspect of an otherwise preposterous attempt to render a vegetable water "metalline." Ripley is thus merely toying with the notion that Raymond's metalline water denotes the resolutive menstruum. If, however, the passage is taken to indicate the *resoluble* menstruum, then further rhetorical quickstepping is uncalled-for, since Raymond has already explained in *De secretis naturae* that this menstruum is a "quintessence of metals."[87] The *Repertorium* may therefore be read as referring only to the metalline resoluble menstruum, in which case its apparent rejection of vegetable ingredients is no longer an issue.

Yet, like the chicken-and-egg paradox, Ripley's explanation seems to lead us to infinite regress. If the resoluble menstruum can be made only by dissolving precious metals, how can it also be the solvent in which they are to be dissolved? Perhaps wary of pushing his patron's patience too far, Ripley here pauses to address him directly. He explains that thus far he has been deliberately diffuse and obscure, since Raymond's true intent is covered by the mantle of philosophy.[88] However, his love for the bishop leads him to speak more plainly, and disclose the true meaning of the two waters.

In Ripley's opinion, Raymond is correct to say that the resolutive menstruum springs from wine, yet he is also right to call it metalline. The secret lies in the fact that resoluble menstrua need not only be made from gold and silver; they provide the potential vapors of *all* metallic bodies. There are, in

85. *Medulla*, fol. 5r: "Si primo modo: sic non est metallica aqua nisi secundum quid, quia est vapor sulphureus et mercurialis, et racione suae sulphureitatis sue ardet cum igne. Videmus etiam in tartaro sole tantum desiccato: mercuriales qualitates ad oculum resplendere. Genus autem metallicum: et sulphur est et argentum viuum."

86. The term is developed from Aristotle's fallacy of "unqualified generalizations" in the *Sophistical Refutations* 5 (167a: 1–20). See William T. Parry and Edward A. Hacker, *Aristotelian Logic* (Albany: State University of New York Press, 1991), 438.

87. *Medulla*, fol. 5r: "Si secundo modo: solutio patet secundum Raymundum in suo questionario."

88. Ibid., fol. 5r-v: "Huc vsque in tractando de hoc lapide diffusi sumus et confusi. Quare ne in hijs que solus amor vestre confert dominacioni scrupulosum quid appareat cum legantur dico quod omnia ‖ ista que Raymundus locutus est cooperiuntur clamide philosophie vult enim vt cum spiritu vini fiat dissolutio. sed in hoc intendit vt habeatur aliud menstruum resolubile quod sine tali resolutiouo haberi nusquam potest."

fact, as many resoluble menstrua as there are metals, and some of these—unlike gold—will dissolve in wine-based solvents. The alchemist therefore needs to find a metal that dissolves either in alcohol or in distilled wine vinegar (that is, the two solvents made from wine). This solution will result in a resoluble menstruum that is in turn used to prepare the *resolutive* menstruum. Although the resolutive still has its ultimate origin in wine, it has thus been "sharpened" by the presence of a metalline ingredient.[89]

Ripley's solution is based on a commonsense knowledge of the behavior of metals and wine-based solvents. For instance, copper dissolved in vinegar produces a vivid green pigment, verdigris. The litharge of lead is also soluble in vinegar, yielding a sweet-tasting white compound, sugar of lead. Such products were well known even in ancient times, appearing, for instance, in Dioscorides's *materia medica*, as well as medieval craft manuals.[90] A fifteenth-century painter, goldsmith, or apothecary would have no need to turn to the enigmatic accounts of philosophical treatises to explicate such processes. Yet in the *Medulla* Ripley approaches this familiar subject matter as a scholastically trained philosopher, rather than an artisan. His aim is not merely to extract a recipe from his source text—although this is still an object with him—but to show how the causes of chemical transformations arise from the nature of matter itself.

By drawing attention to problems in his source text, Ripley here demonstrates his own knowledge of both textual authorities and material substances. His reading shows that he has understood Raymond's true meaning: the resolutive menstruum is not straightforward spirit of wine (which must therefore be a cover name), but a solvent possessed of both vegetable and mineral qualities, drawn from an imperfect metallic body by means of another solvent (distilled vinegar) that springs from wine. Since it is both metalline and vegetable in nature, this menstruum has the power to resolve not only precious metals, but also the contradictions posed by the various strands of the pseudo-Lullian corpus. The Raymond of *De secretis naturae* and the Raymond of the *Repertorium* are shown to be in perfect accord.

---

89. Ibid., fol. 5v: "Illud autem menstruum resolubile est generatum ex solo genere metallico quia est vapor potentialis in quocumque corpore metallico . . . ideo menstruum nostrum resolutiuum est: ad actum perductum vt autem illud habeatur menstruum quod est vnctuosum humidum sulphureum et mercuriale bene cum natura concordans metallorum cum quo artificialiter nostra corpora sint dissoluenda."

90. The making of verdigris is described in Dioscorides, *Materia medica*, 5.91. Processes for verdigris and minium are also included in Theophilus, *De diversis artibus*.

SUBSTITUTION AND EXPERIMENT

Ripley's response to the puzzles of the pseudo-Lullian corpus shows how reading "alchemically" could resolve even contradictory sources. Crucially, however, his solution is also based on the known properties of chemical materials. As we turn to the recipes described in the *Medulla*, it becomes clear that Ripley's hermeneutical groundwork provides a theoretical rationale for his own practice. The resulting process satisfies all the conditions of his authorities in order to yield the true resolutive menstruum: "unctuous, moist, sulphurous, and mercurial, according well with the nature of metals, with which our bodies should be artificially dissolved."[91] This process is the one upon which Ripley's subsequent fame would rest, a fame that seems to rely on Raymond's authority, but also acknowledges the originality of his own formulation.

The practice begins by drawing the first resoluble menstruum from the calx of an imperfect metal—a metallic body that Ripley calls "sericon":

Take the sharpest humidity of grapes, distilled, and in it dissolve the body, well calcined into red (which by masters is called sericon) into crystalline, clear, and heavy water. Of which water let a gum be made, which tastes like alum, which by Raymond is called *vitriolum azoqueus*.[92]

Although the ingredients in the recipe are not clearly defined, they are decodable. "Sericon" is nearly always used in fifteenth-century English texts to denote minium, or red lead—a reddish-orange compound made by careful calcination of litharge.[93] The "sharpest humidity of grapes" is probably distilled wine vinegar, in which the sericon is dissolved. The excess vinegar can then be distilled off, leaving a gum. Ripley equates this gummy substance with the *vitriolum azoqueus* described in the *Testamentum*, thereby substituting a product that is simultaneously mineral and vegetable for Raymond's

---

91. *Medulla*, fol. 5v: "Ideo menstruum nostrum resolutiuum est . . . vnctuosum humidum sulphureum et mercuriale bene cum natura concordans metallorum cum quo artificialiter nostra corpora sint dissoluenda."

92. Ibid.: "Rx acerimam vuarum humiditatem et in ea distillata dissolue corpus optime calcinatum in rubeum (quod a magistris vocatur sericon) in aquam cristillinam limpidam et ponderosam. De qua aqua fiat gummi gustui aluminosum quod vocatur a Raymundo vitriolum azoqueum."

93. I discuss the etymology and alchemical usage of the term in Rampling, "Transmuting Sericon"; see also note 107, below.

minerally derived "green lion azoqueus, which is called vitriol."[94] Even at this stage, Ripley keeps both Lullian strands tightly plaited.

The gum should next be dry distilled until a "faint water" is drawn off. This disposable, watery fraction is merely a precursor to the alchemist's true goal: a white smoke or fume, which should be collected and condensed in a receiver. The resulting "water" provides empirical vindication for Ripley's lengthy exegesis. It is, he exclaims, the true "resolutive menstruum, that before was resoluble"—that is to say, it is a wine-based solvent (and hence a resolutive menstruum according to Raymond's definition of the term), but one that is also drawn from a metalline body (which therefore counts as resoluble). To put it another way, it is the mercury of sericon, drawn out by agency of the vegetable solvent. Eventually, this menstruum will be used to dissolve the bodies of gold and silver, separate their elements, and exalt their calxes into "a marvelous salt."[95] Although Ripley is typically vague concerning the additional steps necessary to achieve this end, such reticence is hardly unusual in this context; his account is, after all, intended to whet the appetite of a patron, not to betray every detail of a complex process.

The resolutive menstruum has other interesting properties, which Ripley describes. When the vapor condenses, it is found to have a sharp taste and a bad smell, earning it another name familiar from the pseudo-Lullian vocabulary: the stinking menstruum. It is also extremely volatile, and if the practitioner wishes to proceed to the elixir, he must do so within an hour of its distillation. When added to its calx (the dregs remaining in the distillation flask), the water begins to boil without the addition of any extraneous heat. For this reason, only just enough liquid should be added to cover the calx.[96]

European secrets literature is rife with meticulously described yet suspect procedures, some of ancient provenance. We might therefore question

---

94. *Testamentum*, 1:198: "Leo viridis azoqueus, qui dicitur vitriolum." Ripley here chooses to reinterpret the Lullian *vitriolum azoqueus*, since earlier in the *Medulla* he uses it to denote vitriol, as we have seen. The boundary between alums and vitriols was not always distinct in medieval alchemy, so Ripley perhaps treats the term as a *Deckname* inspired by the taste of the gum.

95. *Medulla*, fol. 5v: "Cum autem fumus albus inceperit apparere, mutetur receptorium, et lutetur firmissime ne respiret. Et recipiatur nostra aqua ardens et aqua vite Menstruum resolutiuum quod ante erat resolubile vapor potentialis potens corpora dissoluere, putrefacere et purificare elementa diuidere terram quia exaltare in salem mirabilem sua virtute attractiua."

96. Ibid.: "Ista aqua saporem habet acutissimum odorem partim fetidum, ideoque vocatur menstruum fetens et quia est aqua maxime aerea. Ideo infra eandem horam qua distillatur; mittenda est super calcem suam super quam positam incipit bullire que si vas secure obturetur non cessabit ab opere absque extrinseco igne adhibito donec in calcem totaliter desiccetur. Ideo non debet poni de ea in maiori quantitate quam vix sufficiat calcem cooperire."

to what extent an account like Ripley's, deeply embedded within an existing textual tradition, reflects his own experimentation. Throughout the *Medulla*, Ripley provides numerous accounts of processes and observations, suggesting that his careful study of pseudo-Lullian doctrines has indeed been put to the test. His dismissal of rectified spirit of wine as "entirely insufficient for the act of dissolving our body" smacks of personal experience, as do his warnings concerning the volatility of his own vegetable menstruum. More importantly, Ripley's practice is grounded in known chemical properties, such as the solubility of various metallic calxes in distilled vinegar to produce metal acetates. In the case of a lead oxide like litharge, this dissolution would yield sugar of lead, a sweet-tasting, crystalline "gum" that indeed distills to produce a thick, white smoke (as well as a volatile solvent, acetone), although we cannot be certain that this is exactly what Ripley, working with impure materials, would have obtained.[97]

This ambiguity raises important considerations about Ripley's supply of materials and use of language. Whatever the source of Ripley's sericon, it cannot have been pure red lead in our modern sense of the term (i.e., lead tetroxide, Pb_3O_4), since, unlike regular lead monoxide (PbO), this dissolves only with difficulty in normal distilled vinegar.[98] Of course, Ripley seems to have "sharpened" his vinegar, either by further distillations or through the addition of other ingredients, which may have strengthened its capacity to dissolve. But there are also clues that his sericon was impure: thus it "is called the Green Lion because, when dissolved, it is at once decked in a green garment."[99] As it happens, sugar of lead does acquire a greenish color if the "lead" includes a proportion of copper—a feature that raises two intriguing possibilities.

First, Ripley may have unknowingly used an impure source of lead for his work, in which case the success of his practice could have varied depending on where in Europe he was working.[100] Substitutions might be driven by local knowledge or the availability of materials: Yorkshire was a lead-mining area, and Bridlington, Ripley's priory, owned a mine from which lead was

97. The white smoke is an organic material that codistills with acetone. Acetone is highly volatile—a property that accords with Ripley's description of the product of his distillation.

98. I am very grateful to Lawrence Principe for pointing out this property of lead tetroxide, and for his advice on replicating the experiment. On reconstructing the process, see chap. 7, below.

99. *Medulla*, fol. 3r: "Sed proles hec magnifica eo quia dissoluta statim vesti viridi induatur: leo viridis nuncupatur."

100. The composition of local materials can have considerable impact on the outcome of a chemical process. For an early modern example, see Lawrence M. Principe, "Chymical Exotica in the Seventeenth Century, or, How to Make the Bologna Stone," *Ambix* 63 (2016): 118–44.

exported to the Continent.[101] Second, he may have been aware of the impurity, in which case "sericon" offers a particularly ingenious *Deckname* for a mixture of lead and copper, the "red" color of the latter suggesting an analogy with red lead. Such a reading, although speculative, helps explain an anomaly in the *Compound*, written a few years earlier, where Ripley seems to equate the Green Lion with copper. In the poem, he also refers to the metallic body used in his vegetable work as red lead, but qualifies it as "*our* fine Red Lead"—a clue that he indeed intended it as a cover name.[102]

Ripley's treatment of the vegetable stone offers an impressive adaptation of difficult source texts by an alchemist who was evidently familiar with the properties of materials, and willing to look beyond the book for his procedures. What emerges is his willingness to discard or adapt instructions that conflict with his own experience or access to resources. Given the importance of the *Epistola accurtationis* in structuring the *Medulla*, we might also expect to find Ripley borrowing from its practical content. The *Epistola*'s abbreviated process for the vegetable stone begins with repeated distillation of an ingredient described as "black blacker than black" (*nigrum nigrius nigro*).[103] Yet, although Ripley signals in the *Medulla* that he knows of this substance—it is a particular kind of tartar, even "blacker than the tartar of the Catalonian grape"—he sets it aside on the grounds that "this thing is rare in these parts and certain others."[104] Fortunately, a different authority, Guido de Montanor, "has discovered another unctuous humidity, sprung from wine," which provides an adequate substitute.

Ripley's own, lead-based solution is actually rather closer to the final accurtation in the *Epistola*: "From the philosophers' lead is drawn an oil of golden color, or much like it. . . . the hidden oil that makes the medicine pen-

---

101. Colin George Flynn, "The Decline and End of the Lead-Mining Industry in the Northern Pennines, 1865–1914: A Socio-Economic Comparison between Wensleydale, Swaledale, and Teesdale" (PhD diss., Durham University, 1999).

102. Ripley, "Preface," in *TCB*, 126; my emphasis. I discuss the role of copper in the *Compound* on pp. 117–18, below.

103. The recipe for *nigrum nigrius nigro*, one of the most important in pseudo-Lullian alchemy, appears in an extended form in the *Compendium artis alchimiae et naturalis philosophiae*, also known as the *Magica naturalis*; Lull [pseud.], *De alchimia opuscula*, fol. 11r.

104. *Medulla*, fol. 5r: "Sic enim dicit Raymundus. Et tartarum illud nigrius est tartaro vue nigre catalonice, quare vocatur nigrum nigrius nigro. . . . Sed quam res ista in istis et in quibusdam alijs partibus rara est. Guydo de Montanor philosophus grecie invenit aliud vnctuosum humidum quod omnibus liquoribus supernatat a vino ortum." Ripley's rejection of tartar is evidently not based on lack of familiarity with the process, for his earlier, playful comment on tartar's metalline character refers to its appearance when dried in the sun (a procedure necessary to remove excess water prior to distillation).

etrable, friendly, and joining to all bodies."[105] However, Raymond does not specify the kind of lead to be used, while the process is so truncated that a reader, unless already familiar with the procedure for extracting "oil of lead," would not be able to replicate it from this account alone. While Ripley may well have read into this terse reference a confirmation of his own procedure for the vegetable stone, the *Epistola* cannot have been his only source. Ultimately, Ripley passes over Raymond's ambiguous processes in favor of his own metalline water drawn from the enigmatic "red lead": a substance that has no direct parallel in the pseudo-Lullian literature that he cites, and therefore suggests the outcome of his own experimentation and experience.

Yet Ripley's *practicae* are just as likely as his *theoricae* to be shaped by written accounts. In likening his gum to Raymond's *vitriolum azoqueus* and his condensed vapor to a stinking menstruum, Ripley links his observations to two substances reported in the *Testamentum*, thereby generating a Lullian endorsement for a process that he admits has not been taken directly from Raymond. He has also varied the usual recipe for sugar of lead by recommending red lead rather than the other lead compounds, such as ceruse or litharge, more commonly used in its manufacture.

In adjusting his recipes and altering the underlying chemistry, Ripley seems to be acting on information gleaned from his own experience in manipulating a variety of materials. At the same time, he continually seeks to support his use of such innovations through reference to textual sources. Since his red lead is not endorsed by Raymond (except by ingenious reinterpretation of the Lullian resoluble menstruum or oil of lead), he imports a cover name, sericon, of even more authoritative provenance. Sericon is one of the substances mentioned by the philosopher "Mundus" in the *Turba philosophorum*, an early thirteenth-century Latin translation of an Arabic text probably composed around 900 CE.[106] While its early sense and ety-

---

105. *BCC*, 1: 866; translation in Sloane 1091, fol. 101r-v: "off [th]e phi*losoph*orum lede [th]er is drawne out an oyle off goldy color or mych lyke wi*th* the whych || iff [th]ou sublyme iii or iiij tymes [th]e mynerall stone or [th]e ani*m*all aft*er* [th]e fyrst fixyon [th]ou shall be excusyd off all maner off sublymacyons and coag*u*lacyons And [th]e cause is for [th]is is the hydde oyle [th]*a*t makyth [th]e medycyn penetrable frendly and Joyny*n*g to all bodys and [th]e effecte shall be encresyd passing hugely \myghtyly/ So [th]*a*t [th]*er* is no thing so secrete nor more sure in [th]e world."

106. Julius Ruska, ed., *Turba Philosophorum: Ein Beitrag zur Geschichte der Alchemie*, Quellen und Studien zur Geschichte der Naturwissenschaften und der Medizin 1 (Berlin: Springer, 1931), 169: "Oportet igitur, ut plumbum in nigredinem convertatur; tunc decem praedicta in auri fermento apparebunt cum sericon, quod est compositio, quod et decem nuncupatur nominibus." Ruska gives the Arabic name as *sīrīqūn* (30). On the *Turba*, see also Didier Kahn, "The *Turba philosophorum* and Its French Version (15th c.)," in López Pérez et al., *Chymia*, 70–114.

mology are obscure, the term seems to have originally signified a red pigment, which by Ripley's time had come to be identified with red lead.[107] The significance of the color red in alchemical symbolism—associated with both blood and the culminating *rubedo* stage of the philosophers' stone—may also have counted in its favor.[108] By choosing "sericon" as the prime matter of his vegetable work, Ripley locked his novel practice into alchemy's ancient past.

PRACTICAL EXEGESIS

Writing in the 1470s, Ripley was only the latest in a long line of alchemical commentators who sought to understand past authorities by expounding them in light of other books and possible meanings. As a scholar and a churchman, however, he would already have been familiar with the art of interpreting texts in this way, for such techniques were hardly the unique province of alchemy. Since late antiquity, theologians had approached the Bible as a text of almost impossible complexity—a holy book devised by God and intended to be read on many levels, according to the ability and wisdom of the reader. Even seemingly straightforward, narrative accounts encompassed multiple layers of truth, each requiring a different method of interpretetation: literal, allegorical, moral, or anagogical.[109] For readers trained to reflect on meaning beyond the letter of the text, it was a mere step to

107. Isidore of Seville describes *Syricum* as a red pigment used to add the capital letters to books, which he explicitly differentiates from *sericum*, silk; *Isidori Hispalensis episcopi Etymologiarum sive originum libri XX*, ed. W. M. Lindsay (Oxford: Oxford University Press, 1911), bk. 19. Alchemical texts in Byzantine Greek also use σηρικόν to denote a red pigment, although this usage may simply refer back to the Latin *Syricum*. See also Dietlinde Goltz, *Studien zur Geschichte der Mineralnamen in Pharmazie, Chemie und Medizin von den Anfängen bis Paracelsus* (Wiesbaden: Franz Steiner, 1972), 190–91.

108. On the alchemical significance of this color, see Pamela H. Smith, "Vermilion, Mercury, Blood, and Lizards: Matter and Meaning in Metalworking," in *Materials and Expertise in Early Modern Europe: Between Market and Laboratory*, ed. Ursula Klein and E. C. Spary (Chicago: University of Chicago Press, 2010), 29–49, on 41–45.

109. Typically, the literal (or historical) sense concerns past events; the allegorical sense explains one thing through its similarity to another (in scripture, by drawing connections between the Old and New Testaments); the moral (or tropological) sense concerns proper conduct in the present; and the anagogical (or eschatalogical) sense relates to the future, after the end of days. On the techniques of scriptural exegesis, the classic work is Henri de Lubac, *Exégèse médiévale: Les quatre sens de l'Écriture*, 4 vols. (Paris: Aubier, 1959, 1961, 1964), translated as *Medieval Exegesis: The Fourfold Sense of Scripture*, trans. Mark Sebanc (vol. 1), Edward M. Macierowski (vols. 2 and 3), 3 vols. (Grand Rapids, MI: Eerdmans, 1998–2009). See also Beryl Smalley, *The Study of the Bible in the Middle Ages* (Oxford: Clarendon Press, 1941); Lesley Smith, *The Glossa Ordinaria: The Making of a Medieval Bible Commentary* (Leiden: Brill, 2009).

applying the same approach to alchemical writings. No surprise, then, if Ripley's exegetical skills found an alternative outlet in his attempts to resolve the multiplicity of pseudo-Lullian fires, waters, and menstrua—even if, less conventionally, these attempts were assisted by a new kind of philological tool, in the form of his own experimental practice.

We might regard the results as a kind of "practical exegesis," whereby specific processes and products (the resolutive menstruum, the vegetable stone) are forcibly reinterpreted to accommodate such considerations as the availability of local materials and compatibility with the practitioner's own empirical observations. Just as Ripley manipulated conflicting textual sources to obtain consensus, so he modified recipes to fit practical findings, and practical findings to fit established tropes. Thus, while Ripley's practices and theoretical arguments have their origins in recognizable fourteenth- and fifteenth-century exemplars, his *Medulla* may be reduced neither to a compilation of earlier authorities, nor to a straightforward recipe collection. In its consistent elaboration of pseudo-Lullian doctrines, supported by source criticism and applied to material pursuits, it provides both a commentary on a preexisting tradition, and a serious practical engagement with the challenges posed by a confusing and—unknown to Ripley—pseudepigraphic corpus. Between the cracks of Ripley's familiar sources, we catch glimpses of flexibility and innovation in the staging of his own empirical work: a source of knowledge that would feed back into his own writings and those of his later readers.

Once identified, Ripley's sericonian alchemy turns out to be ubiquitous in fifteenth- and sixteenth-century English alchemy. So, too, do his exegetical methods. As later readers sought to recreate the mineral and vegetable stones, they applied the same techniques, modifying their textual sources to accommodate new observations resulting from consequential variations on the original practice—whether these variations were intentional, as in the case of deliberate substitution, or accidental, as might occur through the presence of impure ingredients or modified apparatus. The feedback loop of text and practice offers a tool not just for reading alchemical commentaries, but also for detecting how the underlying chemistry altered over time.

It would be a mistake, however, to imagine this loop as a closed one. The continual circulation between the reading of alchemical texts and the interpretation of experimental findings was not hermetically sealed, but shaped by a host of factors: not just philosophical coherence, but the practitioner's economic circumstances, religious beliefs, and views on correct moral conduct. Any two readers might bring different interpretations to bear on the same source material, or come to different conclusions on the basis of similar practical results. For instance, as we shall see, Ripley was personally concerned with the need to make his science available to poor practitioners of the kind envisaged by John of Rupescissa. His own status as a poor religious man may have shaped his response to the issue of cost, one satisfied by the use of inexpensive base metals like lead and copper.

Sericonian alchemy would continue to shape English alchemical discourse well into the seventeenth century. By the 1650s, a readerly preference for the transmutational goals of Ripley's *Compound* would gradually divert attention from the multipurpose practice outlined in the *Medulla*, which prized the medicinal vegetable stone above the mineral work. Yet the robustness of Ripley's alchemy lay not just in its success as a practical rendering of the prestigious pseudo-Lullian corpus, but also in its adaptability to new interpretations based on differing circumstances. By reading "sericon" not only as red lead, but as any one of a variety of leaden compounds—or as a different metal entirely, or even as a nonmetallic ingredient such as tartar—Ripley's own commentators could substitute new ingredients while still producing interesting chemical outcomes, often with chrysopoeian goals. They could also do so affordably. At once philosophically intelligible, morally unimpeachable, and practically efficacious, Ripley's vegetable stone would become the constant, yet ever-varying, refrain of English alchemy.

CHAPTER THREE

Opinion and Experience

Opinyon is whyle a thyng is in non certayne, and hydde
from mens very knowlegyng.[1]

At the end of the *Compound of Alchemy*, George Ripley confesses that he
is no stranger to failure. A concluding poem, the "Admonition," sets out
his early experimental mishaps, when he was "dyscevyd wyth many falce
Books."[2] The poem goes on to blacklist almost every conceivable alchemical
ingredient, including minerals like antimony, sal ammoniac, and sandiver;
animal products such as urine, hair, and blood; and more exotic matter still,
from "Tarter Egges whyts, and the Oyle of the Snayle" to "The Slyme of
Sterrs that falleth to the grownde." The alchemist concludes with a riddle:
he "never saw true worke treuly but one"—a metalline substance in which
the clearness of gold and silver "be hyd fro thy syght."[3] This is the secret of
Ripley's prime matter; this, he advises, is the work "Of whych in thys tretys
the trewth I have told." But what is the nature of this mysterious prime matter? To answer that question, and unlock the secret of the twelve gates, Ripley's readers must comb once more through the poem. But unless they also
understand how to read like a philosopher, the answer will remain, like the
secret of his gold and silver, hidden from sight.

While Ripley's record of "Many Experyments" testifies to a strong empirical inclination, he was obviously not engaged in an experimental program

---

1. Thomas Usk, "Testament of Love," in Geoffrey Chaucer, *The Workes of Geffray Chaucer Newly
Printed*, ed. William Thynne (London, 1532); cited in *OED*, s.v. "opinion." The text of Usk's *Testament* survives in no original manuscript and is preserved only in Thynne's 1532 edition of Chaucer.

2. Ripley, "An Admonition, Wherein the Author Declareth his Erronious Experiments," in *TCB*,
189–93, on 191.

3. Ibid., 192.

in any modern sense of the term. Ripley is describing his attempts to replicate the processes, products, and effects set down in alchemical recipe collections, which bring together diverse practices from diverse sources, and
using diverse ingredients. He blames his failure not on any lack of skill on his
part, but rather on the negligence of past copyists who allowed practices to
circulate without adequately testing them first. Only after great expenditure
of time and money does he claim to have discovered the errors of these false
experimenta:

> In these I practysyd as in my books I found,
> I wan ryght nought, but lost many a pownde.[4]

By his own account, empirical rigor thus sets his practice apart from that
of compilers who throw together recipes according to fancy rather than
proof of practice. This opposition between proof and opinion is a characteristic feature of Ripley's writing, one that he developed further outside the formal frame of patronage suits like the *Compound*. In particular,
it can be teased out from a fifteenth-century source that has gone unnoticed in modern times: Ripley's own compilation of treatises, recipes, and
poems, gleaned from a manuscript that his early modern readers dubbed the
"Bosome Book."

Despite its previous neglect, the *Bosome Book* offers the key to Ripley's
alchemy. Like the "Admonition," its contents reveal the canon to be first
and foremost a practitioner, who endorsed systematic testing and frankly
admitted his earlier errors in a way that later impressed his seventeenth-
century editor, Elias Ashmole.[5] But although Ashmole saw in these writings
a precursor to the experimental philosophy of his own time, Ripley's trials
were not intended to generate natural knowledge independently of his textual sources. Rather, the canon sought to recover knowledge that had been
known to the authorities all along. Like other practitioners of his time, he
tested alleged "experiences," or *experimenta*, to see whether the processes
described in texts accurately describe what is really observed—whether they
"work."[6] At the simplest level, the activity is one of attempting to replicate
past knowledge and past results through interpreting alchemical texts. The

---

4. Ibid., 191.

5. Elias Ashmole, "Annotations and Discourses upon Some Part of the Preceding Works," in
TCB, 456.

6. On the relationship between *experimenta* and observation, see particularly Park, "Observation in the Margins," in Daston and Lunbeck, *Histories of Scientific Observation*, 15–44; Edward

danger, as Ripley well knew, is that an outcome cannot be reproducible if it was never achieved in the first place: if a source merely records its writer's fancy, or opinion.

"Opinion" is a freighted term in early modern scientific writing. In Middle English, it may express the considered judgment of an individual or group, but also a view or belief based on something other than reason or experience—the product of imagination or groundless supposition.[7] Opinion thus stands in contrast to both natural philosophical reasoning, which seeks certainty through an understanding of universal natures and causes, and the knowledge of individuals gained from extensive experience.

The term begins to creep into English alchemical writing during the fifteenth century, with mixed connotations. Ripley uses it critically and polemically, while confessing that in the past he too has worked according to "opinion" rather than "truth." But for a reader lost in the darkened byways of alchemical literature, opinion might still serve some positive role, if not as a map to the labyrinth, then at least as a lamp to lighten the path. Indeed, in a science whose most authoritative texts demand multiple levels of decipherment, some degree of conjectural interpretation seems impossible to avoid. Whether testing different ingredients to identify the meaning of a tricky cover name, or importing practical knowledge to reconcile an apparent disagreement between authorities, practical exegesis relies on readers forming opinions about their subject matter—opinions that can then be tested. With every theory underdetermined by evidence, the challenge for Ripley and his contemporaries lay in reining in their speculations through the judicious use of observation and experiment.

RECEIPTS AND DECEITS

George Ripley's critique of opinable experiments begins in a familiar place: the philosopher's disdain for the empiric. On the face of it, there is a clear break between the writings of learned philosophers and homelier *experimenta* grounded in individual experiences. The latter, often recorded in English rather than Latin, circulated in ever-increasing numbers throughout the fifteenth century, sometimes prompting a critical response in more

---

Grant, "Medieval Natural Philosophy: Empiricism without Observation," in *The Nature of Natural Philosophy in the Late Middle Ages* (Washington, DC: Catholic University of America Press, 2010), 195–224.

　　7. *OED*, s.v. "opinion." The term derives from Latin *opinor*, which has the sense of imagination or supposition, but probably entered the English language via the French *opinion*.

philosophically oriented writings; even, paradoxically, in treatises that themselves incorporated content from earlier recipe literature.[8] The alchemist thus takes on something of the persona of the medical empiric, a figure reviled by medical authorities from Galen onward, who insisted that medicine should be grounded in reason rather than the nostrums of peddlers and charlatans.[9]

In both alchemy and medicine, this tension reflects the scholastic concern to establish certitude of knowledge within given domains of inquiry. By 1300 physicians had succeeded in accommodating medical theory, and, to a large extent, medical practice, within the natural philosophy of the schools, based on the teachings of Aristotle.[10] Yet physicians still had to accept that the causes of a drug's efficacy could not always be known: a given remedy might simply "work." Empirical findings therefore remained important in practice, while scholastic physicians themselves came under fire from empirical medical practitioners like the Dominican healer Nicholas of Poland (ca. 1235–ca. 1316), who blamed them for relying on the authority of Hippocrates and Galen rather than their own experience.[11]

The danger of dispensing with such evidence in favor of reasoning from universals was obvious to the Montpellier master Arnald of Villanova. In his *De intentione medicorum* (On the Purpose of Physicians), written in the early 1290s, Arnald endeavored to show that the useful knowledge of physicians, acquired over the course of repeated exposure to individual cases, should be viewed on a par with the more certain knowledge of natural philosophers.[12] It was unnecessary for a practicing physician to have complete knowledge of the hidden causes of disease; it was enough that he should recognize the

---

8. On some of the distinctive aspects of English alchemical recipes, see Peter Grund, "The Golden Formulas: Genre Conventions of Alchemical Recipes in the Middle English Period," *Neuphilologische Mitteilungen* 104.4 (2003): 455–75.

9. On the variety of medical practitioners in late medieval England, see Getz, *Medicine in the English Middle Ages*. On medical "charlatans," see David Gentilcore, *Medical Charlatanism in Early Modern Italy* (Oxford: Oxford University Press, 2006); for their alchemical counterparts, Nummedal, *Alchemy and Authority*, chap. 2.

10. On this process, see Roger French, *Canonical Medicine: Gentile da Foligno and Scholasticism* (Leiden: Brill, 2001); French, *Medicine before Science: The Business of Medicine from the Middle Ages to the Enlightenment* (Cambridge: Cambridge University Press, 2003); Siraisi, *Taddeo Alderotti*.

11. William Eamon and Gundolf Keil, "*Plebs amat empirica*: Nicholas of Poland and His Critique of the Medieval Medical Establishment," *Sudhoffs Archiv* 71 (1987): 180–96.

12. Arnald de Villanova, *Opera medica omnia*, vol. 5,1: *Tractatus de intentione medicorum*, ed. Michael R. McVaugh (Barcelona: Publicacions I Edicions de la Univ. de Barcelona, 2000); McVaugh, "The Nature and Limits of Medical Certitude at Early Fourteenth-Century Montpellier," *Osiris*, 2nd ser., 6 (1990): 62–84.

outward signs of the patient's illness and administer remedies known to be effective—an approach that Michael McVaugh has termed "medical instrumentalism."[13]

There is a tempting parallel here with alchemical practice. Arnald's vision of a good "instrumentalist" physician is a practitioner who is hostile neither to reason nor to experience, but who eschews unsupported theorizing in the absence of experiential evidence. Almost two centuries later, Ripley evoked a similar ideal when he urged his own readers to work from both reason and experience, while mistrusting their fancy. In Ripley's view, it was useless for practitioners to stockpile receipts if they were not also willing to test them, and to learn from their results.

Like Arnald, however, he did not dismiss theory. The pseudo-Lullian *Testamentum* and *De secretis naturae* were successful as philosophical authorities in part because they aimed to bridge the chasm between practical experience and the certainty of natural philosophical reasoning, arguing not only that particular processes worked, but that they must of necessity do so. In chapter 1, for instance, we saw how the Magister Testamenti used circular figures and alphabets to package a relatively straightforward recipe for corrosive sublimate made from mercury, vitriol, and saltpeter. For the scholastically minded Magister, knowledge of the operative dimension of alchemy was useless without grounding in general principles: one must "know the practice which is formed by art with theoretical reason." By studying the figures in the *Practica* of the *Testamentum*, a practitioner could memorize the combination of materials necessary to create a given substance, but also, crucially, *understand* why the former must yield the latter. "Unless you know the said alphabet by heart," the Magister warns, "you cannot practice, nor can you even begin."[14]

Recipe collections undermined the authority of such claims, by making explicit the names of substances that were carefully protected in philosophical writings. They also dispensed with causes, presenting alchemical knowledge in a format more closely associated with craft practice than with scholarly learning. Since the circulation of untested receipts threatened to undermine the prestige of alchemy as a science, self-identified philosophers sometimes adopted a critical attitude toward recipes and those who gath-

13. McVaugh, "Nature and Limits of Medical Certitude," 68.

14. *Testamentum*, 2:314: "Fili, istud alphabetum oportet sciri cordetenus a te, si vis scire practicam, que formatur per artem cum ratione / theorice, que venit prompte infra formam memorie omni nobili intellectu[i], si eam studueris. . . . Et nisi scias cordetenus predictum alphabetum, non poteris practicare nec eciam solum incipere."

ered them. Often this involved mythmaking, by inventing or appropriating cautionary tales about wicked alchemists and their dupes—fictional strawmen who gradually acquired a life of their own in successive tellings.

Pseudo-Arnald condemned the harvesting of treatises by practitioners greedy for receipts, underscoring his warning with the exemplum of "a monk who had labored hard in this art for twenty years and still knew nothing."[15] Losing heart, the monk compiled thousands of false receipts in a book called the *Flos paradisi*, which he then deliberately allowed to circulate. In 1477, Thomas Norton silently incorporated this tale of preindustrial sabotage into his own English poem:

> As the monke which a boke dide write
> Of a [thousand and one] receptis in malice for despite;
> which be copied in many a place
> wherebi hath be made pale many a face.

To trust in anonymous recipes was to place oneself in the hands of frauds, and on the path of failure:

> Avoide youre bokis writen of receytis,
> For al such receptis be ful of deceytis.[16]

Philosophers' criticisms were leveled not only at false receipts, but also at false practitioners. Medieval Latin authorities routinely warned against fraudulent or ignorant practice—a tradition that entered the English language not in a practitioner's treatise, but via Chaucer's *Canterbury Tales*. "The Canon's Yeoman's Tale" describes the ignoble contrivances of two canons, one of whom is recognizably an Augustine canon regular.[17] In the early decades of the fifteenth century, vernacular instruction was still sufficiently thin on the ground for Chaucer's poem itself to become established as an alchemical authority. Norton even cited "The Canon's Yeoman's Tale" in the

---

15. Arnald of Villanova [pseud.], *De secretis naturae*, 512: "Vidi autem unum monacum qui bene in ista arte laboraverat per viginti annos et nichil sciebat. Tunc ipse quasi desperatus fecit unum librum et initulavit eum *Flos paradisi*, in quo plus quam 100 000 recepte sunt contente. Et illum librum dabat omnibus ad copiandum. Et sic gentes decipiebat et seipsum quia erat totus desperatus."

16. Norton, *Ordinal*, 7 (ll. 89–100).

17. On the ecclesiastical status of Chaucer's canons, see Marie P. Hamilton, "The Clerical Status of Chaucer's Alchemist," *Speculum* 16 (1941): 103–8. On the persona of the fraudulent alchemist more generally, see Nummedal, *Alchemy and Authority*, chap. 2.

Ordinal to illustrate the use of obscure cover names to denote ingredients: "vnknow bi more vnknow named is she."[18]

Chaucer's tale was probably also the inspiration for Norton's advice to his own fictional interlocutor, the hapless novice Tonsile, who bewails his dealing "In fals Receipts, and in such lewde assayes." His reliance on recipe collections led him to work in "many kinds":

> In heere, in eggis, in merdis, & vryne,
> In Antymonye, arsenek, in hony, wax, & wyne,
> In calce vive, sondyfere [i.e., sandiver], and vitrialle,
> In marchasites, Toties [i.e., tutties], & euery mynerall,
> In malgams, in blaunchers, in citrinacions,
> All fille to nogthe in his operacions.[19]

Norton's rich and varied listing of materials, some of which were undoubtedly used in practice, reflects the ingredients previously cited by Chaucer: "Poudres divers," "salt Peter, and Vitriole," "Sal Tartre, Alkaly, and Sal preparate," and "Tartre, Alym, Glas, Berme, Worte and Argoyle"—to name but a few.[20]

Despite the superficial similarities with his own "Admonition," Ripley may have felt his position more keenly than either Chaucer or Norton. Not only was he a canon himself, but he belonged to the same order as Chaucer's rogue alchemist. In his fifth gate, on "Putrefaction," Ripley turned the tables on Chaucer's portrayal of pilgrims beset by deceitful canons, for the *Compound*'s false alchemists are not religious men, but lay practitioners who

---

18. Norton, *Ordinal*, 38 (ll. 1162–66):

> And chawcer rehersith how titanos is the same,
> In the Canon his tale, saynge: whate is thuse
> But Quod Ignotum per magis ignocius?
> That is to say, whate may this be
> But vnknow bi more vnknow named is she?

19. Norton, *Ordinal*, 35 (ll. 1057–62); my insertions.

20. Chaucer, "The Canon's Yeoman's Tale," in *TCB*, 235–36. On Chaucer's alchemy, see Edgar H. Duncan, "The Literature of Alchemy and Chaucer's Canon's Yeoman's Tale: Framework, Theme, and Characters," *Speculum* 43 (1968): 633–56; Collette and DiMarco, "The Canon's Yeoman's Tale." On the tradition of Chaucerian satire in English alchemy, see Stanton J. Linden, *Darke Hierogliphicks: Alchemy in English Literature from Chaucer to the Restoration* (Lexington: University Press of Kentucky, 1996), chap. 1.

haunt Westminster Abbey and prey on the gullibility of monks and friars.[21] Succumbing to the lure of diverse ingredients, whether soot, dung, urine, wine, blood, or eggs, these charlatans make handsome promises to their patrons and creditors, "But as for Mony yt ys pyssed on the walls."[22]

The false alchemists in Ripley's poem have long been viewed as stock figures of alchemical satire directed against fraud, but his tale also points to a more specific moral—the vice of opinion and the virtue of proof. Thus, although Ripley's characters are rogues, at no point does he claim that their belief in alchemy is insincere: rather, their financial and legal woes arise from the high expenses they incur in the course of unproductive practice, because they are "mevyd to worke after ther fantasy."[23] To tweak McVaugh's term, they are not "alchemical instrumentalists," since they lack the experience to make sense of their book learning.

Such fantastical procedures provide a foil for Ripley's own methodology, which is not only informed by reason, but thoroughly grounded in experience:

> But fyrst examyn, grope and taste;
> And as thou provyst, so put thy confydence,
> And ever beware of grete expence.[24]

Ripley's morality play illustrates two themes that recur throughout his writings: the value of proof over opinion, and the need to avoid high costs in the work. While it underscores the theme of his "Admonition," in which he deplores the circulation of untested *experimenta*, Ripley is also concerned to distinguish such practices from his own work—for, despite his condemnation, the fastidious canon was himself an active compiler of receipts. But unlike the *Compound*'s fictive protagonists, Ripley does not consider his own program to be "opinable," since he has one clear object in view: putting the authorities to work in the service of developing his signature, sericonian practice.

---

21. For instance, Ripley, "Putrefaction," in *TCB*, 157:

> And when they there syt at the wyne,
> These Monkys they sey have many a pound,
> Wolde God (seyth one) that som were myne;
> Hay hoe, careaway, lat the cup go rounde.

22. Ripley, "Putrefaction," 155.

23. Ibid., 153.

24. Ibid., 159.

TRUTH IN PRACTICE

When mathematicians show their working, or artists reveal the preliminary sketches for a finished piece, our assessment of the final result is bound to change. Our perception of Ripley's *Medulla* and *Compound of Alchemy* is transformed by reading them against the *Bosome Book*, a manuscript compendium of treatises and recipes that Ripley seems to have compiled during the 1470s.[25] While I examine the authenticity of the *Book* in greater detail in chapter 7, for now the main point to note is that the original, fifteenth-century manuscript collection, which was almost certainly compiled by Ripley himself or else copied from one of his manuscripts, is no longer extant. Since the *Bosome Book* does not exist in its original form, my reconstruction of its contents therefore relies on the copies and translations made by Ripley's later readers, pieced together primarily from manuscripts held in British archives.[26]

The reconstructed *Bosome Book* shows that although Ripley gathered scores of recipes, he did not do so willy-nilly. His collection includes a much wider variety of ingredients than that represented in the *Medulla*. However, the majority of its content relates to Ripley's twin obsessions, vitriol and sericon: the respective sources of his fire against nature and natural fire.

Ripley himself acknowledges the dominance of his sericonian practice in a brief apologia, appended to a *practica* with the distinctly antimercurialist title of *Praeparatio calcis ovorum* (The Preparation of the Calx of Eggs). In this remarkable passage, which bears more than a surface resemblance to the "Admonition," Ripley confesses to exactly the same failings that he criticizes so harshly in the *Compound*. In his hubris, he has circulated *experimenta* that followed opinion rather than "truth of practice":

> And I pray all men, that wherever they shall find anything concerning my experiments, written by me or titled with my name, that they burn them, or put no faith in them, because I wrote them as I supposed, not as I proved true. That work excepted which was by means of the menstruum, diversely

---

25. Although several components of the *Book* were printed in the early modern period (one under the title of "Bosom-Book"), these have not been examined in the context of the original compilation; CRC, on 132–33, and CRC 3; Rampling, "John Dee and the Alchemists: Practising and Promoting English Alchemy in the Holy Roman Empire," *Studies in History and Philosophy of Science* 43 (2012): 498–508, on 504–5.

26. These are listed under CRC 3. On several copies held in continental archives, see chap. 8, below.

related by me in my writings so that it might be hidden from ill-disposed persons. To that work, let them apply all diligence and they shall find what they desire, God willing. For I myself consigned to the fire many leaves written with experiments that followed opinion not proof, which after proving I found did not accord with the truth. Therefore I crave pardon from God, and from all to whom I was the cause of error through my writings, from the year of our Lord Jesus Christ 1450 to the year 1470, since for so long I sought the stone and did not find it in truth of practice until towards the end of that same year. Then I found him whom my soul loveth, yet not inordinately, as God Himself knows. Herewith G. R.[27]

Undoubtedly there is a performative element to these startling disclosures: Ripley's claim to have found the stone in late 1470 rather too conveniently paves the way for the *Compound*, completed the following year. Since we cannot consult the original document, there is no way of knowing whether the apologia was written at the same time as the *Praeparatio calcis*, or whether it constitutes a late addition. If the former, we can assume that the *Book* was compiled at around the same time as the *Compound*, or shortly after. This conclusion makes sense if we assume the collapse of Ripley's family fortunes in 1471, which may have prompted his search for a patron.

Even if we avoid taking Ripley's alchemical confession at face value, it offers a rare and powerful rhetorical statement on the role of evidence in alchemical practice. In addition to distinguishing between recipes written as he supposed (*opinabar*) and as he proved (*probavi*), Ripley incorporates the same oppositional pairing in one of the self-referential epigrams scattered throughout the *Book*:

---

27. Harley 2411, fol. 64r-v: "Et supplico omnibus hominibus quod vbicumque invenerint aliqua de meis experimentis, scripta per me, vel nomine meo intitulata; comburant ea vel non adhibe||ant eis fidem; quia scripsi sicut opinabar, non vt probaui vera; Excepta operatione illa, quae per menstruum varie traditur per me in scriptis vt occultetur a male dispositis personis, Ibi apponant operam tota diligentia et invenient quod optant, volente Deo. Nam et ipsemet tradidi ignibus multa folia in scripta experimentis secundum existimationem non probationem, quae postea probando non inveni consona veritati. Ideo peto veniam à Deo et ab omnibus quibus causa extiti erroris per scripta mea ab Anno Domini Jesu Christi 1450. vsque ad eiusdem Domini nostri Jesu Christi annum 1470. quia per tantum tempus quaesivi Lapidem, et non inveni illum in veritate practicae, donec eiusdem anni terminus propinquaret, tunc inveni quem diligit anima mea, non tamen inordinate, vt ipse Dominus novit. *Haec G: R.*" In the last line Ripley quotes Song of Songs 3: "In lectulo meo, per noctes, quaesivi quem diligit anima mea" (Vulgate); "In my bed by night, I sought him whom my soul loveth" (Douay-Rheims).

Nec dat opinata, sed vera dat a*tque* probata.[28]

Nor does he offer fancies, but only what is true and tested.

Yet the distinction itself raises the question of why, and how often, alchemical practitioners chose to include a provisional or suppositional element in their recipes. What prompted Ripley—or any alchemical practitioner—to set down untested processes in the first place? How, too, should we evaluate his later practices, which also describe seemingly impossible outcomes, now endorsed by proof of experience? And why is the apologia appended to a recipe for calcining eggs?

Crucially, Ripley does not forswear all of his former practices—his confession is not a rejection of alchemy, of the type later penned by frustrated former practitioners such as Vannoccio Biringuccio or Nicolas Guibert.[29] As in the "Admonition," Ripley urges his readers to disregard all possible works except for one tried-and-tested practice: "the menstruum, diversely related by me in my writings." The apologia thus serves as an advertisement for a single procedure, which we might regard (as Ripley himself seems to have done) as his "signature practice." The mystery menstruum is, of course, the same sericonian solvent that we have already encountered in the *Medulla*, where it serves the ends of both medicine and transmutation. Indeed, Ripley seems to have drawn heavily on the *Book*'s contents while assembling the *Medulla*, which frequently quotes recipes and philosophical dicta from the earlier compilation.

In lashing himself to the sericonian mast, Ripley distances himself from alternative experimental approaches. Although the *Book* retains some diversity (including a number of recipes using arsenic, an ingredient that he later condemns), many of its recipes and short treatises start with imperfect bodies identifiable with sericon. In others, Ripley carefully interprets earlier authorities, including those of the first rank, like Hermes and Aristotle, in accordance with sericonian precepts. Such selectiveness makes the practical content of the *Book* unusually cohesive for a compilation of this period. Certainly this is deliberate: Ripley even notes at the end of one practice that "there are many other branchings of alchemy about which I keep silent," fearing to lead other practitioners to inferior works, and so diminish the

28. Harley 2411, fol. 64v.

29. Vannoccio Biringuccio, *De la pirotechnia. Libri .X.* (Venice: Curtio Navò, 1540); Nicholas Guibert, *Alchymia ratione et experientia ita demum viriliter impugnata* . . . (Strasbourg: Lazarus Zetzner, 1603).

art. He regrets that his scruples are not observed by others of his time, who "circulate fantastical operations, which are had for the deceiving of fools in quicksilver, arsenic, marcasites, salts, and alums, and suchlike, which are all foreign to our most true mastery, as God himself knows."[30]

The one cloud on the horizon of Ripley's sericonian practice is the question of cost. Ripley's menstruum is the key to his work, but it provides only the material principle of the stone; that is, the bulk of its substance. The tenets of mercurialist alchemy required that this base be fermented with something more expensive—the gold and silver that supply the "form" of the elixir, and the source of Raymond's resoluble menstruum. Such a doctrine seems difficult to reconcile with Ripley's warning in the *Compound*, to "medyll wyth nothyng of gret cost."[31]

By a roundabout route, Ripley's dilemma returns us to the same conflict discussed earlier in the book, when mercurialists first objected to pseudo-Avicenna's use of organic ingredients. Such materials might be cheap, but were they effective? This debate was revived by John of Rupescissa, who intended his quintessence to increase poor men's access to medicine. The grafting of John's quintessence into pseudo-Lullian transmutation theory thus resulted in a new source of conflict between authorities, concerning cost: one that Ripley's sericonian alchemy seemed ideally positioned to resolve.

GUIDO VS. RAYMOND

Although mercurialist texts promoted the triad of mercury, gold, and silver on natural philosophical grounds, a moral problem remained. For a poor man, obtaining even a tiny quantity of precious metal still posed a formidable challenge. How was this situation to be squared with the view that God had made the art of alchemy accessible to all—or with the famous aphorism that the stone was so cheap that it was unknowingly trodden underfoot? Philosophical arguments in favor of metallic ingredients did not help to solve the main moral objection to the high cost of doing alchemy: that if the elixir was truly a gift of God then it should be available to all worthy practitioners,

30. Harley 2411, fol. 68r: "Plures alias Alkymicae ramificationes sunt, de quibus taceo, nolens occasionare alios ad operandum inferioribus . . . omnes qui circumeunt fantasticas operationes, quae habentur ad stultorum deceptionem in Argento vivo, Arsenico, sulphure, marcasitis, salibus, et aluminibus et huiusmodi, quae a nostro verissimo magisterio omni moda sunt aliena, vt ipse Dominus novit."

31. Ripley, "Putrefaction," in *TCB*, 158.

whether rich or poor, and regardless of their ability to win the favor of powerful patrons.

Ripley himself was never entirely comfortable with high expenses, including the use of precious metals in the work. Despite his later fame as an expert on fermentation with gold and silver, the cost of the stone is a topic that recurs throughout his writings. He opens the *Medulla* by assuring the bishop that his work does not call for large quantities of precious metals:

> But shall I ask of my lord, for this great treasure, a great sum of gold or silver? Or, to show him in deeds that which I have set out in writing, shall I wish to persuade him to undertake a heavy burden, or put into my hands a great quantity of gold or silver? How should that accord with the philosophers, who say thus: "Purses are not to be loosened for making great expenses, which this art requires not"?[32]

This introduction seems at first to contradict the view of Ripley's mercurialist authorities, including Raymond, that riches are necessary for successful practice.[33] Yet he qualifies his remarks just a few pages later. His "goldless" stone works only for the elixir of metals, and is therefore of no use for making the medicinal vegetable or animal stones. To produce the *elixir vitae* will require a certain quantity of gold, but, according to Raymond, gold made with the help of the mineral stone cannot be used for medicine—for, as we saw in chapter 2, gold tainted by the fire against nature will be deadlier than a basilisk's gaze.[34] For that purpose Ripley must regretfully "require of my Lord not one great pound, but one lesser pound of most fine gold."[35]

Ripley's advance haggling over his expenses may at first glance seem disingenuous. If the ailing George Neville was indeed the intended recipient

---

32. *Medulla*, fol. 1v: "Sed numquid a domino meo pro hoc magno thesauro magnam auri sum[m]am expectam aut argenti? Aut vt factis sibi ostendam que pando literis: numquid volem vt aut sumptuum onus graue subire afficiatur, aut meis manibus auri vel argenti copia exponatur? Quomodo hoc consonum esset philosopho sic dicenti. Non sunt dissoluenda marsupia propter magnas expensas faciendas, quas ars ista non requirit."

33. The need for money as well as book learning is made explicit in the *Testamentum* and the pseudo-Arnaldian *Rosarius philosophorum*, both of which cite an aphorism from *De anima*, warning that "riches, wisdom, and books" are necessary if one is to advance in the art. *De anima*, ed. Moureau, 245: "habet necesse divitias, sapientiam et libros"; Arnald of Villanova [pseud.], *Rosarius philosophorum*, ed. Calvet, 308: "quod quidem requirit divitias, sapientiam et libros"; *Testamentum*, 1:108: "Item hoc requirit sapienciam, divicias et libros."

34. See p. 85, note 74, above.

35. Ibid., fol. 2v: "Requiram a domino meo videlicet auri purissimi non maiorem sed minorem libra vnam."

of the *Medulla*, then he stood in far greater need of medicine than of gold in 1476, and Ripley could have plausibly expected him to place more value on the vegetable stone, particularly since this product, when compounded with the mineral stone, can also transmute metals. Read in the context of Ripley's other writings, however, including the processes gathered in the *Bosome Book*, it appears that the English canon was genuinely committed to keeping the cost of doing alchemy as low as possible. But how could he reconcile such a view with the need for gold and silver apparently insisted upon by his own major authority, Raymond Lull? This does seem to be an instance in which Ripley responds in his practice to concerns that are moral as well as philosophical. The canon's two-tier pricing structure is the outward sign of his deeper engagement with a raft of conflicting moral, philosophical, and practical issues, which turn out to be widely debated not only in his writings, but in those of his English contemporaries.

Ripley's dilemma over cost is grounded in an apparent difference between his two major authorities. Raymond, as we have seen, endorsed the use of gold and silver as ferments for the mineral work. Yet another authority, Guido de Montanor, seemed to argue against their use: a view that Ripley himself took seriously. When Ripley states that "Purses are not to be loosened for making great expenses" in the opening sentences of the *Medulla*, he generically attributes the saying to "the philosophers"—but he is in fact quoting directly from Guido.

Guido de Montanor remains a shadowy figure in English alchemy. Best known as one of Ripley's major authorities, he is cited in both the *Compound* and the *Medulla*. Ripley even used a treatise he believed was written by Guido, the *Scala philosophorum* (Ladder of Philosophers), to provide the structure and much of the content for the *Compound*.[36] Despite Guido's prominence in Ripley's oeuvre no complete copy survives of any of his own writings, although parts of a philosophical treatise, *De arte chymica*, and several alchemical recipes survive in English collections of the fifteenth century.[37] These fragments suggest that Guido was a strong pro-

---

36. I discuss Ripley's adaptation of the *Scala* in Rampling, "Establishing the Canon," 193–200. Other than Ripley's view of the matter, there is no evidence that Guido was in fact the compiler of the *Scala philosophorum*.

37. Guido de Montanor, *De arte chymica*, in *Harmoniae imperscrutabilis Chymico-Philosophicae, sive Philosophorum Antiquorum Consentientium*, ed. Hermann Condeesyanus (Frankfurt, 1625). Outside the *Compound*, references to Guido in fifteenth-century manuscripts include Sloane 3579 (fol. 6r), 3744 (fols. 27r, 31v), and 3747 (fols. 4v, 8r); Ashmole 759 (fols. 87v, 90v); and CCC 136 (fols. 15r, 16v, 42r). Of these five manuscripts, three (Sloane 3579, Sloane 3747, Ashmole 759) are in the hand of a single scribe; see CRC 1 and p. 120, note 52, below. On Guido's writings and relationship

ponent of a lead-based alchemy, which Ripley took as the basis for his own sericonian vegetable stone. As Ripley showed in the *Medulla*, such an approach can be easily reconciled with the alchemy of pseudo-Lull. Yet if Guido's opinion on the substance of the stone is uncontroversial, his views on fermentation are distinctly unorthodox. When he speaks of the "Sol of the Philosophers," he does not intend gold. Instead, he endorses the use of a base metal, identified as Adrop or the Green Lion, suggesting, at least to English readers like Ripley, a serious conflict with the gold-based alchemy of Raymond Lull. But Guido goes farther still, by claiming that alchemists need not worry about what kind of earth they draw their ferment from, "as long as it is fixed"—that is to say, as long as it can withstand the heat of the fire sufficiently to stabilize the volatile component of the stone.[38] Such an argument could potentially be used to justify the use of a nonmetallic ferment, as long as it has the capacity to fix mercury: a view that flies in the face of the doctrine of "mercury alone." It also supplies the authority needed to dispense with gold and silver entirely, and hence to make the work affordable to the poor.

Guido's sayings about the low cost of the work and the use of "fixed" ferments recur throughout the *Bosome Book*. They also appear in the collection of forty-five *notabilia* that Ripley seems to have culled from Guido's writings, which is also one of the few items in the *Book* to survive in a fifteenth-century witness.[39] If we assume that these notes represent the material that Ripley found most interesting in his source text, then they also tell us a great deal about the canon's own priorities. The first seven aphorisms include the following:

1. This science is not given by God except to well disposed persons.
2. Purses are not to be loosened for making great expenses, which this art does not require.
3. All matter which is bought at great price is false, and is unprofitable in our work.

---

to the Ripley Corpus, see Jennifer M. Rampling, "The Alchemy of George Ripley, 1470–1700" (PhD diss., University of Cambridge, 2010), chap. 3; Rampling, "Establishing the Canon."

38. "Notabilia excerpta de Libro Guidonis de Montaynor summi Philosophi in partibus Graeciae," Harley 2411, fols. 50r-53v, on 50v. The *Notabilia*, which also circulated in English translation as the *Notable Rules of Guido*, is CRC 22.

39. Trinity O.8.9, fol. 37v (CRC 22.1). This manuscript was compiled by a scribe possibly connected to Bridlington Priory; see Rampling, "Establishing the Canon," 200.

5. Nature strives and intends towards Sol.

7. The poor as well as the rich may have the stone.[40]

These sayings offer considerable insight into Ripley's self-presentation in the *Medulla*, where he portrays himself not as a lofty philosopher, but as a poor, religious man. However, Guido's precepts address both the right way of living and the right way of practicing. Insofar as we can reconstruct it, the key to his philosophy is the fifth note abstracted by Ripley: "Nature strives and intends towards Sol." If metals are naturally inclined to become gold, then the philosopher need not use precious metals at the start of the work, since any base metal will do. The secret lies in grasping how to activate that natural process by unloosing the vegetative properties of the metals themselves, enabling them to "ripen" into gold.

In the *Bosome Book*, Ripley devotes an entire treatise, the *Concordantia Guidonis et Raimondi* (Concordance between Guido and Raymond), to reconciling this view with the position of his other major authority, Raymond Lull.[41] The disagreement turns on the use of common gold and silver. For Raymond, the vegetable mercury provides the matter of the stone, but cannot sustain the form of the precious metals. Gold and silver are therefore necessary as ferments in the work. Guido, however, claims that the philosophers' gold and silver are not the common metals, but *Decknamen* for red and white tinctures drawn from an imperfect metal, Adrop.[42] Since Adrop contains in itself all the "vegetable" virtue required to precipitate transmutation, common gold and silver are not required at any point. Rather than

---

40. Harley 2411, fol. 50r:

1. Scientia ista non datur a Deo, nisi praecipùe bene dispositis personis.

2. Non sunt dissoluenda marsupia propter magnas expensas faciendas, quas ars ista non requirit.

3. Omnes res quae magno emitur precio mendax et inutilis est in opere nostro. . . .

5. Natura proponit et intendit ad Solem. . . .

7. Tam pauper quam dives lapidem habere potest.

41. Combach printed the text as the *Concordantia Raymundi Lullii & Guidonis philosophi Graeci per Georgium Riplaeum*; *OOC*, 323–26. Like the *Medulla*, this work survives in no fifteenth-century copy: it is first attested in 1557, when its title and incipit were recorded in the *Catalogus* of the antiquarian bishop John Bale; Bale, *Scriptorum illustrium maioris Brytanniae . . . Catalogus*, 623. Bale records two copies of the work in the possession of one John Bushe, of whom no more is known, except that he seems to have been an avid collector of alchemica, owning works by John Dastin and Thomas Norton, as well as several other texts by Ripley.

42. Ripley's source for this position is apparently *De arte chymica*, where Guido discourses on the "Sol of the Philosophers" in his chapter on fermentation; *De arte chymica*, 134.

supplying the form of the precious metals, the job of the ferment in Guido's interpretation is merely—as we have seen—to fix the stone.

At stake for Ripley is the coherence of the philosophy that underwrites his practical program, although the issue also determines whether the work will be available to less well-off practitioners, who may be unable to afford precious metals in their craft. As it happens, he reconciles the two positions using more or less the same argument that he would later employ in the *Medulla*—that is, by dividing the process into two. The first step is to make the "stone": the menstruum extracted from a base metal following its dissolution in a vegetable solvent. On this point, Ripley upholds Guido's authority for a goldless process: "you may know for certain and believe me that the stone can be perfected into white and red, which both spring forth from one root without common gold."[43]

The second step is to prepare ferments from gold and silver, as dictated by Raymond, by using the sericonian menstruum to open up the bodies and draw out their essential mercuries, or natural fire.[44] This position means that everybody's authority is preserved. Raymond was right to recommend common gold and silver, since these metals are the source of the eventual ferment, while Guido is also correct to say that common gold should *not* be used, since the metal must first be altered by the sericonian menstruum.[45] Above all, Ripley upholds his own authority as an agile conciliator, while maintaining the prestige of his vegetable stone as an authentically philosophical product, endorsed by both Raymond and Guido.

Once we grasp the nature of Ripley's signature practice, it turns out to be ubiquitous in his writings. Most importantly, the sericonian menstruum

43. George Ripley, *Concordantia dictorum Guidonis et Raymundi*, in Harley 2411, fols. 47r-49r (hereafter *Concordantia*), on fol. 48r: "scias pro certo, et crede mihi, quod lapis potest perfici in Albo et in rubeo, quae ambo ex vna radice pululant [*sic*] absque auro vulgi." Given considerable differences between various recensions of the text, I have opted to use the version from the *Bosome Book* in Harley 2411, as probably closer to Ripley's original text than Combach's printed edition. Ripley points out that another authority, pseudo-Rāzī, refers to lead as gold and silver *in potentia*, if not in act—a likely reference to the raw, vegetative property of the unfinished metals, which allows them to ripen into gold within the earth; pseudo-Rāzī, *Liber de aluminibus et salibus*, in Robert Steele, "Practical Chemistry in the Twelfth Century: Rasis de aluminibus et salibus," *Isis* 12 (1929): 10–46, on 40–41.

44. Harley 2411, fol. 48r: "sed numquam potest Elixir ex lapide fieri nisi per additionem Auri et Argenti vulgi, quae debent cum [mercu]rio lapidis alterari, et revegetari ac elevari in sulphur crystallinum ac fixari." Ripley here refers to the menstruum as the "mercury of the stone"—that is, the product of the previous stage of the work.

45. *Concordantia*, fol. 49r: "Dixit igitur Raymundus, Cum Auro vulgari, ad innotescundum de quo esset verissimum fermentum. Dixit alius Philosophus [i.e., Guido], Eorum Aurum, non esse Aurum vulgi, ad designandum pro fermento debere accipi Aurum alterandum."

offers the key to unlocking his masterpiece, the *Compound of Alchemy.* In the poem's preface, Ripley warns his readers that his practice is wrought by three "mercuries":

> To understand well *Mercurys* three,
> The keys which of our *Scyens* be.[46]

Even Ripley's early modern commentators were puzzled by the identity of these mercuries, but armed with knowledge gleaned from the *Bosome Book*, turning these keys becomes the work of a moment. The first mercury is used to "naturally calcine" metallic bodies while preserving their natural heat—a process we can recognize as amalgamation with quicksilver, which renders gold and silver into fine calxes without the need for harsh corrosives. The metallic calxes are then dissolved by the second mercury, a "Humydyte Vegetable." This dissolution causes the metals to yield up their "essential" mercuries—the last and most important facet of Ripley's mercurial trinity.[47]

To unlock the poem requires us to understand that Ripley is actually describing four metals, not three. In addition to quicksilver and the essences of gold and silver, Ripley also teases the identity of another metallic mercury, "*Mercury* of other Mettalls essencyall," which is the material principle of the work. Being less complete in nature than gold or silver, this supplies the active virtue needed to catalyze growth in the precious metals. But it cannot be found in them:

> In Soon and Moone our Menstrue ys not sene
> Hyt not appeareth but by effect to syght.[48]

This imperfect metal is also called "Dame Venus," which is "Namyd by *Phylosophers* the *Lyon Greene*."[49] Venus is the planetary analogue of copper,

---

46. Ripley, "Preface," in *TCB*, 124.

47. Ibid., 125. This reading is supported by a passage in the *Scala philosophorum*, Ripley's major source for the *Compound*, where the chapter on calcination opens with a similar (although perhaps more straightforward) account of the first two "waters." *BCC*, 2:138: "Et sic Sol & Luna cum prima aqua calcinantur philosophicè, ut corpora aperiantur: & fiant spongiosa & subtilia; ut aqua secunda melius possit ingredi ad operandum suum opus, quod est exaltare terram in salem mirabilem" ("And thus Sol and Luna are philosophically calcined with the first water, so that the bodies are opened and become spongious and subtle; so that the second water is better able to enter with a view to the operation of its work, which is to exalt the earth into a marvellous salt").

48. Ripley, "Preface," in *TCB*, 124.

49. Ibid., 125.

and frequently denotes this metal in alchemical writing. The most obvious interpretation is that Dame Venus here stands in for copper, although some caution is needed given Ripley's fondness for multiple, overlapping cover names—as we saw in the previous chapter, the *Medulla*'s vegetable stone may have incorporated a mixture of lead and copper. Either way, in describing its dissolution in the vegetable menstruum, Ripley alludes to exactly the same approach that he would later elaborate in the *Bosome Book* and the *Medulla*.

It is now a simple matter to identify the one, true practice that Ripley praises at the close of his "Admonition." This is grounded not in gold and silver, nor a ferment drawn out of them, but "A naturall Mercuryalyte whych cost ryght nought," extracted from the Green Lion by agency of a vegetable solvent. Although this mercury is not itself gold or silver, it contains "The clerenes of the *Moone* and of the *Soone*," hidden from sight. This is Guido's Adrop, the work for which Ripley claims to have abandoned all other practices. Whether or not that is true, we must acknowledge the accuracy of Ripley's claim in the *Bosome Book*, where he promises that readers will find the menstruum "diversely related by me in my writings." In the work of no other English alchemist is the principle of dispersion of knowledge so consistently and deliberately employed.

THREE ALCHEMICAL READINGS

To unpack Ripley's sericonian practice has required us to range across a variety of texts and interpretations, including several new to scholarship. In a sense, we have had to reconstruct Ripley's reading and interpretative habits as much as his practical interests—not always a straightforward process given that the *Medulla* and the *Bosome Book* are preserved only in copies dating from the sixteenth century and later, and the works of one of Ripley's major sources, Guido de Montanor, have not survived complete. We are left with the sketch of a practice based on the dissolution of base metals in a "sharpened" vinegar to yield a powerful mineral-and-vegetable menstruum: a multipurpose elixir that could be adapted both for metallic transmutation and for human medicine.

Within the history of English alchemy, Ripley's process for the vegetable stone is of the first importance. Quite aside from the popularity of the practice over the next century, the circumstances of its production open a window onto Ripley's own vision of nature at work. His choice of materials rests on a specific view of metallic generation, developed from Guido: the idea that metals exist in a continuum between lead (the least perfect) to gold (the most perfect), with the less perfect metals capable of growing or maturing into gold and silver. This process is driven by an active or vegetable virtue present in imperfect metals, which is exhausted only when they achieve their natural termination in the form of gold. Common gold and silver are accordingly of no use in making the stone, since in these metals the vegetative process has come to a halt.

For all its influence, Ripley's sericonian alchemy was not, of course, the only approach to practice in fifteenth-century England, nor was he unique in seeking to reconcile his own experience with the vision of nature revealed in philosophical texts. Ripley's contemporaries also looked for evidence of nature's workings in their practice. Different writers analyzed different materials in the course of their own search for the "attractive power" or "active virtue" capable of animating metals. As they did so, they reflected on the susceptibility of substances to external influences, from the heat of the fire to the celestial influences of stars and planets.[50] All strove to answer the same fundamental question: the identity of the unseen properties in matter, that allowed nature to work transformations without the assistance of human art.[51] Only through practice could these properties be witnessed in action, from their outward effects if not their inner workings; but only reason could help account for what was seen.

The extent to which English readers both experimented with materials and speculated about their properties becomes obvious as we move from Ripleian *theoricae* like the *Concordantia* to the wealth of Middle English *practicae* preserved in manuscript, only a handful of which have so far been published. While these often incorporate commentary from philosophical treatises, their writers sometimes also pause to offer their own opinions on the cause of a given effect, working their speculations back into their own tracts. Some seek to interpret chemical operations—what we might now call reactions—in light of natural philosophical principles. Others turn to alternative sources of authority and explanation, including religious analogies.

50. The influence of celestial virtues is particularly emphasized by John of Rupescissa in *De consideratione*, although, confusingly, he also uses the sun and stars as cover names for gold and the other metals. The effect of celestial influences on the formation of the stone was nonetheless picked up in the pseudo-Lullian corpus (e.g., *De secretis naturae*, fol. 20r), and later developed further in the work of Renaissance Neoplatonists like Marsilio Ficino; concerning which, see chap. 5, below.

51. The relationship between art and nature is discussed in greater detail in Newman, *Promethean Ambitions*. See also Michela Pereira, "L'elixir alchemico fra *artificium* e *natura*," in *Artificialia: La dimensione artificiale della natura umana*, ed. Massimo Negrotti (Bologna: CLUEB, 1995), 255–67; Barbara Obrist, "Art et nature dans l'alchimie médiévale," *Revue d'histoire des sciences* 49 (1996): 215–86.

Taken together, these little-known writings reveal not only the diversity of alchemical practice, but also the diversity of modes of alchemical reading. We can quickly gain a sense of this variety by examining a range of "goldless" elixirs discussed by three fifteenth-century English alchemists, each one drawing on a different kingdom of nature—mineral, vegetable, and animal in turn. As these tracts show, when practitioners read their sources differently, their practice also changed.

Mineral: Opposite Qualities Attract

Our most "traditional" reading is supplied by an anonymous Middle English treatise that I shall here refer to (following the writer's description of the stone) as the *Preciouse Treasure*.[52] This writer attempts to square observation with philosophy by drawing heavily on writings attributed to Geber, Arnald, and Raymond, including the familiar criticism of vegetable and animal ingredients in favor of quicksilver. Like Ripley, the writer recognizes that textual authority must also be tempered by experience—proof is established by "dyuers seiyngges of olde philosophres and also by prouf of manuall practise."[53] He also draws on his medical knowledge of complexional theory to account for the properties of metals, proposing to "naturally prove" his own position "by good reason."

Grounded in Galenic medicine, complexional theory views bodily health as determined by the interplay of the four qualities of heat, cold, moistness, and dryness which together make up an individual's complexion. A healthy disposition of qualities can be maintained by adhering to a proper medical regimen. However, imbalance results in sickness, which can be treated by medicines calibrated to deliver an equal and opposite complexion—the basis of the Galenic precept "Contraries are cured by contraries" (*contraria*

---

52. This Middle English treatise survives in two versions, both dating from the second half of the century, yet different enough to suggest that they represent separate versions of a longer Latin exemplar. Ashmole 1450 gives a slightly fuller rendering of the text (unfortunately truncated before the passage on sublimation cited below), while the three manuscripts of the "Corthop Group," Ashmole 759 and Sloane 3747 and 3579, all include fragments of another, considerably longer version, in a somewhat sparser translation. Of these, Ashmole 759 (fols. 37r-45v) contains what appears to be the most complete version of the text, in which the writer hails the stone as a "preciouse treasure." The same scribe included another version with a divergent ending in Sloane 3747, fols. 56r-60r, and extracts from the earlier part of the text in Sloane 3579, fol. 17r-v. On the Corthop Group, see CRC, on 128; Rampling, "Alchemy of George Ripley," chap. 4.

53. Ashmole 759, fols. 38v-39r.

contrariis curantur). This is not just an aphorism, but a basic philosophical principle that explains why a given simple may counter a given disease.[54] As such, it offers a source of certainty in philosophical reasoning, allowing the writer of the *Preciouse Treasure* to argue that the elixir must be hot and dry in temperament because metals are cold and moist. Thus, when the mercury of iron is parted from its gross substance, it loses its blackness and "apperith white in colour whiche is a tokyn of coldnesse."[55] It also resembles crude quicksilver, suggesting a great abundance of moisture. If metals are indeed cold and moist in nature, then it follows that the elixir cannot be: "wherfore it semyth that the medecyn ought to be hotte and drie. Quia contraria contrarijs curantur [Because contraries are cured by contraries]."[56] To arrive at this conclusion, however, the writer has to defy another medical convention, that metals are actually hot and dry in temperament—a view that he considers to be contradicted by the observable properties of imperfect bodies.

If metals are naturally cold and moist in complexion, how is a hot, dry elixir to be drawn from them? For this writer, the answer lies in vitriol. The nature of quicksilver is cold and moist, but when sublimed with vitriol it becomes imbued with a "working virtue" and "active power" that it lacks on its own. Likewise, vitriol contains a "tincture of redness" that cannot be extracted from its common form, a gross and unclean substance. The two ingredients in fact depend upon one another for the release of their own hidden properties. During sublimation, the quicksilver joins invisibly to the tincture of the vitriol, which in turn causes the quicksilver to congeal.[57] The result is the transformation of common quicksilver into an active "mercury" that also has the capacity to color other metals. For this reason, the philosophers conclude that their "gold and [t]her siluer is no thyng ellys but quyksiluer turnyd into mercury whiche is verey tincture that teynyth all bodies in

---

54. On complexional theory in medieval medicine, see French, *Canonical Medicine*; on its adaptation by medical practitioners in fifteenth-century England (often in light of their own experience), see Peter Murray Jones, "*Complexio* and *Experimentum*: Tensions in Late Medieval English Practice," in *The Body in Balance: Humoral Medicines in Practice*, ed. Peregrine Horden and Elizabeth Hsu (New York: Berghahn, 2013), 107–28.

55. Ashmole 759, fol. 41r.

56. Ibid., fol. 41v.

57. Ibid., fol. 43r: "[T]here is no thyng that can take out that ryall tyncture from his two extremytees but commen ar[gent] vi[ve] by sublymacion where the tyncture of the vitriol in his ascense with the \commen/ ar[gent] vi[ve] doith inuisible ioyne togeder and the tyncture of the vitriol congilith hym."

to very sol or lune."[58] It follows that common gold and silver are unnecessary in the work, since the transformed mercury will suffice to transmute metals on its own.

Like Guido, the writer seems to exclude the precious metals on the grounds that the required tinctures are obtainable elsewhere. His choice of ingredients takes us far from the vegetable stone, to a more old-fashioned process based on corrosive sublimate—one that bears much closer resemblance to Ripley's vitriol-based mineral stone in the *Medulla*. The contribution of the treatise lies not in its choice of substance, but in the writer's explanation of chemical change in terms of both reason and knowledge of substances: a release of active but hidden properties triggered by the combination of two substances under specific conditions, driven by the contrariety of their natures.

Vegetable: The Working of Sweet and Sour

The writer of another "goldless" tract, the *Tractatus brevis sed verus* (A Short Tract but True), uses the doctrine of opposites to even more startling effect in relation to a different pair of contraries—sweet and sour.[59] For this writer, sweetness has an unmistakably moral charge, signifying corruption rather than purity. By this reasoning, it follows that silver is purer than lead because it is less sweet. Lead, visibly the most corrupt of metals, is also sweet-tasting: it "hath an aer off gret suetnesse it is so suete [th]*at* [th]*er* is no thing suett*er*." The writer may have tasted lead himself, but can also point to the experience of leadworkers, who sometimes notice a sweet taste in their mouths as they work: "Also plu*m*ers when thei cast [t]her leed [th]ei p*er*save gret suetnes su*m* tyme."[60]

It follows that in order to purify lead the sweet must be countered with the sour—in this case, by a philosophical vinegar called the Water of Philosophy, which is seven times sharper and sourer than any other vinegar. The resulting opposition of sweet and sour initiates a natural "working," which the writer explains using medicinal and culinary analogies. The sourness of rennet causes sweet milk to curdle into cheese, while in brewing the contrary qualities of sweet wort and sour grout underlie a similar process:

58. Ibid., fol. 43v.

59. This work exists in two fifteenth-century copies: "Tractatus breuis *sed* verus ut opinat*ur*," Sloane 1091, fols. 217v-21v; and a less complete copy in Sloane 3747, fols. 51r-52v. Thomas Robson made a later copy in Glasgow University Library, MS Ferguson 133, fols. 3r-6v.

60. Sloane 1091, fol. 218v.

"Ryght so the suetnes off the leed and [th]e sowrenes off *our* water causyth *our* wyrchyng."[61] In philosophical terms, the contrary natures of sour vinegar and lead are the efficient cause of the metal's dissolution, an interpretation that is also supported by observation. If thin plates of lead are left for seven days in the philosophical vinegar, the sour water will strip away the black coating that is (the writer claims) a by-product of lead's corrupting sweetness, thereby preparing it for further operations.

Once again, personal experience is shaped by reason into a suitable explanation for the otherwise inexplicable "working" of lead and vinegar. But although this idiosyncratic solution begins with a traditionally Aristotelian and Galenic approach to qualities, the author is preoccupied by more than natural philosophy alone. He also perceives chemical operations in light of his own faith, which imbues effects with religious and moral significance. Central to this reading is his association of sweetness with sin, and sourness with contrition: "as the suetnes off syn makyth fowle manys sowle, so [th]e sowrnes off penance doth make hy*m* faer and clene in the syght off god."[62] The writer explicitly develops this dimension of the work by comparing the transformations wrought by natural philosophy to those of moral philosophy. The practical content is, in fact, framed by an extended analogy between the transmutation of lead into silver using the Water of Philosophy, and the conversion of Jews to Christianity through the water of baptism.[63] The tract concludes with the writer's prayer that his own sin might be purged as easily as the corruption of lead: that "I myght haue plente off the wat*er* off compu*n*ccyon to purge.& wash my sowle ffrom the fylthe lyve off syn."[64]

In its fusion of religious and philosophical themes, the *Tractatus brevis* assumes something of the character of a devotional text. But the analogy between moral and natural philosophy breaks down when the writer describes the mechanisms of chemical change. The spiritual transmutation that occurs at the moment of baptism is divine in origin, but in alchemy the "working" of contrary substances is wrought by nature rather than by direct intervention of God. And although the writer cites no authority except scripture, he still perceives transmutation as philosophy, and philosophy as

61. Ibid., fols. 219v-2or.

62. Ibid., fol. 221v.

63. The text is not unusual for its period in criticizing the Jewish faith: the *Testamentum* and other pseudo-Lullian writings also incorporate passages that reveal hostility toward Jews and Muslims; see, e.g., *Testamentum*, 3:444–52.

64. Sloane 1091, fol. 221v.

a key to the "dores and gatys off grace."[65] Even in deeply religiously oriented alchemical texts like the *Tractatus brevis* or John of Rupescissa's *De consideratione*, it is taken for granted that the operations themselves will proceed through natural means—however miraculous they may outwardly appear.

Animal: Fermenting with Eggshells

Our third and final case evokes a Latin *practica* that we have already encountered in the *Bosome Book*: the *Praeparatio calcis*. Here Ripley details his own first attempt at calcining eggshells, when he neglected to thoroughly separate the shells from their skins:

> For as Guido says, "Concerning the earth, it does not matter from what substance it comes, as long as it is fixed." I being less expert once worked with eggshells, the skins not being removed and with the menstruum not rectified, and so my work was infected with indelible blackness. Happy therefore is he whom other men's harms make to beware.[66]

The image of George Ripley, the great mercurialist, stewing eggshells in an attempt to emulate an adept whose works are now all but forgotten casts strange new light on the world of late fifteenth-century alchemical practice. All the ingredients of Ripleian alchemy are present in this terse account: the menstruum, the authority of Guido, the recitation of earlier failings as a warning to future readers. Only the substance itself—an unmistakably "animal" ingredient—seems alien to the practical commitments he records elsewhere. But Ripley offers a rationale for his attempt: he was testing Guido's proposition that even nonmetallic calxes could function as ferments.

Ripley's inclusion of an "animal stone" in the *Bosome Book* makes sense in light of his concern with cost. It also opens up a likely Ripleian connection with one of the most interesting alchemical *theoricae* written in Middle English: the *Accurtations of Raymond* (a work not to be confused with the better-known pseudo-Lullian *Epistola accurtationis*). Of all the Middle English sources we have so far examined, this is the one that engages most

---

65. Ibid.

66. Harley 2411, fol. 64r: "Quia vt dicit Guido. De terra autem non est curandum de qua sit substantia, dummodo sit fixa. Ego semel minus expertus operabar cum testis ovorum, non remotis pelliculis, et cum menstruo non rectificato; et ideo opus meum indelebili nigredine infectum erat. ffelix ergo quem faciunt aliena pericula cautum." The final sentence is a well-known classical aphorism.

directly with the moral dimension of high costs. The writer solves the ferment problem by eliminating the use of gold and silver entirely. His rejection rests on two grounds: first, that the precious metals lie beyond the reach of poor men, which is inequitable; and second, that they simply do not work, resulting in wastage. Rather than making poor men rich, their use will make rich men poor:

[O]lde philosophers did not medle with comyn gold and siluer and therfore they wrote in·[t]her boks that [t]her Work axid no cost and that it myght || be do[ne] of euery powre man as well as of the ryche which were false if it myght not be so without comyn gold and siluer for they be pretiouse and costely and ill for powre men to comeby. [A]nd truly for lak of vnderstondyng in this poynt many on[e] bryngith moche gold and siluer to nought and myspendith myche labour and tyme to hurt and perill bothe of body and sowle which is gret pite.[67]

This popular Middle English treatise later circulated widely under the name of Ripley, with the addition of a lengthy *practica* comprised mainly of processes extracted from pseudo-Lullian sources.[68] There is, however, some reason to doubt Ripley's authorship. Unlike better-attested Ripleian prose works—the *Medulla*, the *Philorcium*, and the contents of the *Bosome Book*—the *Accurtations* was written in Middle English. The writer also diverges from Ripley's compromise in the *Concordantia* by concluding in favor of Guido rather than Raymond. The *Accurtations* thus makes a particularly useful case for testing how far technical content, such as a sericon-based stone and a concern for cost, can be used to attribute authorship to works whose provenance is uncertain.

The author of the *Accurtations* adopts the same two-stage structure as that proposed in Ripley's *Concordantia*: a stone made from Adrop, and an elixir for which a ferment is required, although not necessarily one drawn from gold. The authority he cites for this position is the *Summa* of Guido de Montanor, seemingly a lost work, which, according to the writer of the *Accurtations*, was addressed to a Greek bishop.[69] Guido describes Adrop as a base metal that, unlike gold and silver, is not yet complete by nature; accord-

---

67. Sloane 3747, fols. 3v–4r.

68. On the *Accurtations of Raymond*, see CRC 1; Rampling, "Establishing the Canon," 201–7.

69. The identity of this work is difficult to establish, since the original seems not to have survived in any identifiable form; the cited material does not obviously correspond to Guido's *De arte chymica* as it has come down to us. However, although the original *Summa* seems not to have sur-

ingly, its mercury still possesses a vegetable "attractive power" that the precious metals lack.[70] Once cleansed, this mercury (the Green Lion) will be a thousand times better than gold and silver—it can even be called "Sol," for, like the sun, its attractive power "florisshith and makith grene by attractif power all the worlde."[71] Unlike common gold, however, the body from which the Lion is extracted is neither perfect nor fixed in nature: it is green because it is not yet "ripe."

So far, the process is identical to that we have already encountered in the *Concordantia* and in the *Medulla*'s chapter on the vegetable stone, and hence entirely compatible with Ripley's sericonian alchemy. But rather than attempting to reconcile the two authorities as Ripley did in the *Concordantia*, the writer deliberately pits one against another—this time explicitly invoking the question of cost. Whereas Guido promoted the inexpensive Adrop, Raymond fixed his tinctures "as so plesed hym to do by cost and labour" on common gold and silver. While Raymond's method offers good results, it lies beyond the reach of poor practitioners, since it only "accordith to princes and prelates replete with richesse."[72] For those who lack such lavish resources, a cheaper fix is called for.

For this writer, the use of Adrop satisfies the philosophical requirement that the stone should be made from one thing. To transform this stone into elixir, however, calls for a second stage requiring "dyuers thynggs"—a claim that follows Guido by authorizing the use of nonmetallic ferments. Finally, he reveals his own preferred solution. To "put powre men out of doute," the stone can be fermented with calcined eggshells.[73]

The writer justifies his unexpected use of an animal product by calling upon both practical experience and theoretical argument. When calcined, he explains, eggshells produce a white and subtle calx that can endure the strongest fire better and longer than any other earth. The secret lies in the shells' dryness and lack of "mercury": since all the moisture in the egg devolves to the yolk and albumen, the shell is left without that mercurial humidity possessed by other earths. For that reason, it will "drynk vp our mercury" (the stone) more readily than any metallic calx, fixing it into solid

---

vived intact, we can reconstruct some of its doctrines from passages quoted in the *Accurtations* and, most likely, from Ripley's *Notable Rules of Guido* in the *Bosome Book*.

70. Sloane 3747, fol. 4v.
71. Ibid., fol. 5r.
72. Ibid., fol. 5v.
73. Ibid., fol. 8v.

form.[74] Since this earth is able to receive tincture from the stone, it can also "be turnyd in to the nature of metall by craft"—an unexpected outcome that, the writer admits, "eall [i.e., all] the men that laboren in this science wold neuer trust \except/ hym that hath expert [i.e., experienced] it."

The clinching evidence for the effectiveness of this earth is not authority, but the writer's own experience:

> To prove whether this erth wold drynk vp my mercury or no, onys I put it thervppon, and anon therth [i.e., the earth] become like a fatte crudde and when the mercury was vaporid therfro therth was cytryned with the tyncture therof.[75]

For the writer of the *Accurtations,* the capacity of the eggshell calx to take a citrine color from the mercury, while simultaneously fixing it into a curd, signals its suitability as a ferment. This insight is concretized in the following *practica,* which includes a recipe for an "animal stone" fermented with eggshells, said to be capable of transmuting crude quicksilver to "perfytt gold or Siluer," and for "least cost."[76]

When read in context, we can now see that this *theorica* is uncannily in tune with Ripley's overall practice. If Ripley wrote the *Accurtations,* then at some point in his alchemical career he must have changed his mind on the matter of fermentation—in which case the text may belong among his early recipes, written between 1450 and 1470, that he later encouraged his readers to cast into the fire. In other respects, the theoretical and practical synergies between the Middle English text and Ripley's Latin works are strong enough for us to conclude either that Ripley wrote the *theorica* of the *Accurtations* himself or that the *Accurtations* was the work of one of his own readers. Perhaps it stands as one of those works of opinion that Ripley composed but later disavowed. Either way, when Ripley mentions "Eggs shells I calcenyd twise or thryse" in the "Admonition," it seems that his recollection is not entirely rhetorical.[77] And when he inveighs against the use of diverse ingre-

---

74. Ibid.: "And other erthes which haue mercuriall humydite withyn [t]hem be not so conuenyent to drynk vp our mercury as is this by cause they haue enough of [t]hem self and this hath non at all for the humydite that this shuld haue hadde was multiplied by nature into the white and in to the yolke."

75. Ibid., fols. 8v-9r.

76. London, Wellcome Library, MS 239, 26: "Truly the shortest waye & of least cost that may be is this."

77. Ripley, "Admonition," in *TCB,* 190.

dients, we should perhaps take his words with a pinch of one of the salts he so despised.

THE ALCHEMIST'S OPINION

Our connection with the historical canon of Bridlington ends with the *Medulla alchimiae*, for, although sixteenth-century vitae typically date Ripley's death to 1490, the last contemporary record of his activity remains the cut-off date of 1476 preserved in the colophon of his Latin treatise. George Neville, his putative patron, died in the same year, presumably without having had an opportunity to benefit from the life-prolonging vegetable stone. The *Bosome Book*, from which Ripley teased out so much of his precious *Marrow of Alchemy*, vanished into unknown hands, to emerge only decades later, and in a very different religious and intellectual climate.

As time and circumstance transformed Ripley from Lullian commentator into English authority, both reading and experimentation continued to be guided by the speculations of individual practitioners. Wresting practical information from cryptic and laconic sources required tools, and the ability to formulate opinions allowed readers to keep a record of their qualified responses and speculations, sometimes with a view to later testing.

Such was the approach of one of Ripley's own earliest commentators, who set down his "opyneons" in the course of interpreting the practical content of the *Compound of Alchemy*.[78] This anonymous writer was probably a member of a religious community, as we can infer from the unusual context in which one of his treatises appears. The text has been written on a few blank pages at the end of a late fifteenth-century manuscript that includes a copy of the *Compound*; or, as the writer calls it, the "Book of Ripla" (fig. 2). The commentary seems to have been added a few decades later, as part of a gift—presumably of the entire book—addressed "To the worshipfull Mayster Elles, Prior of Lyse."[79] The opinions of this unknown writer therefore link the alchemical activities of one Augustine canon, Ripley, to those of another: Sir Thomas Ellys (fl. 1493–1557), prior of Little Leighs, who acquired notoriety during the 1530s as an alchemist immediately prior to the dissolution of the monasteries.

While Ripley might have been flattered by the writer's devotion to his

---

78. "Boke conteynyng myne opyneons in the Scyence of this Philozophy," Ashmole 1426, pt. 5.

79. At some point, the two leaves on which the treatise is written became separated. The first leaf remains in situ in Ashmole 1486, pt. 3, fol. 72v; the second is now bound in Ashmole 1492, pt. 8, 125. Since it is addressed to Ellys as prior of Leighs, it must have been written after his appointment in 1527, and presumably before the priory's dissolution in 1538. I discuss Ellys further in chap. 4.

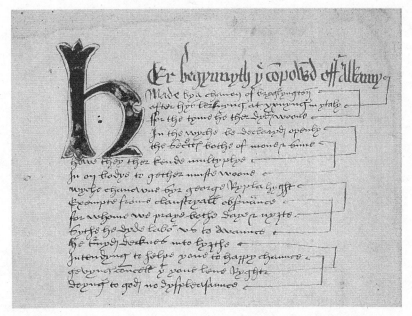

FIGURE 2. George Ripley, *Compound of Alchemy*, alias the "Book of Ripla." This late fifteenth-century copy was later owned by the anonymous Opinator, who in turn presented it to Thomas Ellys, Prior of Little Leighs. Oxford, Bodleian Library, MS Ashmole 1486, pt. 3, fol. 49v. By permission of The Bodleian Libraries, The University of Oxford.

poem, he would surely have been appalled by his methodology. Despite the *Compound*'s injunction against working according to fancy, the writer consistently cites "myne opyneon" and "my fantesye" as tools for interpreting the text. Although he adopts the persona of a master offering counsel to his philosophical son, Ellys, his practice has evidently not yet caught up with his conjectures. The speculative approach is even more marked in a longer treatise written by the same author: an alchemical commentary "aftre the discripcyon of one George Ripla some tyme chanon of Bridlyngton in England," now preserved in Ashmole 1426.[80] Rather than merely paraphrasing Ripley's text, this treatise is embellished with the writer's own conjectural interpretations, as he advertises in the title: "Here begynneth the Boke conteynyng myne opyneons in the Scyence of this Philozophy."[81]

80. Ashmole 1426, pt. 5, 4. The commentary is followed by notes on the *De occulta philosophia liber primus* of Heinrich Cornelius Agrippa von Nettesheim, printed in 1531 and 1533; discussed further in chap. 5, below.

81. Ashmole 1426, pt. 5, 1. It is possible that this work was also written with Ellys in mind. The writer seems to address a clerical audience when describing the obscure language of philosophers,

This treatise, *Myne Opyneons*, helps to fill a vital lacuna in our understanding of alchemical reading: exactly how did reader-practitioners approach the challenge of extracting practical information from an encoded text? Thanks to the unusually frank account of our "Opinator," we can follow the efforts of one eager reader to make sense of Ripley's poem—to do so, furthermore, in the absence of supplementary Latin works such as the *Medulla* or the *Bosome Book*. The writer himself outlines his reasoning:

> For as moche as I do fynde so moche diffyculty in thos workes, becawse I fynd it not playnely set owght. ffor the whiche I wryght thys littil treatyse ·vpon myne owne fantesy Intendyng to declare the most dowbtfull clawsis of the auctor. And therin myne oppyneon.[82]

When this writer speaks of marshaling his opinion, or fantasy, he intends it as yet another exegetical technique. Since Ripley's meaning is obscure, a conjectural reading will help to make sense of the text preparatory to practice. That the Opinator takes the *Compound* completely seriously as a didactic text appears from his struggles to translate the laconic verse into practical instruction. For instance, Ripley's first gate, "Calcination," advises on how to prepare gold and silver for their union with mercury:

> Lat the Body be sotelly fylyd
> With *Mercury*, as much then so subtylyd:
> One of the Sonn, two of the Moone,
> Tyll altogether lyke pap be done.[83]

The Opinator glosses this verse by noting that the metallic bodies, especially gold and silver, must "be Subtylly filed or calcyned *with* Marquery," which will be to them a "naturel calcynacion." Evidently he has had no difficulty interpreting Ripley's "natural calcination" as amalgamation with quicksilver (the first of the *Compound*'s three mercuries). Yet the source text can take him only so far, since Ripley is silent regarding the preparation of his third, imperfect body, Dame Venus. As promised in his title, the Opinator therefore advances his own opinion: "And as I do suppose Ven*u*s showld aulso

---

which has blinded not only the "comen & Rustical or onlerned people," but also "by yowre leaue no smaule sorte of the best lerned clark*es*" (ibid., 6).

82. Ashmole 1426, pt. 5, 30.
83. Ripley, "Calcination," in *TCB*, 130.

be clensed & examyned and subtyly fyled they will the soner dissolue & calcyne."[84]

The tentative conclusion suggests that the Opinator has not yet attempted this stage of the work himself. Nonetheless, in filling Ripley's lacuna he draws upon various sources of information, including both his personal experience of vessels and materials, and his ability to reason through the likely consequences of particular actions. Just as Ripley used his writings to showcase his ability to reconcile Lull and Guido, so the Opinator foregrounds his own skill in deciphering Ripley. His letter to Ellys ends on an optimistic note. If these principles are "discretly consydred," he concludes, "I suppose frute wolbe found."[85] For historians, too, such reflections cast flickering light upon the lost world of alchemical practice, allowing us to pick our way through a profusion of possible meanings, one reader at a time.

84. Ashmole 1426, pt. 5, 4.
85. Ashmole 1492, pt. 8, 125.

PART TWO

The Golden Age of English Alchemy

CHAPTER FOUR

Dissolution and Reformation

Nature shall teach you.[1]

English alchemy in the early decades of the sixteenth century is largely terra incognita. Our lack of knowledge is linked to the relative paucity of manuscript survivals from this period, compared to the flood of transcriptions and translations dating from the second half of the century. While the dissolution of the monasteries, long-established sites for the production of alchemical texts, made many manuscripts available to secular audiences, others perished through neglect or vandalism, prompting Elias Ashmole later to observe, "where a Red letter or a Mathematicall Diagram appeared, they were sufficient to intitle the Booke to be Popish or Diabolicall."[2] The result is a curious gap in the history of English practice, between the vibrant mixed economy of the later fifteenth century and the slew of alchemical petitions, chemical projects, and Paracelsian tracts that distinguish Elizabethan science. Like the space that remains after a building is demolished, this gap does not imply an absence of original material, only that its reconstruction requires much sifting through rubble, and, inevitably, an element of speculation—particularly regarding the elusive alchemical interests of Henry VIII.

In this chapter, I attempt to partly reconstruct how communities of alchemical practitioners pursued their art on the eve of the dissolution, prior to a period of unprecedented upheaval in English social and religious history.

1. Aegidius de Vadis, *Dialogus inter Naturam et Filium Philosophiae, Accedunt Abditarum rerum Chemicarum Tractatus Varii scitu dignissimi ut versa pagina indicabit*, ed. Bernard G. Penot, in *TC* (1602), 2:95–123, on 97: "Natura te docebit."

2. *TCB*, sig. A2v.

During this period, the tradition of monastic practice explored in the first part of this book came to an abrupt end. Between the suppression of smaller houses in the wake of the 1536 Act for the Dissolution of Lesser Monasteries, and the general dissolution of 1537–40, some 20,000 monks, nuns, and lay workers were evicted from their convents.[3] In principle, most religious were eligible to receive pensions from the Crown, proportional to the value of their houses, with larger pensions going to abbots and priors. In practice, large numbers of regulars were presented to livings as secular priests, while many former heads of houses succeeded to bishoprics, deaneries, and canonries: a cost-saving measure for the Crown, which was thereby spared the expense of their pensions. A few chose to live together in small communities in attempts to replicate their former state. Others were denied support on the basis of various transgressions. Some nuns returned to their families, while others married former monks and pooled their pensions. And regulars with interests in science and medicine looked for new ways of supporting their activities in a world that, although untrammeled by the claustral observances that Ripley once sought to avoid, was also open to scrutiny and suspicion.

The dissolution is a matter of concern precisely because the monasteries, as sites of practice, were not restricted to the members of religious orders. The fluidity of boundaries between the religious and secular spheres thus allowed for a vigorous transmission of alchemical knowledge and a greater range of options for patronage and support. For instance, the earlier case of William Morton, the woolman who attempted to establish his own practice at Hatfield Priory in 1415, reveals a complex operation involving monks, merchants, artisans, and local aristocracy. Morton seems to have traversed almost the length of England in order to win support for his project. His partners included a London merchant, Roys, who acted as go-between, perhaps capitalizing on existing networks in order to secure premises and equipment for the investment opportunity afforded by Morton's expertise; and a country prior, Bepsete, who furnished them with laboratory space, in return (we must presume) for a share of the anticipated future profits.[4]

3. Of these, approximately 12,000 were religious. David Knowles and R. Neville Hadcock, *Medieval Religious Houses: England and Wales* (London: Longman, 1971), 494; Martin Heale, *The Abbots and Priors of Late Medieval and Reformation England* (New York: Oxford University Press, 2016), chap. 9; Geoffrey Baskerville, *English Monks and the Suppression of the Monasteries* (London: Jonathan Cape, 1937), chaps. 9–10.

4. KB 27/629, discussed on pp. 61–62, above. Morton also seems to have used Hatfield Peverel as a base for pursuing further patronage opportunities.

Although the role of monks and friars vanished with the suppression, mercantile interest in alchemy not only survived the Reformation, but profited from it. The form that the new order would take was in fact already taking shape before the monasteries closed their doors, mediated by the collection and copying of alchemical books and the honing of practice in secular contexts, such that even monastic alchemists became reliant on outside help. To trace how this happened, we will follow the activities of two communities of reader-practitioners: one centered in the royal library at Richmond, the other in a small Essex monastery. Reconstructing these networks allows us to trace how books and knowledge moved within the mixed economy before the dissolution brought this productive traffic to an end. They reveal the vectors by which pseudo-Lullian writings reached an interested English audience: both through the sharing of books, and, when book learning on its own was not enough, through the search for a master. ·

These activities also bring us into the ambit of the king himself. For much of his life Henry VIII was surrounded and served by men and women with strong alchemical interests. As a child, he learned his French from one of the best-documented alchemists of his reign, Giles Du Wes (d. 1535). In 1547 he appointed another well-known practitioner, Richard Eden (ca. 1520–1576) to serve as his distiller of medicinal waters: an appointment frustrated by his own death a few months later.[5] The life of the king is thus curiously bookended by the activities of two alchemists: men of very different social and educational backgrounds, but with shared humanist leanings and decidedly bookish habits. Their material traces—the signatures, notes, and ownership marks that allow us to connect one book to another, and hence one reader to another—capture, in microcosm, the interactions of wider communities of alchemical practitioners during the tumultuous middle decades of the sixteenth century. To tell this history of books and patronage, there is no better place to start than Henry's own library.

THE LIBRARIAN'S TALE

In a period when English monarchs still styled themselves kings of France, and courtiers kept a close eye on fashions across the English Channel, a grasp of the French tongue was an indispensable attribute for an English prince. As Duke of York, the young Henry received his tuition from the man known at court as Master Giles but better known to posterity by his Latinized name,

5. On Eden, see chap. 6, below.

Aegidius de Vadis. Giles Du Wes was either French or Flemish by birth, and a musician by training; yet, in a way that seems typical of the talented, self-made men of the Tudor court, he rose from the position of court musician to that of tutor to Henry VII's children, whom he taught both lute-playing and the French tongue.[6] He was a "luter" in Prince Henry's household by November 1501, and in 1506 the king placed him in charge of the royal library at Richmond.[7] Although he retained this position on the accession of his former pupil, this did not put an end to his teaching career, as he later tutored Henry VIII's daughter Mary, even accompanying her to Ludlow in August 1525 after she acceded to the duties (if not to the formal title) of Princess of Wales. Later, he incorporated several dialogues based on this experience into his two-part "Introductorie" to the French language, the first French textbook printed in England.[8] By reproducing his "conversations" with Mary—in effect, allowing his readers to practice their French alongside the king's daughter—Du Wes spiced his textbook with a dash of court glamour, consolidating his identity as both scholar and courtier. In the prologue, he reminded his audience that his royal pupils included not only the present sovereign and his late brother, Prince Arthur, but also their sisters Margaret and Mary, respectively, the queens of Scotland and France.

As a native French speaker as well as an enthusiastic reader, Du Wes was well suited to the role of custodian of the royal library, which comprised large numbers of French and Burgundian books, including some acquired during the reign of Edward IV.[9] The royal collections expanded consider-

6. Gordon Kipling, "Duwes [Dewes], Giles [pseud. Aegidius de Vadis] (d. 1535)," *ODNB*. Kipling suggests that he may have been born in the town of Wez (now Le Vay) in Normandy; however, Warner and Gilson propose that he came from Lille in the Duchy of Burgundy, and was thus a Fleming; George F. Warner and Julius P. Gilson, *Catalogue of Western Manuscripts in the Old Royal and King's Collections* (London: Trustees of the British Museum, 1921), 1:xiii.

7. Kipling, "Duwes, Giles."

8. Giles Duwes, "An introductorie for to lerne to rede, to pronounce and to speke French trewly," *L'éclaircissement de la langue française . . . la grammaire de Gilles Du Guez*, ed. F. Génin (Paris: Imprimerie nationale, 1852). See Kathleen Lambley, *The Teaching and Cultivation of the French Language in England during Tudor and Stuart Times* (Manchester: Manchester University Press, 1920).

9. An inventory compiled in 1535, the year of Du Wes's death, includes many books produced in Burgundy that were probably owned by Edward IV. Henry VII also seems to have preferred French works, ordering about twenty books from the Parisian print shop of Antoine Vérard. The inventory is reproduced in H. Omont, "Les manuscrits français des rois d'Angleterre au château de Richmond," in *Études romanes dédiés à Gaston Paris* (Paris: É. Bouillon, 1891), 1–13. See also Janet Backhouse, "The Royal Library from Edward IV to Henry VII," in *The Cambridge History of the Book in Britain*, vol. 3: *1400–1557*, ed. Lotte Hellinga and J. B. Trapp (Cambridge: Cambridge University Press, 1999), 267–73; J. P. Carley, ed., *The Libraries of King Henry VIII* (London: The British

ably under Henry VIII, and not only from the usual gifts and commissions: in the early 1530s, large numbers of monastic books entered the library. Although Du Wes, who died in 1535, did not live to see the dissolution of the monasteries, he was thus extraordinarily well placed to witness the first fruits of the English Reformation.

Du Wes's own collecting and reading practices were, however, formed much earlier. When not occupied with the books and education of the royal family, he read and wrote about alchemy. In 1521, he completed the *Dialogus inter naturam et filium philosophiae* (Dialogue between Nature and a Son of Philosophy), an alchemical treatise that was later widely copied in England and abroad, published by the French Paracelsian Bernard Gilles Penot (ca. 1522–1620), and included in Zetzner's majestic alchemical compendium, the *Theatrum chemicum*.[10] He wrote it in the library at Richmond Palace.[11] Other, shorter works of his, including commentaries on alchemical writings and a letter of alchemical advice to an anonymous friend, appear in several of his manuscripts—although, as we shall see, his contributions are not always easy to detect.

Du Wes's diligence in transcribing alchemical texts, coupled with his confidence in signing his name, means that he is now one of very few identifiable alchemical scribes active in early sixteenth-century England.[12] Several of his compilations survive, such as Harley 3528, a collection of fifteenth-century manuscripts that Du Wes apparently gathered and bound himself, encom-

---

Library in association with The British Academy, 2000). Du Wes's immediate predecessor, who held the role of librarian from 1492, was also a native French speaker, Quentin Poulet, a Flemish priest and scribe from Lille; see Warner and Gilson, *Catalogue of Western Manuscripts*, 1:xiii, 2:336; Green, *Poets and Princepleasers*, 96–97.

10. Aegidius de Vadis, *Dialogus inter Naturam et Filium Philosophiae, Accedunt Abditarum rerum Chemicarum Tractatus Varii scitu dignissimi ut versa pagina indicabit*, ed. Bernard G. Penot (Frankfurt, 1595); reprinted in *TC* (see note 1, above); hereafter *Dialogus*. On Penot, see Eugène Oliver, "Bernard G[illes] Penot (Du Port), médecin et alchimiste," ed. Didier Kahn, *Chrysopoeia* 5 (1992–96): 571–667, on 649–50.

11. *Dialogus*, 96: "Vale, ex bibliotheca regia Richemerum. 17. Idus Iulii Anno 1521. E. D. V." The dialogue also circulated in manuscript in English translation—for instance, in Sloane 3580B (fols. 186v-202v); Ashmole 1487, pt 2 (fols. 100r-106v); Boston, Massachusetts Historical Society, Winthrop 20c (fols. 79r-89v).

12. Aegidius de Vadis's role in compiling the Trinity College manuscripts is mentioned in M. R. James, *The Western Manuscripts in the Library of Trinity College Cambridge: A Descriptive Catalogue* (Cambridge: Cambridge University Press, 1902), 3:414; Rampling, "Alchemy of George Ripley," chap. 5; Anke Timmermann, "Alchemy in Cambridge: An Annotated Catalogue of Alchemical Texts and Illustrations in Cambridge Repositories," *Nuncius* 30 (2015): 345–511, on 423–24. For the place of these manuscripts in a wider tradition of Latin didactic poetry, see Thomas Haye, *Das lateinische Lehrgedicht im Mittelalter: Analyse einer Gattung* (Leiden: Brill, 1997), 321–25.

passing works attributed to such distinguished authorities as Roger Bacon, Hortulanus, John Dastin, and Arnald of Villanova. Besides annotating each item, Du Wes also interpolated short pieces in his own hand, especially copies of Latin alchemical verse. In another manuscript, now part of Ashmole 1441, he copied the two English poems that are his only surviving transcriptions in English, and the only ones signed with the anglicized version of his name, "Giles Duwes."[13]

These manuscripts reveal Du Wes's fascinated preoccupation with the works of medieval alchemists, not only as texts but as material objects. In Harley 3528 he lovingly restored missing or damaged pages by adding paper slips with his transcriptions of the missing text, imitating the original hand and style to produce as seamless a correction as possible (fig. 3).[14] In another of his collections, now Trinity O.8.25, a medieval treatise written on parchment breaks off suddenly, but the text continues uninterrupted on a sheet of paper added in Du Wes's hand. At the same time, the librarian clearly treated these objects as reading copies, dotting the margins with his own annotations, including interpretations of their content and numerous cross-references to other works.

Du Wes was no isolated reader. Manuscript traces allow us to piece together his long and productive association with another alchemist, Sir Robert Greene of Welby (ca. 1467–d. after 1544).[15] Like Du Wes, Greene was a prolific compiler of alchemical texts, as we know from his habit of lavishly signing his name in his books, sometimes transliterated into Greek characters. In 1538 he composed an autobiographical treatise, the *Work of Sir Robert Greene*, in which he reported himself to be seventy-one years of age, and to have devoted forty years of his life to alchemy.[16] The *Work* also purports to describe his own alchemical practice, which is based heavily on pseudo-Lullian alchemy, both medicinal and transmutational. Later, Greene enjoyed the dubious honor of appearing among the foolish alchemists in a famous

13. Ashmole 1441, pt. 2, 89–95; part of a collection of fragments from earlier manuscripts bound together in the seventeenth century by Elias Ashmole.

14. For example, Harley 3528, fol. 94r.

15. On Greene, see Andrew G. Watson, "Robert Green of Welby, Alchemist and Count Palatine, c. 1467–c. 1540," *Notes and Queries*, Sept. 1985, 312–13; Jennifer M. Rampling, "English Alchemy before Newton: An Experimental History," *Circumscribere* 18 (2016): 1–11.

16. Copies of this English tract are found in Ashmole 1415, fols. 85r-96r (seventeenth century, in Elias Ashmole's hand); Ashmole 1426, pt. 9, fols. 3r-17r (mid-sixteenth century); Ashmole 1442, pt. 3 (seventeenth century); Ashmole 1490, fols. 165r-66v (dated 13 August 1592, Simon Forman's hand); Ashmole 1492, pt. 9, 197–205 (dated 23 August 1604, Christopher Taylour's hand). Extracts are found in Sloane 1744, fols. 22v-23v and 58r-v (early seventeenth century, Thomas Robson's hand).

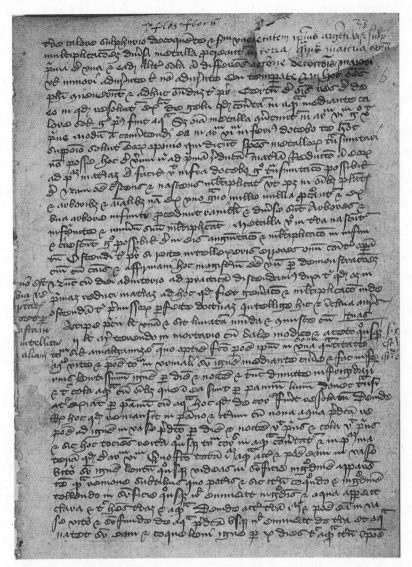

FIGURE 3. Fifteenth-century manuscript repaired by Giles Du Wes. © The British Library Board, MS Harley 3528, fol. 6r.

poem by William Blomfild, a compliment he would share with the man who almost became Henry VIII's distiller, Richard Eden.[17]

Although Greene's relationship with Du Wes has not previously been

17. William Blomfild, "The Compendary of the Noble Science of Alchemy Compiled by Mr Willm Blomefeild Philosopher & Bacheler of Phisick Admitted by King Henry the 8th of Most

noticed, their surviving books suggest a vigorous exchange of alchemical knowledge. The men were certainly acquainted by 1528, the year in which Giles returned with Mary's household from Ludlow. Greene took the opportunity to copy several items from the librarian's medieval compendium, Harley 3528, into his own collection, adding his own name and the date.[18] However, they probably knew one another long before this. When Du Wes completed the *Dialogus* on 17 July 1521 at Richmond, he dedicated the work to a "singular good friend." This may well have been Greene. Although traces of the friend's identity had disappeared by the time the *Dialogus* was printed in 1595, the Latin colophon appended to an earlier English translation concludes, "These, my Sir Robert, were some things that I strove to present to your honor for the sake of love of your virtues."[19]

If the *Dialogus* was indeed dedicated to Greene, it sheds light on the alchemists' relationship. With conventional modesty, Du Wes begs his friend not to be surprised if, despite his ignorance of the sciences, he now approaches a work so far beyond his powers. Virtue, he explains, is of such strength that it inflames even ignorant and lazy men with an ardent desire for "the pure marrow of all the sciences."[20] His friend, needless to say, is already well endowed with this virtue, and Du Wes apologizes for offering such a poor work in exchange—like Achilles in the *Iliad* (Giles here confuses Achilles with Homer's Diomedes), who benefited from trading his own, less valuable armor for the gorgeous trappings of Glaucus. However, he trusts that his friend will accept his gift with a smile, "and that in return you will love me (as you are wont to do)."[21]

More concrete evidence for their relationship dates from 1532, when Du Wes gave Greene a manuscript, now Trinity O.8.24, written in his own hand

Famous Memory," ed. Robert M. Schuler, in "Three Renaissance Scientific Poems," *Studies in Philology* 75 (1978): 21–41; hereafter *Blossoms*. Blomfild and his poem are discussed in chaps. 5–6, below.

18. Cambridge University Library, MS FF.4.12. Greene evidently compiled this collection from multiple sources. Texts likely to have been copied from Harley 3528 include the *Breviloquium Holketti de Serpente* (fols. 333r-44r) and *Expositiones Status Josephe* (fols. 344r-54v).

19. Du Wes, *Dialogue* (English translation of *Dialogus*), in Winthrop 20c, fol. 89v: "Hec erant domine mi roberte que tue dominacioni ob tuarum virtutum amorem offere nitebar."

20. Du Wes, *Dialogus*, 95: "Non mireris super me (virorum optime,) qui omnium scientiarum ignarus, tantum opus super vires aggredior, virtus enim tanti vigoris est, ut non modo ignotos flagrans amore & desiderio conciliet, verum etiam inscios, ignavos, torpentesque instiget, vt meram scientiarum omnium medullam concupiscant."

21. Ibid., 96: "haec quidem Achillis & Glauci permutatio . . . & munus meum (quamvis exiguum) laeta fronte suscipias: mutuoque me amabis (ut soles.) Vale, ex bibliotheca regia Richmerum. 17. Idus Iulii Anno 1521."

and adorned with marginal illuminations. In contrast to his workmanlike transcriptions in Harley 3528, Du Wes seems to have prepared this illustrated volume as a gift, as we can discern from an inscription on the tattered flyleaf: "Ægidius Du Wes, alias De Vadis, gave me to Robert Greene, in the year of our salvation 1532."[22] Above, Greene has added his own distinctive monogram, an intertwined *R* and *G* (fig. 4).[23] A sister manuscript, Trinity O.8.25, includes an alchemical epistle by Du Wes, once more addressed to a "singular good friend," and an informative note: "Experto crede Roberto."[24]

Like Du Wes, Greene collected medieval books, including several copies of the pseudo-Aristotelian *Secretum secretorum*.[25] Unlike his friend, however, Greene seems to have been interested primarily in texts rather than manuscripts: instead of binding old and new books together, he copied out medieval works into large volumes entirely written in his own hand, often signed and dated. Three volumes of his alchemical transcriptions survive, copied between 1528 and 1534 and covering a huge range of material, including works attributed to Greek, Arabic, and Latin authorities.[26] Although most of these are explicitly alchemical in content, others became so only with the aid of alchemical exegesis. For instance, on 30 April 1528 Greene copied a version of *De ave phoenice* (Concerning the Phoenix Bird) by the theologian Lactantius. The late antique poem is accompanied by a long commentary that interprets the life and death of the phoenix as an allegory for the philosophers' stone.[27] The author was probably Du Wes, who wrote out both the poem and the commentary in his own hand in Trinity O.8.24, suggesting that Greene may have copied the work from his friend's manuscript even before it came to him as a gift.[28]

These traces allow us to partially reconstruct a previously unknown relationship between two alchemical readers connected to English courtly

22. Trinity O.8.24, front flyleaf: "Ægidius du Wes a*lias* De Vadis | Me dedit Robe[. . . . . .]eene | An*n*o salu*a*toris 153[2]."

23. The monogram and the date 1532 appear again at the base of the page, in Greene's hand. The same monogram is found in several of Greene's manuscripts, including the end flyleaf of CCC 118.

24. Trinity O.8.25, fol. 3v. The note follows Du Wes's own commentary on the first text in the compilation: the *Visio mystica* of pseudo-Arnald of Villanova. While this manuscript includes no definitive signs of Greene's ownership, it seems likely that he is the Robert addressed.

25. Now Sloane 2413; and Bodleian Library, MS Laud Misc. 708.

26. Cambridge University Library, MSS FF.4.12 and FF.4.13 (copied between 1528–1529); Ashmole 1467 (copied between 1531 and 1534). In 1544 he copied a volume of medical material: Glasgow University Library, MS Hunter 403.

27. Cambridge University Library FF.4.13, fol. 322v: "Per me Robertum greene De Welbe A*n*no 1528. *c*urrente vltima aprilis."

28. Trinity O.8.24, fols. 5r-16r.

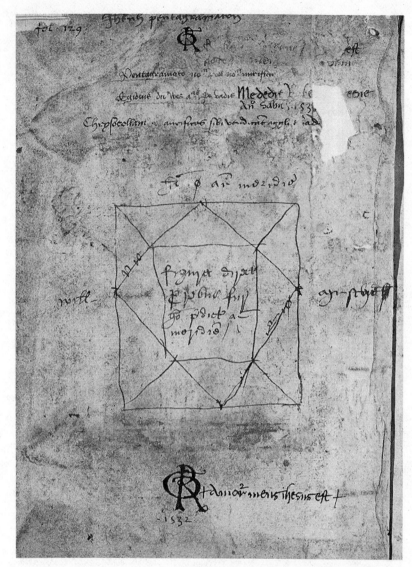

FIGURE 4. Ownership marks of Giles Du Wes and Robert Greene of Welby in Cambridge, Trinity College Library, MS O.8.24, flyleaf. By permission of the Master and Fellows of Trinity College, Cambridge.

circles during an otherwise poorly documented period of English practice. They reveal the importance that both Du Wes and Greene placed on owning alchemical books, whether original manuscripts or transcriptions copied from friends' collections. They also show how these men read the medieval authorities that were so necessary to their own work, both as sources of practical instruction and as models for composing their own treatises. The authority on whom both overwhelmingly relied was Raymond Lull, whose writings served as the foundation of their collections and the keystone of their own alchemical practice. Whether collecting books, reconciling theoretical positions, or devising experiments, Du Wes's and Greene's activities share the quintessentially Lullian character that defines the alchemy of Henrician England.

COLLECTING RAYMOND

Despite the wide European influence of Lullian alchemy, most readers in the first half of the sixteenth century would have known this material only through manuscript. The only pseudo-Lullian text in print, the first two books of *De secretis naturae*, was published with a collection of medical *consilia* in 1514—a context that downplayed the work's alchemical associations, and omitted the third, chrysopoetic book entirely.[29] The dearth of printed copies prompted interest in comprehensive manuscript collections, particularly those of early provenance. In 1541, the Nuremberg printer Johannes Petreius published a volume of pseudo-Geberian texts. At the end of the book, he included a list of manuscripts already in his possession, including numerous alchemical works of Lull, which he hoped to publish in full. He invited readers to send him additional unpublished manuscripts to print, promising to return them together with free copies of their printed incarnations.[30]

29. Giovanni Matteo Ferrari da Grado [Gradi], *Consilia . . . cum tabula Consiliorum ecundum viam Avicenne ordinatorum utile repertorium* ([Venice]: [Mandato et impensis heredum Octaviani Scoti & sociorum, impressa per Georgium Arrivabenum], [1514]), fol. 103r: "Incipit liber prime distinctionis secretorum nature seu quinte essentie sacri Doctoris Magistri Raymundi Lulij de insula maioricarum."

30. Johannes Petreius, *In hoc volumine de Alchemia continentur haec. Gebri Arabis* (Nuremberg: Johannes Petreius, 1451), 374–75. See Carlos Gilly, "On the Genesis of L. Zetzner's *Theatrum Chemicum* in Strasbourg," in *Magia, alchimia, scienza dal '400 al '700: L'influsso di Ermete Trismegisto,* ed. Carlos Gilly and Cis van Heertum (Florence: Centro Di, 2002), 1:451–67, on 452; Kahn, *Alchimie et Paracelsianisme,* 100–102. On Petreius's scientific publications, see Joseph C. Shipman, "Johannes Petreius, Nuremberg Publisher of Scientific Works 1524–1580, with a Short-Title List

In England, the scribal collections of Du Wes and Greene anticipated the kind of medieval compendium that Petreius hoped to publish. Between them they owned at least three large collections of pseudo-Lullian alchemical writings that date from the early to mid-fifteenth century, and that today comprise some of the earliest and most influential witnesses of Lullian doctrines in England. They and their circle also recognized the importance of these documents. The manuscripts were annotated and copied throughout the sixteenth century.

We have already encountered one of these volumes in Oxford, Corpus Christi College, MS 244. Partially transcribed in 1456 by Henry VI's chaplain John Kirkeby, it offers one of the most important witnesses to interest in pseudo-Lull in English court circles. It opens with a fine copy of the *Testamentum* in both Latin and Catalan, now edited by Michela Pereira, and which, as we have seen, may have been associated with Kirkeby's successful petition to Henry VI in 1456. Other major Lullian works follow: *De secretis naturae*, *Compendium animae artis transmutationis metallorum*, and *Liber lapidarii*, all in Kirkeby's hand.[31] Whether or not this luxury volume was ever seen by Henry VI, a generation later it had fallen into the hands of Robert Greene. A few short notes in his hand testify to his later ownership, although we do not know exactly how or when he obtained it.[32]

Beinecke Library MS Mellon 12 is another mid-fifteenth-century collection that once comprised just the *Testamentum*, *Codicillus*, and *De secretis naturae*.[33] This manuscript has a convoluted history. Like Kirkeby in CCC 244, its scribe records that the *Testamentum* was written at St. Katharine's Hospital in 1332, and translated from Catalan into Latin in 1443.[34] By 1506 the

---

of His Imprints," in *Homage to a Bookman: Essays on Manuscripts, Books, and Printing Written for Hans P. Kraus on His 60th Birthday Oct. 12, 1967*, ed. H. Lehmann-Haupt (Berlin: Mann, 1967), 147–16.

31. For detailed descriptions of the manuscript, see Pereira, "Descrizione del Manoscritto Oxford, Corpus Christi College, 244," in *Testamentum*, 591–600; R. M. Thomson, *A Descriptive Catalogue of the Medieval Manuscripts of Corpus Christi College, Oxford* (Oxford: D. S. Brewer, 2011).

32. Greene's notes are in CCC 244, fols. 4r and 37r.

33. Description in Laurence C. Witten II and Richard Pachella, comps., *Alchemy and the Occult: A Catalogue of Books and Manuscripts from the Collection of Paul and Mary Mellon Given to Yale University Library*, vol. 3, *Manuscripts: 1225–1671* (New Haven: Yale University Library, 1977), 79–93.

34. In New Haven, Beinecke Library, MS Mellon 12, the date of 1443 applies only to the *Practica*; the translation of the *Theorica* from Catalan (*de hispana lingua*) is dated 1446: "Explicit Theorica testamenti translata de hispana Lingua in Latinam. Anno 1446" (fol. 87r). However, probably owing to a missing page in the original, Giles Du Wes has added this colophon in his own hand,

book was in the hands of Giles Du Wes, who exercised his powers of librarianship to the full, by extensively annotating and correcting the original contents, then rebinding them with additional transcriptions of pseudo-Lullian writings in his own hand. To many of these he appended his name and the date, 1506—the same year in which he took up his post at Richmond.

The earliest of the three collections, now Cambridge, Corpus Christi College Library, MS 395, was probably written abroad in the first half of the fifteenth century and then brought to England.[35] It is the most diverse in content, including the Lullian *Tertia distinctio* but also a fine copy of John of Rupescissa's *De consideratione*. Once again, Du Wes supplemented the original material with additional works in his own hand, signed and dated to 1506. None of these additions is attributed to Raymond, suggesting that Du Wes viewed Mellon 12 rather than CCCC 395 as his main repository of Lullian writings. However, several of the interpolated texts may be his own compositions: among them, an alchemical dream poem (*Pulchrum somnium*); a treatise on fame (*De fama*); a practical alchemical work (*Secretum meum mihi*); and, intriguingly, an alchemical dialogue with necromantic overtones, the *Dialogus inter Hilardum necromanticum et quendam spiritum* (Dialogue between the Necromancer Hilardus and a Certain Spirit).[36]

Mellon 12 and CCCC 395 also eventually came into the possession of Robert Greene. Possibly he received them as gifts or bequests directly from Du Wes, although he had no compunction in erasing all signs of his friend's involvement. Greene seems to have been jealous of his books, overwriting earlier marks of ownership, and possibly even of authorship, in a way that shows little interest in their provenance. For instance, he confidently overwrote the name of a former owner—John Dunstable (d. 1453), the influential English composer—from another of his manuscripts, a copy of

---

raising the possibility that the colophon was not included in the original manuscript but was copied from another exemplar that included the variant dating. Unlike CCC 244, the Mellon 12 colophon omits reference to the translator's name (Lambert) or the supposed site of translation at St. Bartholomew's priory.

35. Description in Paul Binksi and Stella Panayotova, *The Cambridge Illuminations: Ten Centuries of Book Production in the Medieval West* (London: Harvey Miller, 2005), 323–24 (with information on dating); Timmermann, "Alchemy in Cambridge," 470–74.

36. Of these, the *Dialogus inter Hilardum necromanticum et quendam spiritum* enjoyed the widest subsequent circulation, appearing in Trinity R.14.56, fols. 22r-v (hand of Richard Eden); Trinity O.8.5, fols. 132v-33v (ex libris John Dee); and a lost copy previously owned by Sir Thomas Smith, detailed in "An Inventarie of suche thinges a[s] were in the Stilhowse," Cambridge, Queens' College, MS 49, fol. 117v. All were thus owned by Cambridge men; see chap. 6, below.

Boethius's *De musica*.[37] When Du Wes's two medieval compendia fell into his hands, Greene erased his name at every point where it appeared and replaced it with his own, although he left the date, 1506, intact. Only the edges of Du Wes's distinctive signature now remain, still just visible beneath Greene's own.[38]

The fact that these three manuscripts were at one point held in the same collection, perhaps even side by side on the same shelf, gives Greene an unexpectedly significant place in the genealogy of English Lullianism. Certainly both he and Du Wes understood the importance of books like these, which preserved the medieval authorities so necessary to their own work. Du Wes's careful attention to the material conservation of his manuscripts suggests that he also regarded them as antiquities to be cherished and preserved, although this was not merely antiquarian interest on his part. Du Wes did more than collect and repair old texts; he also read them with close attention and a critical eye. His *Dialogus* is filled with signs of his engagement with these manuscripts, suggesting that he wrote the treatise with his books close to hand. Indeed, since the *Dialogus* provides a commentary of sorts on the *Testamentum*, our access to Du Wes's exemplary manuscripts offers unparalleled insight into the process of alchemical composition in the early decades of the sixteenth century.

READING RAYMOND

As the title suggests, the *Dialogus* of Giles Du Wes is presented as a conversation between an alchemist and the personification of Nature, who, like Philosophy appearing to Boethius in his hour of need, appears in order to lead her bewildered disciple back onto the true path of philosophy.[39] As the dialogue develops, Nature leads the Disciple through the basic alchemical doc-

---

37. CCC 118, end flyleaf. On Dunstable's ownership, and Greene's erasure of his name, see Rodney M. Thomson, "John Dunstable and His Books," *Musical Times* 150 (2009): 3–16.

38. Mellon 12 does include one instance of a later date, 1522, in Du Wes's hand, suggesting that Greene acquired it after that time, but prior to his friend's death in 1535.

39. Nature was a familiar protagonist in medieval dialogues, most famously in *De planctu naturae* of Alain of Lille (d. 1202); see Willemien Otten, "The Return to Paradise: Role and Function of Early Medieval Allegories of Nature," in *The Book of Nature in Antiquity and the Middle Ages*, ed. A. Vanderjagt and K. VanBerkel (Leuven: Peeters, 2005), 97–121. On Nature's personification in alchemical contexts, see Michela Pereira, "Natura naturam vincit," in *De natura: La naturaleza en la Edad Media*, ed. José Luis Fuertes Herreros and Ángel Poncela González (Porto: Húmus, 2015), 1:101–20; Newman, *Promethean Ambitions*, 77–82; Barbara Obrist, "Nude Nature and the Art of Alchemy in Jean Perréal's Early Sixteenth-Century Miniature," in *Chymists and Chymistry*, ed. Lawrence M. Principe (Sagamore Beach, MA: Science History Publications, 2006), 113–24. On

trines, including the ubiquity of mercury, the role of fermentation, and the importance of establishing the correct proportion of ingredients (even if the actual answer is never disclosed). Throughout, Nature voices the alchemy of the *Testamentum* and other medieval authorities, while sometimes admitting newer intellectual trends, including the writings of Giovanni Pico della Mirandola (1463–1494) and Johannes Reuchlin (1455–1522).

In this respect, the *Dialogus* offers an extended concordance, in which Du Wes seeks to show that all the authorities, including those of more recent times, ultimately speak with one voice. While Du Wes's manuscripts testify to his skill at gathering texts in physical form, his treatise reveals that this ability also extended to interpreting and reconciling their alchemical contents. As Nature answers the Disciple's questions, she ventriloquizes the answers that Du Wes has himself arrived at through his painstaking study and annotation of philosophical treatises, particularly pseudo-Lullian writings. Since we now have access to Du Wes's manuscript sources, it is possible to partially reconstruct this process.

For instance, at one point the Disciple asks Nature to explicate an alchemical conundrum: why do the philosophers say that the Sun ought to be exalted in Aries?[40] Should alchemists actually track the movement of the heavens—that is, should they take account of astrological factors when preparing their work? Or do the philosophers merely use "Sol" to denote gold—in which case, how is it possible for the most perfect and stable of metals to be "exalted," or made volatile?

Du Wes's annotations in Mellon 12 show that this was a question he asked himself in his own reading. Coming to a passage in the *Testamentum* that mentions the exaltation of the Sun in Aries, he may have been reminded of the opening lines of a famous English poem, the *Mystery of Alchemists*: "When Sol in Aries and Phoebus shines bright."[41] He added a marginal note: "Here is the exposition of that text which is called that work because Sol in the first work is exalted in Aries."[42] In the *Dialogus*, Nature voices his eventual solution: there is no need to watch the heavens for an astrologically propitious moment, because (paraphrasing pseudo-Geber) every moment is

---

Boethius's personification of Philosophy, see Seth Lerer, *Boethius and Dialogue: Literary Method in "The Consolation of Philosophy"* (Princeton: Princeton University Press, 1985).

40. *Dialogus*, 106.

41. "The Mistery of Alchymists, Composed by Sir George Ripley Chanon of Bridlington," in *TCB*, 380 (CRC 19). Despite Ashmole's title, there is no evidence that Ripley composed the poem.

42. Mellon 12, fol. 145r: "Hic expositionem illius textus qui dicitur quod opus quod sol in primo operis sit exaltatur in ariete."

apt for her to bring about generation and corruption.[43] The philosophers' saying therefore refers neither to the sun nor to common gold, but to the "Sol of the Philosophers": the imperfect metal used in the early stages of the work. This reading suggests that Giles was familiar with the argument that George Ripley had outlined in his *Compound* and *Concordantia* some fifty years earlier to justify the use of base metals rather than gold. "Aries," in the meantime, alludes to the low-grade heat of the springtime sun, suitable for operations that require gentle heating, such as putrefaction.[44]

This puzzle was one that Du Wes had evidently solved to his own satisfaction in the course of his reading. Not all problems were so easily resolved. An issue that he wrestled with over the course of several decades was the correct proportion of alchemical ingredients. We can first trace this preoccupation to 1506 or thereabouts, during the period when Du Wes was compiling and annotating the Lullian contents of Mellon 12, particularly the first book of *De secretis naturae*. Here, Raymond justified the multipurpose nature of his art by explaining that the same alchemical "water" could be congealed into either elixir or precious stones, "according as the matter is proportioned [to] either."[45] Du Wes underlined this passage, noting in the margin that these proportions were hard and difficult to determine, since none of the philosophers had fully explained them.[46]

Hints of Du Wes's frustration linger in the *Dialogus*, written fifteen years later, where Nature and the Disciple debate exactly this problem. Nature, citing scripture, concedes "that all things are created by weight, measure, and number," observing that the philosophers speak more obscurely on this topic than any other.[47] Probably Du Wes still had Raymond's protean, multi-

43. *Dialogus*, 107; *Summa perfectionis*, 649. See also William R. Newman and Anthony Grafton, "Introduction: The Problematic Status of Astrology and Alchemy in Early Modern Europe," in *Secrets of Nature: Astrology and Alchemy in Early Modern Europe*, ed. Newman and Grafton (Cambridge, MA: MIT Press, 2001), 1–37, on 21–22.

44. The three "astrological" grades of alchemical heat (associated with, respectively, Aries, Leo, and Sagittarius) are set out in Geber [pseud.], *De alchimia. Libri tres* (Strasbourg: Johann Grüninger, 1529), fol. 57r; see Peter J. Forshaw, "'Chemistry, that Starry Science': Early Modern Conjunctions of Astrology and Alchemy," in *Sky and Symbol*, ed. Nicholas Campion and Liz Greene (Lampeter: Sophia Centre Press, 2013), 143–84, on 156.

45. Mellon 12, fol. 222v: "Et aqu[a]e aere[a]e habent *potestatem* indurari et coagulari, ta*m* in elixir q*uasi in* lapides *p*reciosos *secundu*m q*uod* proporcionat*ur* mat*eria* uel ad elixir uel ad lapides preciosos ia*m* dictos."

46. Marginal note by Giles Du Wes, Mellon 12, fol. 222v: "quia duru*m* & diffic*ile* cu*m* nemo ph*i*losoph*orum sit qui de ist*is* proporcionibus* tractet."

47. A well-known verse from Wisdom 11:21 (Douay-Rheims): "Thou has ordered all things in measure, and number, and weight."

purpose matter in mind, for Nature explains that changing the proportio[n] of matter causes the different metals to vary in weight, even though all ste[m] from this single root. Likewise, precious stones differ from metals only in proportion, rather than in qualities such as color, softness, or fusibility. Put more simply, the same matter, governed differently by Nature (or by human craft ministering to Nature), will produce different results, depending on the proportion of ingredients used.[48]

Yet this brings the reader no closer to identifying the correct proportion. In the *Dialogus*, Nature cites an impressive range of alchemical authorities on this topic, proving that Du Wes had searched for an answer beyond the pseudo-Lullian corpus, and possibly beyond the royal library. Roving outside the canon of alchemy, he turned to authorities on other topics—Neoplatonic philosophy, and even the newly Christianized art of Cabala—in the hope of grasping the fundamental proportions of matter teased by medieval adepts. In doing so, he offers a glimpse of an English intellectual culture that is very different from that of Ripley and Norton half a century earlier.

During the period between Du Wes's original note in 1506 and the completion of his treatise in 1521, the English court had embraced humanist learning under the impetus of its precocious new king. Henry VIII sought to attract prominent humanists to England, most notably Erasmus of Rotterdam, Lady Margaret's Professor of Divinity at Cambridge from 1510 to 1515.[49] As keeper of the library, Du Wes was on the front line of Henry's book-buying practices, although—perhaps because of a lack of scholarly credentials—he seems not to have gravitated toward the circles of highly educated humanists like Erasmus, Thomas More, John Fisher, and John Colet, the dean of St. Paul's. On the other hand, he would surely have been fascinated by one of Colet's guests: the young German doctor of divinity Heinrich Cornelius Agrippa von Nettesheim (1486–1535), who visited London as part of the imperial embassy of Maximilian I in the fall of 1510. Agrippa arrived fresh from his controver-

48. *Dialogus*, 109: "eo quod ista proportio est clauis omnium secretorum; quia metalla quae ab vna radice originem sumpserunt, in pondere diuersificantur, per istam proportionem dumtaxat, atque lapides pretiosi ab ipsis metallis non differunt qualitate, calore vel leuitate, fusioneve, nisi ista mensura siue proportione mediante: eo quod illud, quod producit metalla in esse, ea gubernatione mediante: similiter preciosos lapillos in esse producit; secundum tamen vniuersa instrumenta, atque informationem diuersam, quam a me accipit, vel ab ipso artifice mihi ministrante."

49. The influence of humanism had already been felt during the previous century, particularly in the circle of Henry V's brother Humphrey, Duke of Gloucester: Roberto Weiss, *Humanism in England during the Fifteenth Century*, 3rd ed. (Oxford: Blackwell, 1967); Alessandra Petrina, *Cultural Politics in Fifteenth-Century England: The Case of Humphrey, Duke of Gloucester* (Leiden: Brill, 2004).

sial lectures in Paris on Reuchlin's work of Christian Cabala, *De verbo mirifico* (On the Wonder-Working Word), which had prompted one critic, Jean Catilinet, to assail him as a "judaizing heretic."[50] Agrippa composed his dignified reply to Catilinet while staying in Colet's house in Stepney.[51]

While there is no evidence that Agrippa and Du Wes ever met, we must wonder whether Agrippa's timely visit contributed to the latter's interest in Cabala. A decade later, Du Wes invoked Cabala in the context of his ongoing struggle to understand the philosophers' proportion. In the *Dialogus*, Nature warns her disciple that the secret of proportion should be disclosed only to wise men, and not set down fully in writing:

> Wherefore some say this science to be part of Cabala, which reception (*receptio*) is explained through speaking together (*per colloquiam*). For the philosophers, treating of these things, wrap them in such enigmas, allegorical writings, glyphs, and riddles, that Pythagoras teaches as much with his silence as they do in their writings.[52]

Nature's admonition invokes Cabala in a purely analogical sense—like Cabala, alchemy can be regarded as an oral tradition, suggesting that Du Wes conceived of it as a tradition of profound knowledge conveyed by word of mouth, rather than through writing alone. Cabala thus had a clear affinity with alchemical secrets, which were likewise passed from master to disciple.[53] Du Wes may also have studied Reuchlin by this time. Certainly he was familiar with Pico della Mirandola's Cabalistic reflections in the *900 Conclu-*

50. Christopher I. Lehrich, *The Language of Demons and Angels: Cornelius Agrippa's Occult Philosophy* (Leiden: Brill, 2003), 40. On the controversial reception of Reuchlin's work on Cabala, see Franz Posset, *Johann Reuchlin (1455–1522): A Theological Biography* (Berlin: De Gruyter, 2015), esp. chap. 3; Charles Zika, *Reuchlin und die okkulte Tradition der Renaissance* (Sigmaringen: Thorbecke, 1998); Zika, "Reuchlin's *De Verbo Mirifico* and the Magic Debate of the Late Fifteenth Century," *Journal of the Warburg and Courtauld Institutes* 39 (1976): 104–38.

51. This was eventually published as Agrippa, "Expostulatio super Expositione sua in librum de Verbo Mirifico cum Joanne Catilineti fratrum Franciscanorum per Burgundiam provinciali ministro sacrae Theologiae doctori," in Agrippa, *De Nobilitate et Praecellentia Foeminei Sexus* (Cologne, 1532).

52. *Dialogus*, 109: "[Q]uare aliqui dixerunt, istam scientiam esse partem cabalae, quae receptio interpretatur per colloquium scilicet. Nam Philosophi de ea tractantes tantis aenigmatibus, tropicis scirpis [*sic*: scriptis], gryphis, atque problematibus inuoluunt, quod tantum docet Pythagoras suo silentio, quantum ipsi scripturis suis." Du Wes's discussion of proportion is omitted from several English versions of the *Dialogus* (e.g., in Sloane 3580B), possibly because it was seen as offering little practical value.

53. On the connections between alchemy and Cabala, see Peter J. Forshaw, "Cabala Chymica or Chemia Cabalistica—Early Modern Alchemists and Cabala," *Ambix* 60 (2013): 361–89.

sions (1486), since the Disciple cites this work on the usefulness of number for philosophizing: "Moreover, the philosophers say marvelous things about numbers. For Pico della Mirandola extolls them thus, that he is not afraid to say that by them he can answer to everything knowable, which certainly I think to be most true."[54]

Du Wes's interest in Cabala resurfaces in his French textbook, the "Introductorie." Although the book was first printed after 1533, Du Wes probably wrote the sample dialogues in the late 1520s, before the Princess Mary fell out of favor, and some years before the king's marriage to Anne Boleyn. In one, he tells Mary that God granted Moses the wisdom to understand his own works, "of the whiche knowlege, the cabalystes doth make fyftie gates that they name of intelligence"—an allusion to the "fifty gates of understanding" discussed in the third book of Reuchlin's *De arte cabalistica*.[55] In the Cabalistic tradition, Moses achieved knowledge of forty-nine of those gates, with only the final one, signifying comprehension of God himself, being denied him. Du Wes also read *De verbo mirifico*, and his inscription on the flyleaf of Trinity O.8.24 includes a note of the pentagrammaton, Reuchlin's "wonder-working word" (fig. 4).

These references show that, for all his love of manuscripts, Du Wes did not view alchemy as locked in the medieval past. As befitted the upwardly mobile librarian of a humanist prince, he kept himself up to date with continental literature, including the work of Italian Neoplatonists like Pico and Marsilio Ficino, as well as Reuchlin's controversial Christian Cabala.[56] He also applied this knowledge to understanding alchemical doctrine.

54. *Dialogus*, 331: "praeterea Philosophi mira dicunt de numeris. Nam Picus Mirandulanus ita eos extollit, vt non vereatur dicere, per ipsos responderi posse, ad omne scibile, quod profecto verissimum arbitror." Du Wes here glosses Pico's eleventh conclusion on mathematics from the *Conclusiones nongentae*: "Per numeros habetur uia ad omnis scibilis inuestigationem et in/tellectionem" ("Through numbers a method exists to the investigation and understanding of everything knowable"); Giovanni Pico della Mirandola, *Syncretism in the West: Pico's 900 Theses (1486); The Evolution of Traditional Religious and Philosophical Systems*, ed. and trans. S. A. Farmer (Tempe, AZ: Medieval & Renaissance Texts & Studies, 1998), 468–69.

55. Du Wes, "Introductorie," 1058. See Reuchlin, *De arte cabalistica libri tres* (Anshelm, 1517); Posset, *Johann Reuchlin*, 703. That Du Wes must have read Reuchlin is also noted by François Secret, *Les Kabbalistes chrétiens de la Renaissance* (Paris: Dunod, 1964), 229; Kahn, *Alchemie et Paracelsisme*, 65.

56. Du Wes copies a passage from Ficino's *De triplici vita* in a marginal note on Augurelli's 1515 poem, *Chrysopoeia*, in Trinity O.8.24, fol. 60r; Marsilio Ficino, *Three Books on Life: A Critical Edition and Translation*, ed. and trans. Carol V. Kaske and John R. Clark (Binghamton, NY: Medieval & Renaissance Texts & Studies in Conjunction with the Renaissance Society of America, 1989), 257. This note is mentioned in Zweder von Martels, "Augurello's 'Chrysopoeia' (1515)—A Turning Point in the Literary Tradition of Alchemical Texts," *Early Science and Medicine* 5 (2000): 178–95, on 186.

Although the *Testamentum* provides the bedrock of Du Wes's *Dialogus*, the librarian did not restrict himself to pseudo-Lullian material when interpreting alchemical texts. Rather, with all the skill of a royal teacher of language, Nature leads her disciple, and Du Wes's readers, through the interpretative labyrinth of texts, pausing where needed to work through examples selected from the books of the philosophers, then applying what is learned to the solution of new puzzles. The book is ultimately more precious as a guide to thinking and reading like a philosopher than it is as a manual of practice. As Du Wes's own experience illustrates, the process of learning to read alchemically is, like learning a language, a lifelong pursuit.

ALCHEMICAL WITNESSING

It is only thanks to the fortuitousness of manuscript survivals that we know of Du Wes's relationship with Robert Greene. Hitherto these men have led separate bibliographical existences, each a known collector of books and author of an alchemical treatise, but essentially disassociated from other practitioners. Yet, although they shared books and interests, including a mutual respect for Raymond, it does not follow that they approached their materials in the same way. As we see when we turn to Greene's own treatise, methods and priorities could vary widely even within a single alchemical circle concerned with broadly the same body of textual sources.

If Master Giles was preoccupied with the philosophical problems of alchemy, then his fellow Robert Greene was at least as concerned with the particulars of its manufacture. Greene's writing reveals that he placed greater emphasis than Du Wes on practice and experience—specifically, Greene's own success at recreating the marvelous effects described by earlier authorities. The *Work of Sir Robert Greene* is a shorter text than the *Dialogus*, written in English and in Greene's own voice. Finished in 1538, it purports to summarize the fruits of the forty years that Greene has spent studying alchemical texts, seeking to wrest practical results from the enigmatic procedures they describe.

For instance, it is Greene's emphasis on reconstruction that sets his approach to reading apart from Du Wes's interest in textual problem-solving. Like Du Wes, he frequently cites Raymond, but his aim in doing so, his language and tone, belong to another world: less the library than the workshop of the practicing alchemist. Although he throws out doctrinal points in passing, Greene does not plunge into their exposition, nor does he call upon Nature as an intermediary. And while acknowledging that skill in alchemical reading is necessary to understand "the misticall writing of the most noble philosophers," he usually links textual passages to descriptions of his own practical findings.[57] Thus, he admits that he was unable to grasp the nature of the volatile matter drawn from gold, silver, and mercury "till I had studied the second Chapter of Raymond Lullyes testament. For I never could fynde in all the authours that over I red the preparation so plainly declared."[58]

He goes on to illustrate what he means. The philosophers speak of "calcination," but only experience has taught Greene what they intend by it—the amalgamation of gold and mercury "which rather increaseth moisture radicall then otherwise"—because he has himself "seene and done [it] all manner of wayes."[59] He knows that Raymond's stinking menstruum contains the "secret and life of the stone," because he has seen and proved it in his own practice—in its first corruption it smells like brass, but gradually it changes, from savor to savor, until it becomes sweet-smelling, and "this truely have I proved by experience in making."[60] He offers information on the color of matter as it changes; how it smells; what it feels like to the touch; even its sound as it "cries" in the glass. In recording these observations, Green relates his mishaps as well as his triumphs. He describes his experience with a volatile calx that escaped from a poorly sealed flask, and his satisfaction at making another, ruby-red calx that retains its radical moisture: "vpon the which calce I have my whole trust and confidence vnder God therein."[61] Whether relating success or failure, Greene embellishes his *Work* with accounts of observations that testify to his own empirical, sensory knowledge.

Greene offers, in effect, a string of significant products, observations, and effects that he has achieved in his experimental practice, and that serve as a record of his practical skill in the absence of living witnesses. That such testimony mattered to him appears from his approval of a passage in the *Testamentum*. There, the Magister describes carrying out the effects he wrought in the presence of several high-status witnesses:

In the presence of certain of my fellows, in my practice I mortified common argent vive with its menstrual. And another time, in the presence of one of the fellows in whose company we were, two leagues from Naples and in the presence of John of Rhodes and Bernard de la Brett and others, we caused

57. Ashmole 1492, pt. 9, 197.
58. Ibid., 199.
59. Ibid., 197.
60. Ibid., 199.
61. Ibid., 198.

argent vive to be congealed by its menstrual; and although this was done openly in their presence, to sight and touch, still they did not know how it could be, or how made, except only simply, in a rustical way.[62]

Readers took note of the fact that Raymond's experiments were conducted before witnesses. In Greene's own copy in CCC 244, the names of John of Rhodes and Bernard de la Brett have been added, in elaborate script, in the margin beside this passage.[63] And Greene cites the passage approvingly in his *Work*: Raymond "openly certifieth to all friendes that he and other hath done and also proved these artes."[64] This passage supports not only Raymond's veracity but also, in a roundabout way, that of his reader Greene, who claims to have followed Raymond's experiments in practice. In the *Work*, Greene admits that he can call upon no other witness than God for the wonderful effects he claims to have reproduced, and therefore urges his probity on his readers: "I assure and also Certify you that I have seene with myne eyes as I have before declared." When describing his red calx, he even imagines its likely effect on observers: for "whatsoever man had seene this calce that is learned would have Judged that it was the true calce of the wise philosophers."[65]

For all his emphasis on supplying evidence and citing testimony, Greene's *Work* failed to convince sixteenth-century readers of his expertise. Although it was later transcribed by the two great alchemical copyists of early modern England, Christopher Taylour and Thomas Robson, Greene's apologia did not achieve wide circulation in manuscript, and was published relatively late, and in an anonymous, truncated form.[66] In this regard, he was outstripped by his social inferior, Giles Du Wes. The *Dialogus* published under the name of

---

62. *Testamentum*, 1:282: "Ego in presencia aliquorum sociorum meorum in mea practica mortificavi argentum vivum vulgare cum suo menstruali. Et alia vice coram uno sociorum, in societate cuius eramus, duabus leucis prope Neapolim et in presencia Johannis de Rodes et Bernardi de la Brett et aliorum fecimus congelari argentum vivum per suum menstruale; et quamvis hoc factum erat in presencia manifeste, visum et palpatum, nesciverunt tamen quomodo hoc erat aut fiebat, nisi tantummodo simpliciter ad modum rusticalem." "Bernard de la Brett" is presumably the Gascon noble Bernard Ezi V, sieur de l'Abret from 1324 to 1358, a period that overlaps with the production of the *Testamentum*. "John of Rhodes" is more difficult to place, since Rhodes itself was in the hands of the Knights Hospitaller at this time. The John Gabalas who governed Rhodes on behalf of the Byzantine emperor prior to the Genoese invasion of the island in 1248 had died long before the meeting described.

63. CCC 244, fol. 42r.

64. Ashmole 1492, pt. 9, 200.

65. Ibid., 198.

66. Anon., *The Practice of Lights: or An Excellent and Ancient Treatise of the Philosophers Stone*, in Eirenaeus Philalethes, *The Secret of the Immortal Liquor called Alkahest, or Ignis-Aqua* (London: for William Cooper, 1683); reprinted in *Collectanea Chymica: A Collection of Ten Several Treatises in*

Aegidius de Vadis is at once more sophisticated in content, more learned in citation, and more elegant in style than the *Work of Sir Robert Greene*. The Latin text also benefited from being translated into English at a relatively early date, making it accessible to both Latin and vernacular audiences.[67] Above all, it presents Du Wes as a philosophical author in the approved mode. Rather than describing his own particular array of practices, Du Wes presents the science in universal terms, as a single practice intended to yield the great elixir—in the process, demonstrating both his own mastery of alchemical reading, and his ability to solve philosophical problems.

The difference between the two writers is manifest in how each handles failure. When Greene makes a poor choice of lute for his flask, allowing the mercury to escape and fall into the ashes of his furnace, he admits to his error, before patiently scraping the precious droplets out of the ashes ("but they would never meddle nor yet joyne togither no more") and retrying the experiment, this time with a new proportion of ingredients.[68] When Du Wes fails to settle on the right proportion, his avatar, Nature, merely becomes obscure. Her strategic retreat maintains the illusion of a coy philosopher, whereas Greene openly acknowledges that he has obtained promising effects, but failed overall—an ending that proved too much for one later redactor, who tweaked the *Work* into a more palatable format as the anonymous *Practice of Lights*, published by William Cooper in 1683.[69] Greene, whose account of repeated testing comes closer to what we would now expect of an experimental program, is in the end a less convincing alchemical philosopher than Du Wes, who never permits the stench of the laboratory to disturb his communion with Nature.

THE PRIOR'S TALE

While Du Wes pored over Lullian manuscripts in the royal library, another alchemist, this time in holy orders, was vainly seeking insight from his own books in an Augustinian priory some fifty miles to the northeast. Sir Thomas

---

Chymistry, concerning The Liquor Alkahest, the Mercury of Philosophers, and other Curiosities worthy the Perusal ... (London: for William Cooper, 1684), 27–44. On Taylour and Robson see chap. 9, below.

67. The *Dialogue*'s inclusion in Thomas Potter's 1580 compilation in Sloane 3580B shows that it had been translated by that date: "A dialoge betwene nature & [th]e disciple of Philosophye," fols. 186v–202v.

68. Ashmole 1492, 197–98.

69. Anon., *The Practice of Lights*. I discuss the alterations to Greene's text in Rampling, "English Alchemy before Newton."

Ellys was the last prior of Little Leighs, or Leez, near Chelmsford in Essex. Like Du Wes, he collected alchemical manuscripts and also viewed alchemy as a collaborative endeavor. He also sought to transmute his theoretical knowledge of alchemy into practical skill, at first through his reading, and later through paying for expertise—an ultimately ill-fated enterprise that marks the last documented case of English monastic alchemy, on the very cusp of the dissolution.

Despite his spiritual calling Ellys seems to have been concerned with material outcomes: specifically, the making of gold and silver for his personal benefit. The earliest records of his career show him to have been ambitious in the cause of his advancement, if not always entirely scrupulous in obtaining it. While still a clerk in orders he attempted to win a living by suing the prior of Prittlewell, a Cluniac priory in Essex. Ellys persuaded the prior to endorse a document that named him as the next incumbent, but which he had craftily dated three years earlier, giving him priority over two men who had previously been granted the presentation of the benefice (and on whose behalf he claimed to be acting).[70] After this scheme was foiled, Ellys joined the Augustinian order, initially as a canon at the house of Little Dunmow.[71] In 1527, he was elected prior of Little Leighs.[72]

The prior later claimed that his engagement with alchemy started not in the laboratory but in the library, where his interest was provoked by "the redyng of my bokys."[73] However, Ellys did not stumble across all of these books by accident. As we saw in the previous chapter, one of his manuscripts survives, which certainly originated outside the house, and is packed with English texts, including one of the earliest known copies of Ripley's *Compound of Alchemy*. It was sent to Ellys by an anonymous correspondent who inscribed a letter on a blank page inside, sharing his opinions on how to interpret Ripley's work.[74]

For Sir Thomas, however, expounding textual authorities in light of "opyneon" was apparently not enough. Lacking the necessary expertise to

70. National Archives, STAC 2/14/111, 112. The case is summarized in James Edwin Oxley, *The Reformation in Essex to the Death of Mary* (Manchester: Manchester University Press, 1965), 36–37.

71. Presumably by coincidence, this house had been connected with dubious metallurgical practices over a century earlier, when William de Stoke was charged with counterfeiting in 1369; see chap. 1, note 19.

72. "Houses of Austin Canons: Priory of Little Leighs," in *A History of the County of Essex*, vol. 2, ed. William Page and J. Horace Round (London: Constable, 1907), 155–157n56; *British History Online*, http://www.british-history.ac.uk/vch/essex/vol2/pp155–157 (accessed 28 December 2012).

73. National Archives, STAC 3/7/85, fol. 1r.

74. Ashmole 1486, pt. 3, discussed on pp. 128–29, above.

put his bookish knowledge into practice, he began to search outside his priory for a master. These attempts eventually brought him back before the Court of Star Chamber, in proceedings that, although embarrassing for the prior, offer an unusually detailed overview of his alchemical career. A series of archival coincidences has preserved not only the prior's deposition and that of his assistant, but also treatises written by his master, Thomas Peter, and by his unknown correspondent, the Opinator. Together, these sources present Ellys as one actor within a diverse and previously unrecognized circle of alchemists, whose activities exemplify the mixed economy of secular and religious practice even immediately prior to the dissolution.

As the head of a religious house in the late 1520s, Ellys's first problem was locating a master alchemist. Given the illegality of multiplication and its close relationship with counterfeiting, practitioners seem not to have advertised their expertise widely, although they may have approached potential patrons directly. It was discreet word of mouth, however, that led Thomas Ellys to his own master, following a conversation with a goldsmith named Crawthorne in Lombard Street. The men may already have had dealings in the course of normal priory business, but Lombard Street, as the center of London's goldworking district, was in any case an excellent starting point for a budding adept. Of all London's craftsmen, the goldsmiths were the most likely to have traffic with alchemists, as both assayers and potential purchasers of alchemically produced metal. This connection was well enough known by the late fourteenth century for the Guild of Goldsmiths to express concern over its members' ability to duplicate gold and silver, and, rather later, for Thomas Norton to excuse smiths for succumbing to the lure of alchemy, "For sightis in theire Craft movith hem to bileve."[75]

It is unclear whether Crawthorne personally practiced alchemy, but he knew a man who did: a priest named Sir George, who had "made hymselfe cunnyng in suche mater."[76] Crawthorne agreed to introduce Ellys to Sir George, and the priest in turn introduced Ellys to the alchemist in this case, Master Thomas Peter. Like Crawthorne, Peter came from a background that was artisanal rather than clerical—in Ellys's words, he was a "clothe worker in london that sayd he hade the scyens of Alkemy as well as eny man in yngland."[77] As we will see, he also seems to have had some knowledge

75. Norton, *Ordinal*, 6 (l. 32).

76. STAC 3/7/85, fol. IIr.

77. Ibid., fol. Ir. The case is also summarized, with a transcription of the documents, in William Chapman Waller, "An Essex Alchemist," *Essex Review* 13 (1904): 19–23. To retain the pecularities of the original spelling, I use my own transcriptions throughout.

of goldsmithing techniques, possibly through his acquaintance with Craw-thorne. Whatever the source of Peter's skill, Ellys must have esteemed it highly, since in return for Peter's instruction he agreed to pay a substantial fee of £20, comprising a down payment of twenty-two nobles and the prom-ise of a further twenty marks "to be payd by a byll of my hande when [the work] was fynysthyd."[78]

Back at Leighs Priory, Ellys set up his laboratory under Peter's super-vision. The alchemist supplied him with starting materials, including an amalgam made from quicksilver and an ounce of silver filings. In addition to his own labors, the prior requisitioned the services of a young canon, Edmund Freake (ca. 1516–1591), whose job it was—Freake later recalled—to keep the fire burning in his furnace "bothe day & nyght."[79] Eventually, how-ever, Ellys became disillusioned and resolved to break the deal. As he later confessed, "within a certeyne tyme I perceyuyd yt was but a falce crafte [and] I wolde not paye hym no more mony."[80] He broke the vessel to retrieve the silver, which he then sold, together with those vessels that remained intact.

Three or four years passed, and Ellys may have believed that he had successfully extricated himself from his unproductive engagement with alchemy. This state was rudely interrupted when he received a subpoena from a London attorney, Hugh Oldcastle, suing Ellys for the outstanding sum of twenty marks. Given the legal situation with regard to multiplication, this seems an audacious move on Peter's part, and one that suggests that he, at least, regarded their arrangement as a business transaction subject to the normal rules of recovery. Perhaps Peter knew Ellys well enough to guess that the prior would not wish his financial dealings to come to the attention of his brethren. Soon after Ellys received Oldcastle's letter, he was approached by one Master Richard Lynsell, who warned him that it was illegal for him to retain the money, and advised him to settle the debt. Ellys asked Lynsell to lend him the twenty marks, and Lynsell agreed, in return for the lease of the parsonage of Matching, which lay within the prior's gift.

It seems that Lynsell's appearance was no coincidence. Later, Ellys reported that Lynsell became aware of the debt from "the men as I haue writyne afore"—that is, from the alchemists themselves. Lynsell and his henchman Thomas Wysman put pressure on Ellys, sending him "dyvers letters" and threatening him with "great wurdes."[81] To complicate Ellys's

78. STAC 3/7/85, fol. Ir.
79. Ibid., fol. Vr.
80. Ibid., fol. Ir.
81. Ibid., fol. IVr.

position further, Matching parsonage had been offered to another tenant, Sprenger, so the prior risked a breach of promise suit if he reneged on the agreement. The lease was also worth more than the twenty marks offered by Lynsell, so the priory stood to make a loss. Yet, if the threatened litigation was not damaging enough, hanging in the background was the specter of criminal indictment. Ellys cracked. In his own words, "I wolde not abyde the tryall of the lawe."[82] He signed over the parsonage, Lynsell paid off the alchemists, and the incriminating bill was returned.

Soon after, Ellys confessed his actions to the brothers of his priory. The matter apparently came to the attention of the authorities only after the dissolution of Leighs, although it is not clear whether the case arose from a civil suit regarding Sprenger's title to the parsonage, a criminal charge related to multiplication, or, most likely of all, an investigation instigated by the royal commissioners in the course of the visitation of 1535–36.[83] So far, all the details have been taken from Ellys's and Freake's undated depositions to the Court of Star Chamber, at this time under the direction of Henry's Lord Privy Seal, Thomas Cromwell.

Freake's testimony is valuable because it is rare to have the reports of multiple witnesses in cases of English alchemical practice. By Freake's own account he was twelve years old when the prior co-opted his services, although he would obtain distinction in later life, serving as chaplain to Elizabeth I and holding three successive bishoprics during her reign.[84] At the time of his deposition, however, Freake was a canon at Waltham Priory, having moved there after the dissolution of Leighs. This move allows us to date the court proceedings to between 1538 and 1540, when Waltham, then the last abbey standing in England, was finally dissolved. The timing of the case therefore coincides with Cromwell's campaign against superstitious practices in monasteries, suggesting that the hearing was intended to discredit the management of Leighs rather than to indict the prior personally.

For Ellys, Freake must have been an awkward witness, since the young man's account did not always tally with that of his master, detailing a longer timescale and a more complex series of practices than that deposed by the former prior. For instance, Ellys reported that he followed Peter's instruc-

---

82. Ibid., fol. 1r.

83. During the visitation, regulars were encouraged to report bad behavior on the part of their heads of houses. On the background to the visitation, and the kind of complaint elicited, see Heale, *Abbots and Priors*, 291–94.

84. Freake was bishop first of Rochester, then of Norwich, and finally of Worcester. C. S. Knighton, "Freake, Edmund (c.1516–1591)," *ODNB*.

tions over a ten-week period, during which time the alchemist came to visit him twice.[85] However, young Edmund recollects stoking the fire "the tyme of eight monthys or more to my remembrans," during which time Ellys was frequently visited not only by Peter, but also by a priest—presumably the mysterious Sir George.[86] Ellys's abbreviated version of events may reflect his concern over self-incrimination, an anxiety certainly justified by the questions posed to him, which sought to establish whether he had treasonably coined money. "I neuer coynyd," Ellys retorted, "nor neuer thowght to do nor neuer wyll god wyllyng."[87]

From the point of view of reconstructing Ellys's practice, Freake's deposition thus provides an important foil to the prior's laconic reporting. For instance, when interrogated about the role of precious metals used in his work, Ellys admits that he sealed a silver amalgam "that Peter put together" inside a glass, placed it in a water bath, and "so kept yt in a heate." According to this account, Ellys did little more than warm up Peter's prepared amalgam. Freake expands on his testimony, remarking that Ellys sometimes had as many as three or four glasses on the furnace at once, containing an unidentified metal—"but what metall yt was, yt ys to me vnknowen." Before subjecting it to the heat, however, the prior tempered this unknown metal in his hands, with quicksilver:

I dyd se hym sett thes glassys in th[e] fyer. & after when they had takyn a very grete hete, with a payer of pynchyus [i.e., pincers] he wold wrest the mowthys of them to gether, thys forsayd metall was temperyd in hys handes euer [i.e., ere] yt was putt in to thes glassys, wyche parte of yt was quycke syluer, & the rest of yt was in thynne platys lyke tyne or whyght lede.[88]

Freake's testimony describes a very different kind of alchemical practice from that carried out over a century earlier at Hatfield Priory, which involved a range of ingredients, including alums and salts. Eschewing such diversity, the prior instead prepared an imperfect body (seemingly a tin-mercury amalgam) in a manner far more compatible with the mercurialist alchemy

85. STAC 3/7/85, fol. IIv: "I dyd folowe thys peters Instruccions vppon a ten wekys & falleyd yt in a glasse & put it in a pot of erthe with water & so kept yt in a heate & no body with me but my selfe & my lade & peter came thyther twysse in the mene season."

86. Ibid., fol. Vr: "wyche often tymes came with the sayd master Peter."

87. Ibid., fol. IIv. During the 1530s, Henry VIII clamped down on clipping and coining by clerics. His commissioners also recognized priories as sites of illegal multiplication, as suggested by the discovery, in 1536, of chemical apparatus at Walsingham: Peter Marshall, "Forgery and Miracles in the Reign of Henry VIII," *Past and Present* 178 (2003): 39–73, on 69–71.

88. Ibid., fol. Vr.

that we have already encountered in the works of Arnald, Raymond, and Ripley. As we shall see, there is good evidence that this practice was developed by Thomas Peter directly from pseudo-Lullian sources, which, by the 1530s, were circulating in London in courtly and mercantile circles.

Although the prior's encounter with alchemy ultimately brought him nothing but trouble, he seems to have survived it with his living intact. Ellys was the last prior of Little Leighs. On 6 July 1534, he swore the oath of supremacy to Henry VIII, two years before the dissolution of the priory. He accepted his pension, and in 1538 finally entered the secular priesthood as vicar of Blackmore, a parish some fifteen miles from his former house.[89] This situation seems not to have been affected by Cromwell's investigation, and in fact Ellys's fortunes would later rise during the reign of Mary. He was made a canon of Norwich Cathedral in 1557 and, almost uniquely among Marian appointments, retained that position under Elizabeth, after assenting to royal supremacy for a second time.[90] His former laboratory assistant, Freake, went farther still, arriving at Norwich in 1575 as its bishop. In the absence of an alchemical elixir, however, it seems unlikely that the elderly prior lived to see the boy who once stoked his furnace enthroned as his ecclesiastical superior.

MASTER PETER'S TALE

While working conditions for alchemical religious like Ellys changed dramatically with the dissolution, the shift was less extreme for laymen and secular clergy. One of the most serious outcomes was the loss of a significant source of patronage, as well as potential sites of practice, that the monasteries had afforded even to secular practitioners like Morton and Peter. Yet heads of houses were not the only source of investment for lay practitioners, who also established premises for their work in other locales. Henry VIII and members of his council displayed interest in the potential of alchemical processes, while alchemists like Thomas Peter were not remiss in offering their services to such illustrious prospective patrons. Religious and laymen studied essentially the same books and sought to master the same practical techniques, including those enshrined in pseudo-Lullian texts. As Peter's case illustrates, by the 1530s knowledge based on learned Latin sources was

89. "Henry VIII: July 1534, 26–31," *Letters and Papers, Foreign and Domestic, Henry VIII*, vol. 7, *1534 [1883]*, ed. James Gairdner (London: Her Majesty's Stationery Office, 1883), 385–401; *British History Online*, http://www.british-history.ac.uk/report.aspx?compid=79328&strquery=leez priory (accessed 10 May 2009); "Houses of Austin Canons: Priory of Little Leighs," 155–57n57.

90. Heale, *Abbots and Priors*, 370, 374.

as apt to travel from artisanal workshops to religious houses as the other way round.

So far, the clothworker Peter has remained a shadowy figure in the background of the prior's tale. Clearly he had acquired a reputation as an alchemist by the late 1520s, but Thomas Ellys's deposition offers no clue to his alchemical activities beyond the bounds of their ill-fated partnership. Happily, Peter seems to have written a treatise of his own, preserved in a rare copy in the library of Trinity College, Cambridge. Here the elusive alchemist records his views on the theory and practice of alchemy, shedding light on the reputation that first convinced Ellys to hire him, while suggesting that he may have set his sights considerably higher than the patronage of a country prior.

This neglected text, which lacks a formal title, is an English treatise on transmutation written in eight chapters, now part of Trinity O.4.39. The manuscript is written in a hand of the first half of the sixteenth century, although a later owner appended a new front page in 1571, which confusingly attributes the work to Raymond Lull. However, Lull was not the author but the authority of the ensuing treatise, as appears from a colophon written in the original hand: "Here endeth thole [i.e., the whole] practise playn set forth without intricate words of all thole science lawde be to god, vnto the moost noble king Henry the viijth. by me Thomas Petre."[91]

While Peter/Petre is not an uncommon name, the practical content of the treatise strongly evokes young Edmund Freake's description of Ellys at work, making it likely that the author was indeed Ellys's master and nemesis. For instance, "Petre" begins by describing the sublimation of mercury. For every pound of this sublimed mercury, one should take a pound of tin and a pound of "rawe" quicksilver, then amalgamate these materials "as the goldsmithes do"—that is to say, by adding the hot quicksilver to molten tin and then, once the amalgam has cooled, rubbing it between the fingers: "in a yron morter small that no knobbes may be felte betwixt your fingers, then it is well beaten."[92] The beaten amalgam is then placed in a vessel and gently heated. As we know from Freake's report, Ellys also tempered an amalgam of thin plates "lyke tyne" in his hands before confining it to the glass. If this *Work of Petre* was indeed written by Ellys's master, it seems that Peter instructed the prior in the same technique that he set down in writing for Henry VIII—

91. Trinity O.4.39, fol. 250r. The text was later copied by Ashmole into Ashmole 1507, fols. 121r-25r, quite possibly from this manuscript; however, he does not refer to the fact that the opening page of the Trinity version was added in a later hand.

92. Trinity O.4.39, fol. 246r.

although, since the treatise is undated, we have no way of knowing whether the *Work* was composed before or after the enterprise at Leighs.

The survival of the *Work* allows us to expand on Freake's testimony, and investigate Peter's own approach to alchemy. For instance, one striking feature of the *Work* is its absence of philosophical mystification. Rather, keeping his promise to set out his practice clearly and "without intricate words," Peter describes his procedures in resolutely practical language that is closer to that of recipe collections than philosophical treatises. There are also hints that he has gained his knowledge not just from reading, but from craft practice. Peter repeatedly compares his procedures to goldsmithing techniques, for instance when describing the manufacture of *aqua fortis*, using a long-necked vessel designed for drawing strong waters, "as the goldfyners do," or when preparing the ferment, by amalgamating gold and silver with quicksilver, "as the goldsmithes do when they guilte plate."[93]

At the same time, book learning was a necessary attribute of the true adept, and Peter is careful to acknowledge an appropriate spectrum of authorities. A reference to the key revealed by a divinely inspired "gardener" suggests that he has studied the *Rosarium parvum* of Hortulanus, and identified its three key *Decknamen*, a set of three "herbs," as mercury, vitriol, and saltpeter.[94] His main source, though, is Raymond. His process begins with a chapter translated directly from the practical alphabet in the *Testamentum*—specifically, the pseudo-Lullian process for corrosive sublimate, including the signification of letters *B* (quicksilver), *C* (saltpeter), *D* (vitriol), and *E* (the menstrual water). The treatise is larded with pseudo-Lullian terminology: thus the purpose of the tin amalgam is to yield a red oil, which is, Peter tells us, the "inward bodie of quicksilver," "aqua fetence" (stinking water), and a "preciouse quinta essence." The recipe is followed by a list of chapters from the *Testamentum* that appear to support this process.

The result is a curious amalgam of traditional Latin authority and knowledge related to craft practice. Peter's perfunctory nods to Raymond and Hortulanus are a far cry from the sophisticated exegesis we have already encountered in Ripley's Latin *Medulla*. At the same time, there are moments when Peter reflects on the chemical processes that underlie his practice. He notes that quicksilver contains a "virtue," and that "when he is sublimyd with

---

93. Ibid., fols. 246v, 249v.

94. It is not clear whether Peter was reading his authorities in the original Latin. An English translation of the *Rosarium parvum* was circulating by the end of the fifteenth century (e.g., in Sloane 1091, fols. 125r-32v), although Peter may also have relied on his priestly partner, Sir George, for advice on Latin works.

his virtue, he congelith other mercury into heigh medicine"—a remark that, for all its terseness, may explain his decision to compound the sublimed mercury with crude quicksilver. Vitriol and salt seem little more than instruments in a process that is ultimately mercurialist in outlook, relying on the properties of metals rather than salts.

Peter's inner mercury, drawn from the bodies of quicksilver and an imperfect body, in fact sounds a lot like Raymond's natural fire: a metallic solvent "*with* the which ye maye dissolue all bodies at your pleasour," including gold and silver. The process concludes with Peter using this mercury to draw out quintessences from the precious metals in a process almost identical to that described in Ripley's *Medulla*. The major difference is that whereas Ripley seems to advocate copper and lead as base metals, Peter prefers tin; unless, as happens so often in alchemical writing, he employs "tin" merely as a cover name.

Thanks to the depositions of Ellys and Freake, Peter's treatise sheds light not just on the activities of a solitary alchemist, but on a whole network of practitioners with shared alchemical interests. In other respects they are a motley group, comprising, at the least, a senior cleric, a priest, a clothworker, and a goldsmith. To these we may even add an attorney-at-law, if Peter's lawyer Hugh Oldcastle is the same man later outed as a fraudulent alchemist in William Blomfild's alchemical poem, the *Blossoms*.[95]

The *Work of Petre* also introduces a new figure into our narrative: the presumed recipient of Peter's treatise, Henry VIII. The king's appearance reminds us of the importance that patronage has played in attested instances of English alchemical practice, from courtly poems addressed to English monarchs to agreements struck between priors and laymen in Essex monasteries. It also shows that diversity was not restricted to the matter of alchemical writing, but was also manifest in its form. Ripley and Norton crafted their verses for the edification and possibly the education of kings, although in practice they may well have presented copies of the same work to multiple audiences.[96] Their poems are, however, very different productions from the *Work of Petre*. The elaborate features of Ripley's *Compound*, with its dedicatory verses, satirical passages on the antics of fraudulent alchemists, and the

---

95. Blomfild, *Blossoms*, 26 (l. 138).

96. We can speculate as much on the basis of the number of early, presentation-quality copies of Norton's *Ordinal*, one of which survives as British Library, MS Add. 10302, while two others are mentioned by Ashmole, *TCB*, 455 (one of which Ashmole thought to have been "Henry the seventh's own Booke"). John Reidy, "Introduction," in Norton, *Ordinal*, xiv.

imaginative conceit of the twelve-gated castle, are devices that might have rendered the Englished doctrines of obscure alchemical philosophers more attractive to a royal palate.

Thomas Peter dispenses with such trappings in his own pragmatic *Work*. His only concession to his presumed royal audience is to present the scale and aspiration of his project on a princely scale: "you must haue a hundred pounde weight [of sublimed mercury] which wolbe litle ynoughe, yf ye woll haue a riche worke for a prince, to transmute all vnparfite metal into golde || and silver."[97] Possibly Peter wrote his treatise to accompany a petition for a license to practice—a context in which he may have felt that practical content would be valued over rhetorical flourishing.

Yet there is no evidence that the *Work* was ever sent to Henry, let alone read by him. We are on firmer ground in stating that, by the late 1530s, Henry VIII stood in little need of alchemy as a source of bullion. His fortune lay with an even more profitable kind of dissolution, albeit one less helpful to alchemists—the suppression of the religious houses that had formerly offered shelter and funding for their practice.

THE END OF THE MIXED ECONOMY

Of the practitioners discussed in this chapter, none except possibly Peter was still involved in alchemy after the dissolution. Thomas Ellys, no doubt a sadder and wiser man, had abandoned his practice even before the suppression of his house. Giles Du Wes did not live to see the end of English monasticism. When he died on 12 April 1535, he left Robert Greene in possession of both his pseudo-Lullian manuscripts, as well as Trinity O.8.24 and possibly its sister manuscript, O.8.25. But Greene's once passionate interest in alchemy was already on the wane. From 1528 he copied large numbers of Latin alchemical treatises into a series of folio volumes, but his last dated transcription was made on 2 July 1534. By this time, Du Wes's health may already have been declining. On 11 March 1534, William Tyldysley was awarded the patent to succeed him as keeper of the royal library at Richmond. On 20 December, Du Wes made his will.[98]

Although Greene's own decision to abandon alchemy was made just a few years after the loss of his long-term collaborator, he blamed it more straight-

---

97. Trinity O.4.39, fol. 245r-v.
98. Kipling, "Duwes, Giles."

forwardly on shortage of funds. In 1538 he wrote the *Work* as an apologia for his career. He had, he explains, "laboured the space of 40 yeeres and more in the Theoricke." During that time, he had grown a mercurial tree and made a calx of many colors, and "in this I have done and proved which I could never see man that could shew me so much." He still believes that the science of alchemy is true, "surely without doubt or fable," but he no longer has the will or the means to continue:

> And so left I of my busines in this art and never since I did this art. I laboured
> no more for lacke and want of worldly goods, and for envy of the perverse
> and false world. And now of this litle treatise I have made an end vnder
> Gods power in my old yeeres and age of 71 Anno 1538 and in the reigne of
> our most sovereigne Lord King Henry the 8. the 30th yeere.[99]

Only one manuscript in Greene's hand survives from after this date: a compendium of medical treatises and recipes dated to 1544 that contains few hints of his earlier alchemical interests. Only toward the end do a few references slip in: a recipe for a compound water, and a "miraculous water of Raymond."[100] Greene continued to collect and parse medical receipts, but his alchemical career now lay behind him.

The year 1538 marks a pivot point in the history of English alchemy, not just as the year in which Greene abandoned his practice, but as the start of the great dissolution of the monasteries. Yet the activities of men like Du Wes and Greene show that the preservation of alchemical books was underway long before the suppression of religious houses was even anticipated— before Henry VIII even came to the throne. Medieval manuscripts were already prized by readers from outside monastic settings, providing the basis for further copies that disseminated rapidly in the second half of the century. The boom in Elizabethan alchemy was fed not just by materials retrieved from former religious houses at the moment of dissolution, but also by the bookish outputs of a mixed economy that had long since permeated secular spaces. There was thus no grand *translatio studii* of alchemical books from monastic to private hands, although large numbers of the former undoubtedly did end up in the libraries of the latter, partly through the keen efforts of collectors like John Dee who sought to salvage as much as possible from the dispersal of monastic collections. But several of Greene's books also came

---

99. Ashmole 1492, 205.
100. MS Hunter 403, 292: "aqua vite composita"; 297: "Aqua mirabilis a Reimundo."

into Dee's hands, including two of the great Lullian compendia, CCC 244 and Mellon 12, and the copy of Boethius's *De musica*.[101] In the end, it was on the shelves of scholarly collectors like Dee, rather than in a monastic workroom, that the mixed economy of English alchemy finally came to rest, the books of Thomas Ellys and his peers now lying cover to cover with the secular productions of merchants, artisans, and librarians.

101. Mellon 12 is DM94; CCC 244 is DM148; and CCC 118 (Boethius) is M142 in Julian Roberts and Andrew G. Watson, eds., *John Dee's Library Catalogue* (Cambridge: Bibliographical Society, 1990).

CHAPTER FIVE
Nature and Magic

Our vitrioll, our sulphur, our lunary most of price.
Put the key in the locke, & it will open with a trice.[1]

The material consequences of the English Reformation are still underexplored in histories of science. Studies of Henrician science have long focused on the court, where the young king sought to set an example for Europe through his patronage of humanist scholarship, or on the circles of university graduates whose futures lay in royal administration and mercantile ventures rather than the contemplative life.[2] These settings were undeniably important loci for natural knowledge, including knowledge of alchemy. The foregrounding of lay practice should not, however, blind us to the impact of the dissolution. The mixed economy of religious and lay practice depended on monasteries as sites of practice and sources of patronage, and on the religious themselves as collectors, compilers, and translators of scientific and medical texts. Yet in the 1530s, the mixed economy collapsed into a single stream. What, then, became of the monks, their knowledge, and their books? And what options remained for alchemical practitioners in need of support?

Undoubtedly the single greatest factor in blurring our picture of alchemy

1. Blomfild, *Blossoms*, on 24 (ll. 90–91).
2. Nicholas Kratzer (1487?–1550), the Munich mathematician and instrument-maker appointed as astronomer by Henry VIII, exemplifies both sites: in addition to his court position, he was appointed by Wolsey to lecture at Corpus Christi College, Oxford, an important site for humanist learning. On humanism and scientific knowledge during Henry's reign, see Antonia McLean, *Humanism and the Rise of Science in Tudor England* (New York: Heinemann, 1972); Kenneth Charlton, "Holbein's 'Ambassadors' and Sixteenth-Century Education," *Journal of the History of Ideas* 21 (1960): 99–109; John North, *The Ambassadors' Secret: Holbein and the World of the Renaissance* (London: Hambledon, 2002).

in the first half of the sixteenth century has been the dearth of contem
rary documents. By patchworking together surviving Henrician sou
with precious Elizabethan copies, we learn that transmutation and medicinal alchemy not only flourished during Henry's reign, but received at least
the tentative support of the king and his privy council. At the same time,
this interest was tempered by concern about a rise in conjuring practices,
fed by the popularity of Agrippa von Nettesheim's *De occulta philosophia
libri tres* (Three Books Concerning Occult Philosophy, first printed in 1533),
which linked alchemy to magic beneath the umbrella of "occult philosophy."
Such associations led to concern over implications of unregulated alchemical
practice, not just for the economy, but for the spiritual safety of the realm.

It has long been assumed that Henry VIII, unlike his Plantagenet forebears, was uninterested in transmutation—a view dating back to 1884, when
J. S. Brewer, editor of the state papers, crushingly observed, "I do not find
that Henry ever dabbled in alchemy, the royal amusement of the Scotch
kings."[3] The perception of Henrician indifference is bolstered by a dearth of
licenses granted to alchemical practitioners, suggesting that fewer of them
were encouraged to petition the king directly for support.

Yet an absence of alchemical petitioners was not the perception of one
aggrieved practitioner, whose own suit to the king survives in a later Elizabethan copy. Writing from his prison cell, this former monk denounces
the "dyuerse persones . . . which have promysed vnto your moste excellent
Maiestie to enforse themselves, to worke in the excellent work of Alkymye.[4]
Despite their boasts, these men are not adepts, since they cannot properly
construe "the occult and hidden books of the ancient and old philosophers."
Instead, lacking a true understanding of the causes of things, they occupy
themselves in useless and sundry practices, trapped like blind men in "the
thickets and briars of ignorance." Warming to his metaphor, the writer imagines the books of the authorities as a dense forest, which the ignorant seek
blindly to traverse:

> Groping for the pathway to the large Camp of Philosophy, wherein they be
> entered a little way in the thickets of the philosophers' books, whose path-
> way is darkened and hidden with many leaves and bypaths, whereby they
> stand amazed, knowing not whither to turn unto their journey's end.[5]

3. J. S. Brewer, *The Reign of Henry VIII from His Accession to the Death of Wolsey*, ed. James
Gairdner (London: John Murray, 1884), 1:233n1.

4. Sloane 2170, fol. 56r.

5. Ibid.

To modern readers of Ashmole's *Theatrum*, this allegory may have a familiar ring. The same conceit was employed in 1557 by William Blomfild (fl. 1529–1574) in his famous English poem, *The Compendiary of the Noble Science of Alchemy*, printed by Ashmole as *Bloomefield's Blossoms, or The Campe of Philosophy*.[6] In this verse treatise, Blomfild seized the opportunity to criticize the false alchemists of Henry's reign, who sought the Camp of Philosophy but lost themselves within "A thicket haueing by pathes many [a one]."[7] Unable to construe the enigmas of the philosophers, such men failed to penetrate the intended meaning of their texts—or, in consequence, the garden of Lady Philosophy, who waits beyond the wilderness to welcome her true disciples.

The coincidence between the two works suggests that Henry's petitioner may have been Blomfild himself, and that he was exasperated by the existence of potential rivals for the king's attention and support. The image of alchemists clamoring for royal patronage is not one typically associated with the Henrician age. In an unsettled political and religious climate, however, practitioners were eager to obtain preferment. Just as the profusion of unlicensed healers prompted Thomas Linacre (ca. 1460–1524) and his peers to petition Henry to establish a College of Physicians to regulate medical practice, so self-styled alchemical philosophers like Blomfild urged the king to dispense with the services of ignorant chemists in favor of more reliable practitioners.[8] Such support may have had particular value for former religious like Blomfild, who now sought to make their own way in post-Reformation England.

If the dissolution marks the pivot point in English alchemical history, it makes sense to focus on a pivotal figure. In this chapter we take up the enterprise of William Blomfild, a former Benedictine monk who was himself employed as an alchemist in London in the wake of the dissolution. In personality, affiliation, and context, Blomfild is a very different character from his contemporary Thomas Ellys, prior of Little Leighs. Ellys remained Catholic and received a cathedral canonry under Mary; Blomfild embraced the Reformation and gained a parish under Elizabeth—only to lose it within a year

6. Blomfild, *Blossoms*, first published by Ashmole in *TCB*, 305–22. Robert Schuler proposes the "Compendiary" as the original title in his edition of the work; Schuler, "Three Renaissance Scientific Poems."

7. Blomfild, *Blossoms*, 26 (l. 149).

8. Henry's letters patent to the petitioners were confirmed by a statute of 1523: 15 Henry VIII, c.5. On Thomas Linacre, Henry VIII's physician and first president of the college, see Francis Maddison, Margaret Pelling, and Charles Webster, *Essays on the Life and Work of Thomas Linacre, c. 1460–1524* (New York: Oxford University Press, 1977).

when his evangelism proved too potent a brew for his moderate congrega-
tion. The prior quickly abjured alchemy, while Blomfild was still haranguing
his sovereign on the topic some thirty years later, having narrowly avoided
prosecution for conjuring along the way. For all the differences, however,
the two cases are united by alchemical doctrine and practical commitments,
based on the quintessential alchemy of pseudo-Lull and his English commentator, Ripley. They also reveal the extent to which both transmutation
and alchemical medicine continued to be practiced and funded in the early
years of the Reformation. And they show how practicing alchemists continued to engage with the books of their authorities every step of the way.

HENRY VIII AS PROSPECTIVE PATRON

Although Henry was temporarily enriched by the proceeds of the dissolution of the monasteries and the suppression of chantries, the supply of bullion was soon under pressure. Prices in England had been rising since the
beginning of the century, and by the 1540s, despite the influx of monastic
plate into royal coffers, England faced a serious financial crisis. Between
1543 and 1546 Henry prosecuted an expensive series of wars against France,
financed by a tax hike, as well as forced loans, the sale of Crown lands, and—
most significantly for our purposes—the debasement of the coinage. Henry's
response to rampant inflation was to devalue English money by reducing the
proportion of silver in the alloy, which led, predictably, to a loss of confidence in English coin.[9]

It is often at times of currency crisis that alchemists come into their own,
and we might expect the same of Henrician England. Indeed, the 1530s and
40s saw a new generation of English practitioners petition the king for support, testifying to continuing engagement with alchemy among the mercantile community, even prior to the dissolution. The substantive content of
their books and petitions confirms the trend we have already detected in the
manuscript tradition—namely, that pseudo-Lullian alchemy was gathering
pace in mercantile and artisanal circles. The mercer Robert Freelove prepared an elaborate presentation volume of pseudo-Lullian transcriptions in
1536, which may have been intended to attract the king's attention.[10] As we
have seen, the clothworker Thomas Peter also addressed a treatise to Henry

9. On the "great debasement," see C. E. Challis, *The Tudor Coinage* (New York: Manchester
University Press, 1978), 81–112; J. D. Gould, *The Great Debasement: Currency and the Economy in
Mid-Tudor England* (Oxford: Clarendon Press, 1970).

10. Sloane 3604; discussed in chap. 6, below.

that cites Raymond at length, although there is scant evidence that the king or his councillors licensed him, or any alchemical practitioner.

In fact, the only warrant associated with Henry VIII is almost certainly misdated. The text survives in a later copy, in which the king grants the merchant John Misselden and his son Robert permission to practice transmutation, on the grounds that Misselden, during his time abroad, "hathe learned and vsed by Crafte or scyence of Philosophie vnperfecte mettall to bringe & Transpos vnto perfecte Mettaylle and at alsayes [i.e., at all assays] to abyde the hamere bothe gold & Syluere aswelle at the vre [i.e., as the ore] that growethe in any myne."[11] An ambiguity in the dating—the king seals the letter on 13 February, in the thirtieth year of his reign, at Westminster Palace—led the editors of Henry VIII's *Letters and Papers* to date the grant to 1539. Yet the primary recipient is surely the same John Mistelden licensed in 1452, the thirtieth regnal year of Henry VI.[12] The manuscript thus preserves a lost document of Henry VI rather than the sole warrant of Henry VIII.

The text itself is typical of fifteenth-century licenses, underscoring the requirement for the transmuted metal to pass assay, and hence to provide reliable coin, to the same standard as naturally occurring gold and silver. It also reflects Henry VI's positive view of alchemy—the Misseldens are to execute their project not only for their own benefit, but also "for the greate avayle that maye therby within breefe tyme growe vnto vs and vnto our leege people."[13] However, it also differs from contemporary licenses in significant (and prescient) ways. Unlike other petitioners, the Misseldens had the advantage of having learned their skill outside the realm, and hence could claim expertise in practice without also having to admit to flouting the law by multiplying metals on English soil. The same pattern would be repeated later during the reign of Elizabeth I, as practitioners with experience gained abroad, like Cornelius de Lannoy and Giovanni Battista Agnello, enjoyed

11. Harley 660, fol. 85v; summarized in "Letters and Papers: February 1539, 11–15," in *Letters and Papers, Foreign and Domestic, Henry VIII*, vol. 14, pt. 1, *January–July 1539*, ed. James Gairdner and R. H. Brodie (London: Her Majesty's Stationery Office, 1894), 108; *British History Online*, http://www.british-history.ac.uk/letters-papers-hen8/vol14/no1/pp107–117 (accessed 5 August 2017). Harley 660 comprises numerous copies of patents and other royal administrative documents.

12. On the Latin license, which mentions Mistelden's three servants but not his son Robert, see p. 67, note 18, above. The warrant in Harley 660, fol. 85v, concludes: "vnder our sygnete at our pallace of Westemenstere the xiij daye of ffever ffeveriere the yeare of our Reigne xxxt"; misleadingly summarized in the *Letters and Papers*: "Under our signet at Westminster, 13 Feb. 30 Hen. VIII." The fact that Westminster was damaged by fire in 1512, ending its term as a royal residence, reinforces the earlier date.

13. Harley 660, fol. 85v.

greater success in winning royal and noble support than the majority of tl.
disgruntled English rivals.[14]

The warrant differs in another way, one that suggests a shift in England
alchemical culture. As ever, the king directs that his officers and liege people
should not disturb the Misseldens' operations, but he adds a caveat: they will
escape harassment provided that they proceed "without any Crafte of Nec-
romansye but onely by playne science of Philossophie."[15] This qualification,
unique among surviving documents, reflects concern over magical practices. This concern would be realized during the reign of a later Henry, when
several alchemists in search of patronage, including Blomfild, did become
embroiled in conjuring.

In Henry VIII's England, such apprehensions encompassed more than
alchemy. The English Reformation oversaw a period of tremendous spiritual and temporal upheaval, unleashing fears over the influence of the devil
in the world, as revealed through superstitious practices—particularly those
that Reformed propaganda tended to associate with Catholicism, and with
monks in particular. The anxieties of the period also manifested in political
prophecy and visions, which sometimes accrued a popular following that
posed a threat to order, both spiritual and political.[16] Agrippa's books contributed to an increased interest in learned magic, which had a more prosaic sequel in the pursuit of power and riches through magical means, from
crafting rings of invisibility to seeking buried treasure. Such practices were
legislated against. In 1542, political prophecies were banned by Act of Parliament.[17] Another statute passed that year (although first drafted in 1533) made
witchcraft a felony, forbidding "Invocacons or co[n]juracons of Sprites wit-
checraftes enchauntementes or sorceries" for any "unlawfull intente or purpose" regardless of whether harm was actually caused.[18]

Despite a modern tendency to conflate alchemy and magic under the
rubric of occult philosophy, the relationship between them is difficult to pin-

14. Discussed in chap. 6, below.

15. Harley 660, fol. 85v.

16. Madeleine Hope Dodds, "Political Prophecies in the Reign of Henry VIII," *Modern Language Review* 11 (1916): 276–84; Keith Thomas, *Religion and the Decline of Magic* (London: Weidenfeld & Nicolson, 1971; repr., London: Penguin, 1991), 471–77; G. R. Elton, *Policy and Police: The
Enforcement of the Reformation in the Age of Thomas Cromwell* (Cambridge: Cambridge University
Press, 1972), chap. 2.

17. 33 Henry VIII, c.14.

18. 33 Henry VIII, c.8. On the associations between magic and treason, see Francis Young,
Magic as a Political Crime in Medieval and Early Modern England: A History of Sorcery and Treason
(London: I. B. Tauris, 2018); Jonathan K. van Patten, "Magic, Prophecy, and the Law of Treason in
Reformation England," *American Journal of Legal History* 27 (1983): 1–32; Malcolm Gaskill, "Witch-
craft and Evidence in Early Modern England," *Past and Present* 198 (2008): 33–70.

point, and in practice varied in line with the opinions and interests of particular practitioners. In early modern Europe, the distinction is most obvious between alchemy and ritual or ceremonial magic. Whereas alchemy utilizes the operations of nature, achieving change through manipulation of properties intrinsic to matter, ritual magic relies (implicitly or explicitly) upon the agency of spirits. Although alchemists overstepped the law when they attempted counterfeiting, their practices were otherwise no more dangerous than any other metallurgical process. Conjuring, on the other hand, achieved its effects through angelic or demonic intervention, and was therefore spiritually dangerous in and of itself, however benevolent the practitioner's intention.[19] The difference was not always obvious to nonspecialists, and fourteenth-century writers had to vehemently reject intimations that their work was effected by demons.[20] In this sense alchemy had its analogue in astrology, whose defenders also sought to decouple their art from illicit divinatory practices: for instance, by distinguishing what they considered to be the useful, pragmatic aspects of their work from overly determinist readings of the stars, which threatened to impinge on both divine authority and human free will.[21]

For a self-identified alchemical philosopher keen to avoid guilt by association, it was just as important to distinguish conjuring activities from alchemy, which (practitioners claimed) was wholly grounded in nature. This contrast could even be used to score an argumentative point. Writing to Elizabeth I in 1565, the alchemist Thomas Charnock later recalled his own youthful follies during this period, when he studied magical books and dabbled in such "vayne sciencis" as geomantia, hydromantia, aeromantia, and pyromantia—

19. On ritual magic in its English context, see Frank Klaassen, *The Transformations of Magic: Illicit Learned Magic in the Later Middle Ages and Renaissance* (University Park: Pennsylvania State University Press, 2013). Legal and moral condemnations of "diabolic" magic are more generally surveyed in Richard Kieckhefer, *Magic in the Middle Ages* (New York: Cambridge University Press, 1989), chap. 8; Michael D. Bailey, "Diabolic Magic," in *The Cambridge History of Magic and Witchcraft in the West: From Antiquity to the Present*, ed. David J. Collins (New York: Cambridge University Press, 2015), 361–92. Major studies include Stuart Clark, *Thinking with Demons: The Idea of Witchcraft in Early Modern Europe* (Oxford: Oxford University Press, 1997); Thomas, *Religion and the Decline of Magic.*

20. These medieval objections against transmutation are detailed in Newman, *Promethean Ambitions*, 44–49.

21. John Dee, for instance, famously distinguished "true" from "false" astrological practice in his defense of the former: Dee, *The Mathematicall Praeface to the Elements of Geometrie of Euclid of Megara (1570)* (New York: Science History Publications, 1975), sigs. [A.i.]v–[A.ij.]v; [b.iii]r–v. On the risks and benefits of astrology in English courtly settings, see Hilary M. Carey, *Courting Disaster: Astrology at the English Court and University in the Later Middle Ages* (London: Macmillan, 1992).

means of prophesying by manipulating the various elements. He
from reading Michael Scot how to divine the future by plotting th
birds, and scrambled up to a jackdaw's nest on the advice of Albe
nus, "to have that stone named Aldronicus to go invisible."[22] His eaɪɪy ⌐ᵣ
rience with the less reputable aspects of "that ryche science off Alchimie,"
when still guided by opinion rather than proof, also belonged to this period
of misguided application. Only when he began to study astronomy, cosmography, medicine, and natural philosophy did he gain an inkling of the truth,
and so, he jokes, "by gods grace I euer fell from the worst, to the better."

The distinction between change wrought by nature and that accomplished by spiritual intervention is already present in Henry VI's license to
John Misselden. Yet by seeming to approve alchemy conducted according to
the "playne science of Philossophie," the Plantagenet king created a loophole
for practitioners under suspicion of conjuring. Evidence of alchemical skill
might be used to defuse concerns about other, more suspicious activities if
the accused could only demonstrate that this work, at least, was grounded in
the principles and practices of conventional natural philosophy, rather than
illicit magic. The stone also offered a potentially mitigating source of profit.
Blomfild was not the only practitioner implicated in conjuring who saw the
philosophers' stone as valuable not just in its own right, but also as a means
of escaping the law.

ALCHEMY AND MAGIC

The blurred boundary between alchemy and magic presents a demarcation
problem, one that modern historians share with early modern practitioners
and the authorities who sought to regulate their activities. If alchemy was
readily distinguished from conjuring, it was less clearly delineated from *natural* magic, an art grounded in the hidden correspondences and influences
presumed to exist in nature, rather than in the action of spiritual entities. The
existence of occult properties, such as the ability of a lodestone to attract
iron, was a scholarly and popular commonplace in early modern Europe,
and underpinned a vast array of practices from medicine to navigation.[23]
Ambiguities could arise, however, even in the case of such "natural" work-

22. Lansdowne 703, fols. 44v–45r.

23. The literature on occult properties in early modern Europe is extensive. See, in particular,
the pioneering studies of Keith Hutchison, "What Happened to Occult Qualities in the Scientific
Revolution?," *Isis* 73 (1982): 233–53; John Henry, "Occult Qualities and Experimental Philosophy,"
History of Science 24 (1986): 335–81.

ings: notably in relation to the making of amulets, whose effects could be variously explained by natural sympathies between materials, celestial virtues, or demonic agency.[24]

Since alchemical change was mediated by hidden actions and properties, it is not surprising that even the alchemists themselves sometimes viewed their work as contiguous with magic, particularly if they were already interested in the latter for its own sake. The connection was made explicit by the Florentine priest and physician Marsilio Ficino, who saw a correspondence between the Rupescissan and pseudo-Lullian "quintessence" and the universal spirit (or "World-Soul") that he discussed in his own Neoplatonist philosophy.[25] For Ficino, all generation, whether animal, vegetable, or mineral, is achieved through this spirit, which is not specific to metals, but exists in all matter. The ubiquity of the spirit allows careful operators to harness a sympathetic connection between given substances and their celestial analogues. This, for Ficino, is natural magic. The spirit is, furthermore, inhibited by the presence of the grosser material in which it is embedded. Only "diligent natural philosophers" can artificially separate it from this denser matter—and this is alchemy. "When they separate this sort of spirit from gold by sublimation over fire," Ficino states, they can "employ it on any of the metals and will make it gold."[26]

For readers already steeped in pseudo-Lullian alchemical doctrine, this view will have struck a familiar chord. The third chapter of the *Testamentum* describes just such a universal power, albeit expressed as an attractive force, or virtue, rather than as a spirit. This virtue is further invested with a religious and apocalyptic character in keeping with its fourteenth-century context. In the Magister's account, all matter originally shared in the purity of its parent substance "until the time of sin," when nature was corrupted by

24. On the problems related to amulets and image magic, see Klaassen, *Transformations of Magic*. D. P. Walker, *Spiritual and Demonic Magic from Ficino to Campanella* (London: Warburg Institute, 1958; repr., University Park: Pennsylvania State University Press, 2000), still provides a good overview.

25. The link is made in the third chapter of *De vita coelita comparanda* (On Obtaining Life from the Heavens), the famous third book of Ficino's *De triplici vita* (Three Books on Life), first printed in Florence, in 1489.

26. Translation in Ficino, *Three Books on Life*, 257. As Sylvain Matton has shown, Ficino's reading of the quintessence in turn had an enormous influence on sixteenth-century alchemical theorizing; Sylvain Matton, "Marsile Ficin et l'alchimie: Sa position, son influence," in *Alchimie et philosophie à la Renaissance, Actes du colloque international de Tours (4–7 décembre 1991)*, ed. Sylvain Matton and Jean-Claude Margolin (Paris: Vrin, 1993), 123–92.

the Fall of Man.[27] From that time on, human intervention has been required to recover the original matter from its corrupt accretions: "For, by reason of her gross and corrupt matter, Nature cannot make a thing as perfect as she did at the beginning."[28] Only at the end of days will this pure matter emerge, unsullied, from the refining fires of the Last Judgment.

Ficino's own reading of the quintessence achieved wide dissemination when Agrippa included it in the first book of his *De occulta philosophia*, concerning natural magic, written in 1510.[29] Although Agrippa says little directly on the subject of alchemy, he does repeat Ficino's view that the quintessence diffuses the World-Soul through all things (including metals), and that this spirit might be beneficially extracted, "if any one knew how to separate it from the Elements."[30] The influence of this synthesis was soon felt in England, although its reception was not uniform. While Agrippa's book, loaded with evidence of his humanist credentials and saturated with Ficinian Neoplatonism, arguably reflects a new current in alchemical thinking, some readers treated the text much as they would any other "philosophical" source—that is to say, they studied it alongside their other books in the hope of gaining insight into practice.

Hints in English manuscripts suggest that some readers viewed the "spirit" of Ficino and Agrippa as another reference to the quintessence already familiar from the writings of Raymond and Ripley. Thus Thomas Ellys's anonymous correspondent, the author of the *Book Concerning Myne Opyneons*,

27. *Testamentum*, 1:14: "Ista quattuor elementa sic creata remanserunt pura et clara racione clare partis nature ex qua erant creata usque ad tempus peccati, quod exivit a natura et adhuc est ad tempus indulgencie post peccatum."

28. Ibid.: "[Q]uoniam natura non potest facere rem tam perfectam, racione sue materie grosse et corrupte, sicut fecerat in suo principio. Sed natura in operando imperfeccionis participat cum magna corrupcione propter materiam elementorum minus purorum, quam quotidie ipsa invenit."

29. A revised version of the first book was printed in Paris, Cologne, and Antwerp in 1531, while the complete work—including the contentious third book on ritual magic—appeared in Cologne in 1533. However, the first version of the book circulated widely in manuscript before this time. On Agrippa's use of Ficino, see Matton, "Marsile Ficin et l'alchimie."

30. Heinrich Cornelius Agrippa, *Three Books of Occult Philosophy*, trans. J. F. (London: R. W. for Gregory Moule, 1651 [i.e., 1650]), 33. Agrippa is silent regarding his own alchemical practice, although Sylvain Matton argues that he was in fact the author of an alchemical treatise later attributed to Ficino; Matton, "Introduction," in Heinrich Cornelius Agrippa (attr. to), *De arte chimica (The Art of Alchemy): A Critical Edition of the Latin Text with a Seventeenth-Century English Translation*, ed. Sylvain Matton (Paris: S.É.H.A.; Milan: Archè, 2014). Elsewhere, Agrippa included a dismissive passage on alchemy in *De incertitudine et vanitate scientiarum et artium atque de excellentia verbi Dei declamatio invectiva* (Cologne, 1527); translated as *The Vanity of Arts and Sciences* (London: Samuel Speed, 1676), 312–16.

rs a brief commentary on the fifth chapter of Agrippa's first book, "Con-
ing the Wonderful Natures of Fire and Earth," which interprets Agrippa's
discussion of the elements almost entirely in light of Ripley's *Compound of
Alchemy*.[31] Ripley's doctrine of four fires is cited to show that the "fire" in
Agrippa's chapter heading is not elemental fire "that brennyth the brand" (a
direct quotation from the *Compound*) but rather a "pure essencyaule fyre of
nature clere & bright counfortable & Nutratyue." It is, in fact, the natural fire
of pseudo-Lullian alchemy, whose "wonderfull essencyall power" lies buried
within the common metals. Initially hindered by the gross, earthy dregs of
their material substance, the "most lyuely acytyue powers" of the metals can
still be "porged & clensid from theyre origenal synnes" through alchemical
operations.[32] This passage blends old and new sources: Ficino's theory of the
universal spirit (mediated via Agrippa) and the *Testamentum*'s notion of matter corrupted by original sin (mediated via Ripley).[33] This, however, is as far
as the Opinator goes. Whereas Du Wes treats the writings of Ficino, Pico,
and Reuchlin as new sources of insight into alchemical processes, the writer
of *Myne Opyneons* is apparently interested in Agrippa's book only insofar as
it helps him decipher Ripley's alchemy.

But Agrippa was also studied by readers whose interests went beyond
transmutation, drawing alchemy into murkier waters. In 1532, the Oxford
scholar Richard Jones was arrested for his role in a deception involving the
nobleman and poet William Neville, brother of Lord Latimer (and hence,
coincidentally, brother-in-law of Latimer's wife Catherine Parr, the future
queen). Neville was himself interested in magic: he had attempted to make
a cloak of invisibility, and later sought a ring that would win him royal favor.
As with Thomas Ellys, lack of expertise sent him in search of specialists—in
this case, diviners such as Jones, who prophesied that Neville would succeed
to the earldom of Warwick following the death of the king. Neville's indis-
creet reaction to this good news troubled his chaplain to the extent that he
disclosed the whole affair to Cromwell, who had the main actors arrested
and confined to the Tower on suspicion of treason.[34]

31. Anon., "Opus henrici Cornelij Agrippe de occulta philozofia liber primus capitulo .[quint]o.
de mirabilibus ignis at terre natures," Ashmole 1426, pt. 5, 37–42.

32. Ibid., 42.

33. Ripley alludes to the *Testamentum*'s apocalyptic analogy particularly in the verses to his
wheel (a favorite source of the Opinator); Rampling, "Depicting the Medieval Cosmos," 63–64.

34. "Henry VIII: December 1532, 16–31," in *Letters and Papers, Foreign and Domestic, Henry VIII*,
vol. 5, *1531–1532*, ed. James Gairdner (London: Her Majesty's Stationery Office, 1880), 681–700;
British History Online, http://www.british-history.ac.uk/letters-papers-hen8/vol5/pp681–700
(accessed 5 August 2017); Elton, *Policy and Police*, 50–56.

Necromancy rather than alchemy presented the chief danger to Jones, although he was involved in both. Neville later deposed that he saw various books and magical accoutrements in Jones's rooms, in addition to the more conventional stills and alembics of the practicing alchemist. These books presumably included a copy of Agrippa's *De occulta philosophia*, which, Neville reported, Jones had discussed with him, specifically in the context of magical images. For Geoffrey Elton, who reconstructed the Neville affair from the state papers, there was little to distinguish Jones's magical interests from his alchemy; yet, as we have seen, transmutation was clearly demarcated from ritual magic not only by philosophical and technical content but also by legal status. While multiplication was still a felony, it did not alarm Henry and his ministers to nearly the same extent as necromantic and divinatory practices, or the manipulation of political prophecy—for instance, when predicting the death of the king or Anne Boleyn, or the downfall of favorites like Cromwell.

Jones was certainly well aware of the difference when he wrote to Cromwell in an appeal for mercy, understanding, and employment.[35] Imprisoned in the Tower, he seems to have concluded that his best hope lay in convincing Henry's powerful counselor not just of his innocence in the Neville affair, but also of his usefulness as an alchemist. In a petition that conspicuously makes no reference to magical practice, Jones offers his chemical services, asking Cromwell to convince the king that he and his friends are willing to be bound, to the sum of a pound or more, "to make the phylosopher stone" in twelve and a half months on gold and twelve months on silver. Jones is sure of his ability to bring this about—indeed, if Cromwell only shared his confidence, he would regret the time he had wasted in keeping Jones incarcerated, "for hyt is a pressyus thyng." As for the charges against him, Jones implies that his dealings in prophecy amounted to little more than a prank on the gullible Neville, who was already a laughing stock in the county.[36]

As an example of an alchemical petition, this is both cruder and more urgent than the suit of Thomas Peter. Even so, it offers glimpses of familiar tropes, such as the injunction against false practitioners and the use of authoritative precepts to shore up the writer's own expertise. The stone, Jones explains, "ys to many men dowtfull," for most have been deceived by it—and not without cause, "for they had not the knowlege of the Ryght." To

35. National Archives, SP 1/73, fol. 1r-v.

36. Ibid., fol. 1v: "the most that I haue offendyd was yn \laughyng/ at hys contynawnce [i.e., countenance] as one wold \do/ at one of hys behauyng."

demonstrate his own competence, he cites scripture: "Nisi granum frumenti cadens in terra mortuum fuerit ipsum solum manet" ("Unless the grain of wheat falling into the ground die, itself remaineth alone").[37] Whatever the phrase signified to Cromwell, any reader familiar with alchemy would instantly have recognized the allusion to the "death" of the stone, which must endure blackening and putrefaction in order to release its vegetative virtue. Jones may have hoped that the phrase would also advertise his fitness to any experts whom Cromwell might care to consult, while remaining within the respectable bounds of scriptural quotation. Possibly he had good grounds for optimism: his offer of sureties was taken up, and he was released later that year, although it is not known whether Cromwell actually took advantage of his offered service. The letter is now filed among Cromwell's papers, under a laconic heading: "Richard Jones would make the Philosophers stonne."[38]

Jones's case hints at two intriguing, if unproven, conclusions. First, it confirms the picture of Henrician alchemy as an art that was tolerated by the king, to the extent that its practitioners felt able to tout their prowess as a means of escaping more serious charges. Second, it opens up the possibility that Jones may actually have been employed on Henry's behalf. Both conclusions return us to the open question of whether the king was personally interested in sponsoring alchemical practice.

Despite the presence of an alchemist in his library and the importunities of alchemical petitioners seeking licenses to practice, there is still little hard evidence for this. The best evidence we have that an alchemist was actually employed on Henry's business rests on the torn remnants of a letter filed in Cromwell's papers alongside Jones's petition. This unsigned and undated epistle is addressed to a priest—possibly Jones himself—who has been engaging in "this science," a common euphemism for alchemy. The unknown writer has been accused of trying to entice the priest away from this work by claiming that the king would spend the proceeds "wantonly and in tyranny."[39] He denies ever having used such an argument, and urges the alchemist to back him up, "for hyt can not helpe to do me hurt." Rather, the priest should tell the council that he intended only to set aside some of the proceeds of his work to secure himself a bishopric. The writer has kept a copy of the same words, so that "yours and myn may agre."

37. John 12:24–25 (Douay-Rheims).
38. SP 1/73, fol. 2v.
39. Ibid., fol. 3r.

It is surely the writer's bad luck that this naive and impolitic letter fell into Cromwell's hands rather than those of the alchemist—indeed, the fact that it has been torn into small pieces (some still missing) and then carefully reassembled suggests a foiled attempt to prevent just this eventuality. This may be the very same letter that was found sewn into the coat of one Roger Tyler, a "simple person" who had asked after Jones in Oxford after the latter's arrest, and who was subsequently apprehended by the commissar of the university.[40] It transpired that Tyler had brought two horses with him to Oxford, intending that he and Jones would ride to the "said Monk at St. Albones."[41]

Since Neville's testimony records that Jones was already engaged in alchemy when he first visited his apartments in Oxford, the letter raises the intriguing possibility that the conjuror was also working for the king or his officers, and that Tyler was employed to lure him away from this service— possibly acting on behalf of an unidentified member of the great abbey of St. Albans. The house was not dissolved until 1539—an action hotly opposed by its brethren, who may have had good cause to fear that the king would spend his ill-gotten gains "in tyranny," whether from alchemy or the disposal of monastic property.[42] Beyond these slim connections, however, there is no evidence that Jones was engaged in any kind of officially sanctioned project, although the shredded letter shows that at least one English alchemist was thought to be.

The repercussions of the Jones affair, with its illicit brew of alchemy, prophecy, and ritual magic, suggest a connection between conjuring and transmutation similar to that already encountered in Henry VI's addendum to the Misseldens' license to practice. Jones himself seems to have been conscious that his best hope lay in emphasizing the natural philosophical dimension of his work, linking it to the "playne science of Philossophie" rather than conjuring. This, at least, is the ploy attempted in his letter to Cromwell,

---

40. John Longland, Bishop of Lincoln, to Thomas Cromwell, SP 1/69, fols. 12r and 15r; summarized in "Henry VIII: January 1532, 11–20," in *Letters and Papers*, 5:339–49; *British History Online*, http://www.british-history.ac.uk/letters-papers-hen8/vol5/pp339–349 (accessed 5 August 2017). The commissar handed Tyler over to Longland, the bishop of Lincoln, who forwarded both the letter and a report on Tyler's examination to Cromwell. It is plausible that this is the letter now filed alongside Jones's petition to Cromwell in SP 1/73. The letter was probably reassembled soon after the original attempt at destruction, although details of its provenance had evidently been lost by the time it was labeled: "A *lettre* pasted vppon paper but whose & for whom not to be knowen. only it concerneth Alchumy" (fol. 3v).

41. SP 1/73, fol. 3r.

42. On the abbey's resistance to the dissolution, see James G. Clark, "Reformation and Reaction at St Albans Abbey, 1530–58," *English Historical Review* 115 (2000): 297–328.

ough we cannot be certain that it helped his cause, it certainly seems ve damaged it further.

stratagems acquired increasing importance as the stakes of magical practice rose with the passing of the Witchcraft Act in 1542. The statute removed benefit of clergy, thereby withholding protection from literate practitioners, including any learned readers of Agrippa and other writers on magic who might overstep the bounds of caution by attempting to put their illicit operations into practice. In this respect, popular misconceptions that linked alchemy and magic (not to mention Agrippa's learned and quite explicit connection of the two) might pose a danger even for alchemists without an interest in magic. Over a decade after Jones's embroilment in necromancy, the question of whether alchemy conducted according to "playne science" could trump allegations of conjuring would again be tested, this time in a case involving one of England's most famous alchemists, William Blomfild.

THE REFORMATION OF WILLIAM BLOMFILD

If Ellys, Peter, and the Misseldens have so far featured little in histories of English alchemy, the same cannot be said of William Blomfild, a Janus figure whose life spans the "before" and "after" of the English Reformation.[43] Although Blomfild was formerly a Benedictine monk of Bury St. Edmunds, his Reformed commitments later helped him secure a parish in Norwich under John Parkhurst, Edmund Freake's predecessor in that diocese.[44] A brief encomium by his kinsman Miles Blomfild encapsulates his turbulent and paradoxical career:

> beyng in his Lyf tyme a A Monke, a preyst A preacher a physicion A phylosopher A Alchimiste A good Latinist partly a Gretian, & an hebritian havyng also [th]e tonge of dyverse languages as, Dutch & French But in Alchimistri & Distillation he hath not left his lyke in this Nation.[45]

43. On Blomfild, see R. M. Schuler, "William Blomfild, Elizabethan Alchemist," *Ambix* 20 (1973): 75–87; Schuler, "An Alchemical Poem: Authorship and Manuscripts," *The Library*, 5th ser., 28 (1973): 240–43; Schuler, "Hermetic and Alchemical Traditions of the English Renaissance and Seventeenth Century, with an Essay on Their Relation to Alchemical Poetry, as Illustrated by an Edition of 'Blomfild's Blossoms,' 1557" (PhD diss., University of Colorado, 1971); Lawrence M. Principe, "Blomfild, William (fl. 1529–1574)," *ODNB*.

44. On Parkhurst, see Ralph Houlbrooke, "Parkhurst, John (1511?–1575)," *ODNB*. As Houlbrooke notes, "Parkhurst faced at first a severe shortage of adequately qualified clergy," a shortage that points to how Blomfild, surely an idiosyncratic candidate, was able to secure a parish.

45. Cambridge University Library, MS DD.3.83, art. 6, front cover.

Despite his alleged linguistic prowess, all of the alchemist's surviving works are written in English. Blomfild (also called Blundefielde or Blundeville) is best known for the poem that, following Ashmole, I shall simply refer to as Blomfild's *Blossoms*. Dated to 1557 and containing four acrostics of Blomfild's name, the poem circulated widely in Elizabethan England, cementing the writer's reputation as a major English adept.[46] A second work, *Blomfild's Quintessence, or The Regiment of Life*, an English treatise on alchemical medicine dedicated to Elizabeth I, was written around 1575 and survives in a single manuscript copy.[47] As with Ripley, several shorter, practically oriented works later came to be associated with him, including a late copy of a *practica*, transcribed by Ashmole, provocatively titled "Blundefielde his worke to King Henry 8: Anglie" (hereafter *Blundefield*).[48]

The details of Blomfild's Henrician exploits have so far been known almost entirely through the studies of Robert Schuler, which show that Blomfild enjoyed few periods of quiet during his turbulent career. He first came to the attention of the authorities in 1529, while still a monk, when he was required to recant several heretical positions. In this respect his predicament contrasts with Thomas Ellys's legal difficulties in the late 1530s. However questionable Ellys's financial dealings, there is no evidence to suggest that the Essex prior was anything other than orthodox in matters of religion, and his alchemical dabblings seem not to have hampered his later ecclesiastical preferment. Blomfild, on the other hand, showed early leanings toward Lutheranism, and his recantation later secured him a mention in John Foxe's *Book of Martyrs*.[49] He may have left the abbey of Bury before its formal dissolution, since he does not appear among the monks assigned pensions on 4 November 1539.[50]

Blomfild thus presents a very different picture of the monkish alchemist to career churchmen like Ellys and Freake, reminding us that not all regulars with alchemical interests opposed the Reformation or resisted a return to secular life. At the same time, his monastic background and clerical status distinguish him from contemporaries in the merchant community, like Rob-

46. For the manuscript witnesses, see Schuler, "Three Renaissance Scientific Poems," 17; Schuler, "Hermetic and Alchemical Traditions," 352–77.

47. MS DD.3.83, art. 6; hereafter *Regiment of Life*.

48. Ashmole 1415, fols. 96v–100r; printed in Schuler, "Hermetic and Alchemical Traditions," 488–90. To my knowledge this work does not survive in earlier manuscripts.

49. Schuler, "William Blomfild, Elizabethan Alchemist," 76–77.

50. "Letters and Papers: November 1539, 1–5," in *Letters and Papers, Foreign and Domestic, Henry VIII*, vol. 14, pt. 2, *August–December 1539*, ed. James Gairdner and R. H. Brodie (London: Her Majesty's Stationery Office, 1895), 168; *British History Online*, http://www.british-history.ac.uk/letters-papers-hen8/vol14/no2/pp160–170 (accessed 22 April 2018).

ert Freelove, who were interested in alchemy at around the same time. We do not know how Blomfild supported himself after leaving Bury, although it seems likely that he practiced medicine in some capacity. The subtitle of the *Blossoms* later claimed that he was admitted a bachelor of physick by Henry VIII, suggesting that Blomfild was able to obtain Henry's endorsement as a physician even in the absence of a university degree.

Like many of his contemporaries, both Catholic and Protestant, Blomfild acquired an interest in ritual magic, and consequently spent a period of imprisonment in London's Marshalsea prison during the 1540s. The privy council minutes of 22 April 1543 record that all documents relating to Blomfild should be delivered at that time to the attorney and solicitor general, "to reporte what matter they sholde fynde therein, and how the law wolde way [i.e., weigh] in the same."[51] The next official record is dated two and a half years later, in October 1545, when the council requested the knight marshall, keeper of the Marshalsea, "to deliver oon William Blomvile, a prisoner, to such oon as your Majesty shall send for him."[52] What happened after he came before the council is uncertain. Although Blomfild was arraigned for conjuring in 1546, he was apparently not convicted, and quite possibly never tried.

Blomfild did not spend the period of his imprisonment in idleness, but used it to petition members of the privy council, and Henry himself, on the basis of his alchemical prowess. Like the *Work of Petre*, his suits have escaped notice owing to the peculiar circumstances of their preservation. Two treatises survive, both in later copies made by Elizabethan compilers who did not identify them as works of Blomfild. The first was copied from an unknown source in or around 1604 by the Elizabethan clergyman Christopher Taylour, who misleadingly attributed it to "Raymundus" on the basis of its predominantly Lullian content.[53] The second treatise, which refers to the "Camp of Philosophy" already cited above, survives in a later sixteenth-century collection, now Sloane 2170, copied from an older manuscript, pos-

---

51. National Archives, SP 1/222 (29 July 1546), fol. 132; Schuler, "William Blomfild, Elizabethan Alchemist," 78.

52. National Archives, SP 4/1 (October 1545); Schuler, "William Blomfild, Elizabethan Alchemist," 78.

53. Ashmole 1492, pt. 9, 152 (hand of Christopher Taylour). Taylour dated an earlier work in the manuscript "1604 Octob[er]" (fol. 138v). Black mistakenly described the treatise as a translation of the *Testamentum practica*, which it little resembles; William Henry Black, *A Descriptive, Analytical, and Critical Catalogue of the Manuscripts Bequeathed unto the University of Oxford by Elias Ashmole, Esq., M.D., F.R.S.* (Oxford: Oxford University Press, 1845), 1374.

sibly by the compiler Edward Dekyngstone.[54] More cautious than Taylour in attributing the text, the scribe added an intriguing provenance note: "This Copie hearafter, I tooke fowrthe of a very olde litle Booke of Mr Blagbornes: But what was the fyrste Author therof. I know not: But (as I suppose) It was Dedicate. to Kinge Henrie, the eight."[55]

In fact both treatises are addressed to Henry and his councillors, and both were written from prison by alchemists with strong pseudo-Lullian leanings. On the grounds of content, style, and circumstance it seems highly probable that both were penned by Blomfild himself during his documented period of incarceration, possibly as part of a whole sequence of tracts written to aid his release. For instance, the writer of "Raymundus" alludes to a previous work, "the former Alphabet which we call our Theoricke," which he intended to serve as "an elucidary toward the royall worke of Alchemy."[56] In that work he had promised to complete a "Practicke" at a later date, a promise now to be redeemed with the present treatise (which I will accordingly refer to as the *Practicke*).[57] The writer of the Sloane 2170 treatise (hereafter *An Incomparable Work*) offers a more straightforward dedication, begging Henry to pardon his presumption in offering this "peculier gyfte to so excelent a kynge," while defending his decision to do so: for "An incomparable worke I have dedycate."[58] Although the writer opens by addressing the king directly, he concludes with an appeal to one of the members of his council. Since there is no obvious transition between the two modes of address, this may be an instance where the scribe has transcribed only part of the work, or else his exemplar may itself have been incomplete, suggesting a draft rather than a finished copy.

While both works allude to the author's imprisonment, the *Practicke* unfolds a remarkable history. The writer claims to have made the philosophers' stone some ten years earlier, producing thirty-eight ounces of the marvelous substance on which he lived for five years (he does not say how). A man named Goddard, living in Shoreditch, then took the remainder of the

---

54. On Dekyngstone, see Grund, *"Misticall Wordes and Names Infinite,"* 112–14; Peter Grund, "A Previously Unrecorded Fragment of the Middle English Short Metrical Chronicle in Bibliotheca Philosophica Hermetica M199," *English Studies* 87 (2006): 277–93, on 278–79.

55. Sloane 2170, fols. 56r-59v (hand of Edward Dekyngstone?).

56. Ashmole 1492, pt. 9, 152.

57. While we have no reason to doubt his word, this style of opening also evokes such pseudo-Lullian classics as the *Epistola accurtationis*, which starts with Raymond's recollection of his previous works written for his patron, the king of Sicily, or the Ripleian *Epistle to Edward IV*, which refers to earlier correspondence with the English king.

58. Sloane 2170, fols. 56r-59v.

stone and handed it over to the lord chancellor, Sir Thomas Audley, who in turn delivered it "into the handes of one Cowper then being his secretary, the which Cowper never would deliver it agayne."[59] Since that time he has struggled to replicate the stone, apparently while working under strict and unwelcome supervision:

> Now last of all I have diligently wrought this whole yeere past in this worke and through an evill pay maister and overseer my worke 3 tymes had evill successe. But God be praised the 4 tyme is well come to passe towardes the perfection vnto the performaunce wherof I shall gladly do my diligence, most humbly desyring the Kinges highnes and your honours to be meanes for me, that my goodwill may the rather be accepted, the evill report of myne enemyes notwithstanding.[60]

Tales of stolen elixirs, conniving officials, and traduced practitioners are familiar staples of alchemical narratives, including a memorable account in the *Ordinal of Alchemy* where Norton laments the theft of his elixir by the wife of a Bristol merchant. What stands out in this version is the reference to Audley, who as lord chancellor actually presided over the council that Blomfild now addresses—suggesting that the *Practicke* dates from after Audley's death in April 1544, when he was no longer alive to counter the writer's claims.[61] This period happens to coincide with Blomfild's known spell in the Marshalsea, increasing the likelihood that he is the author of the *Practicke*, who complains bitterly of "this imprisonment and oppressing with Irons, by reason whereof I feare the success to be death or perpetuall Impotency of my body."[62]

Further details are supplied in *An Incomparable Work*, where the writer records that he has been imprisoned for two and a half years. He pleads for an opportunity to present his case before the council:

59. Ashmole 1492, pt. 9, 151.

60. Ibid.

61. Audley was knighted in 1532, appointed lord chancellor on 26 January 1533, and created Baron Audley of Walden on 29 November 1538. Since he is named "Sir Thomas" rather than "Lord Audley," we can possibly narrow the date further; Audley was ennobled in November 1538, and the *Practicke* is supposed to have been written five years after his servant took the stone. Allowing for some aberration in the writer's memory, we can estimate that the *Practicke* was written between 1537 and 1545. On Audley, see L. L. Ford, "Audley, Thomas, Baron Audley of Walden (1487/8–1544)," *ODNB*.

62. Ashmole 1492, pt. 9, 151.

Desyryng yow of your goodnes, at the leaste to obtayne, of the Kyngs maj-
estie, and his moste honorable Cownsell to be callyd byfore theym to hear
mee speak. and as they shall thynck by their discreassion to do vnto mee
with mercy althowghe I have deservyd litle, or none.[63]

Blomfild was in fact summoned before the privy council in October 1545,
exactly two and a half years after his arrest, suggesting that *An Incomparable Work* may be the remnant of a successful petition for an audience. If
so, he cannot have entirely succeeded in persuading the councillors of his
innocence, since evidence was still being gathered against him nine months
later. Whether Blomfild's lost petitions helped or hindered his case, they
nonetheless tell us a great deal about how alchemical expertise functioned
in Henrician England. The alchemist offers his audience a trove of valuable
knowledge, philosophical and practical, that can be made available not only
in return for financial patronage, but also, in Blomfild's case, for freedom. In
their close attention to pseudo-Lullian doctrine, these writings show how
the kind of texts preserved by Giles Du Wes and Robert Greene could be
adapted for use in urgently pragmatic contexts. They also show that Thomas
Peter was not the only alchemist to offer a practice based on Lullian precepts
to Henry VIII. Blomfild's writings elucidate not just his aspirations as an
alchemical philosopher, but the methods by which he used pseudo-Lullian
and sericonian alchemy as evidence to distance himself from suspicion of
conjuring.

BLOMFILD IN PRACTICE

There is an episode in Blomfild's *Blossoms* in which the dreamer, guided by
Father Time to the Camp of Philosophy, gains an audience with Lady Philosophy herself. In a scene reminiscent of the Tudor hierarchy of patronage,
Time intercedes with the lady on the dreamer's behalf, and she accepts him
as her disciple. She then commits his further instruction to the care of Raymond Lull, who guides him through the alchemical work:

First into a towre, most beautifull constructe,
Father Ramond brought me, & thence immediately
He led me to her garden, planted most deliciously.[64]

---

63. Sloane 2170, fol. 59r.
64. Blomfild, *Blossoms*, 29 (ll. 222–24).

In the poem, Blomfild uses his passage through tower and garden as an allegory for his own introduction to philosophy through the writings of Raymond. The features of the garden have additional meaning. For instance, the well-constructed tower refers to Lull's famous, tower-like philosophical furnaces, which are illustrated in medieval copies of *De secretis naturae*. Through correct use of Raymond's furnace, Blomfild's dreamer enters a garden blooming with the alchemical ingredients that philosophers so often chose to represent as flowers: the "Blossoms" that give the poem its name.

As we learn from the newly discovered prison writings, Blomfild in fact took Raymond as his philosophical guide some twenty years prior to the composition of the *Blossoms*. Even in *An Incomparable Work*, the least technical of his prose works, he includes two diagrams of Lullian furnaces that have been copied directly from *De secretis naturae*. It is the *Practicke*, however, that most closely evokes the quintessential alchemy of *De secretis naturae*, in both theoretical framing and technical content. In it Blomfild, like Raymond, and Ripley before him, distinguishes between natural and contra-natural processes: the chrysopoetic Lesser Work that dissolves gold using "Corrosive waters and poysons," and the Greater Work or *Opus regale*, "done by dissolving the gold in waters medecinable." He chooses the latter in preference to mere gold-making, expressing a public-spirited desire that the council might hand over his treatise to some "learned or expert" practitioner, "to worke vnto the Commodity of our sovereigne and the succour of his commonwealth."[65] Yet the actual instructions are so heavily abbreviated that replication would surely have been impossible to achieve without further assistance from Blomfild himself—as the alchemist no doubt intended.

The *Practicke* includes, for instance, a brief summary of the manufacture of "lunary," or the quintessence, based heavily on *De secretis naturae*, followed by advice on how each of the seven metals may be reduced to a calx, or powder. This advice is accompanied by a drawing of the double circulation flask, or "gemissaries," copied from the *Epistola accurtationis* (although the recipe as set down here does not call for this apparatus), and a rather straightforward example of a Lullian wheel, in which each of the metals has been assigned a letter (although the insertion of this alphabet has no obvious practical implications) (fig. 5). It seems that the author is withholding crucial information, as, indeed, he partly confesses: "In this operation only I might conclude \all/ the whole worke, but so highe a secrett is rather to be wrought of vs then written." The treatise thus ends on a cliff-hanger with

65. Ashmole 1492, pt. 9, 151.

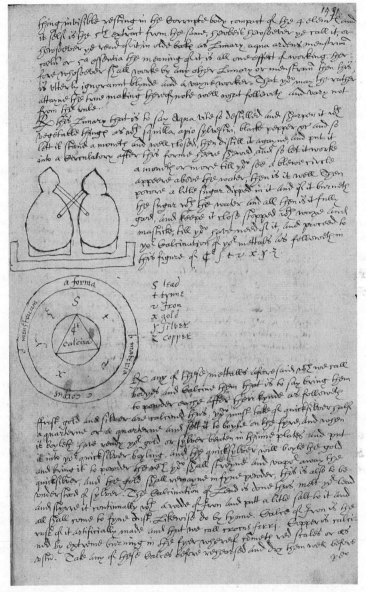

FIGURE 5. Pseudo-Lullian wheel and gemissaries. William Blomfild, *Practicke*, in a copy made by Christopher Taylour. Oxford, Bodleian Library, MS Ashmole 1492, 153. By permission of The Bodleian Libraries, The University of Oxford.

the dissolution of the golden calx into a water—which, the writer observes, "shall be sufficient for vs at this tyme."[66]

For all this deliberate obscurity it is still possible to reconstruct, at least in broad strokes, the practice of the *Practicke*. To this we owe the fact that the alchemist, having apparently hit upon a process that satisfied him, retained it as the centerpiece of all his subsequent work. The essentials of Blomfild's signature practice remain the same across his various writings, cleverly adapted to suit the conventions of different genres—whether a technical *practica* (*Blundefielde*), a petition framed as a philosophical treatise (*Practicke*), a patronage suit laid out as a recipe collection (*Regiment of Life*), or an allegorical poem (*Blossoms*).[67] The information necessary for replication is diffused across different texts, requiring careful reassembly on the part of readers. It would have been impossible, for instance, for a nonspecialist to complete the process using the guidance of the *Practicke* alone, in which Blomfild frequently omits or abbreviates technical information in favor of philosophical exposition—an approach wholly in keeping with the text's function as a petition. On the other hand, *Blundefielde His Worke* abandons alphabets and philosophical explanations entirely to give a more thorough account of the practice.[68]

We can triangulate textual clues from across three of Blomfild's works to gain an outline of a single practice, strongly Lullian and sericonian in character. It proceeds in three stages: first, the distillation of a vegetable menstruum called "lunary" (*aqua vitae* sharpened with herbs), followed by the preparation of precious and base metals, respectively. Gold is calcined by first casting thin plates into "boyling hott" quicksilver, then straining the amalgam through a cloth to yield a "whight lumpe" (*Regiment*) or "round white masse" (*Blundefielde*).[69] An incombustible oil is drawn from this calx by means of

<hr />

66. Ibid., 154.

67. To these can be added several recipes that may have been excerpted at a later time from any of these writings, not necessarily by Blomfild himself—for instance, the process titled "An Excellent worke of W. B." in Lincoln's Inn Library, MS Hale 90, fol. 82v; printed in Schuler, "Hermetic and Alchemical Traditions," 506.

68. It is possible that Blomfild wrote *Blundefielde* in response to further inquiries occasioned by the *Practicke*, to plug some of the omissions in the earlier text. However, given the late date of the copy, the practical focus may be the outcome of posthumous editing of a once-longer work by a copyist concerned to preserve only technical information.

69. *Regiment of Life*, fol. 11r: "Then hete your gold & put yt into the quiksiluer & yt wil eate up the gold then cast yt al into a dishe of water & strayne away the quyksiluer from [th]e golde & then yt wyll be lyke whight lumpe"; Ashmole 1415, fol. 97v: "putt itt into [th]e [mercury]. fumeinge; & boyling, & [th]e [mercury]. shall eate vp your [gold] then streine itt through a thicke Cloath as hard as you may wringe itt, then the [gold]. remayninge behind in a round white masse."

lunary, "by whose operation we dissolve the mettall."[70] The oil is then heated with the lunary in a circulatory vessel for about twenty days, and reduced until it becomes "lyke a ruby ston" (*Regiment*), a "thicke substance" (*Blundefielde*), or a "gummy substaunce like a stone" (*Practicke*).[71] To make the transmuting elixir, this oil of gold is next combined with a stone drawn from a base metal, which Blomfild calls "sulphur of nature." Any base metal will do, depending on how the practitioner intends to proceed at a later stage. For instance, to transmute lead into gold requires that the sulphur of nature also be drawn from lead, which is prepared by heating the metal in a crucible with a little salt to make a "poulder" (*Regiment*), "fyne dust" (*Practicke*), or "yellowe dust" (*Blundefielde*). This can be readily identified with litharge, the bright yellow oxide of lead.

From this point on, *Blundefielde* continues to supply detailed instructions, while the practical content of the *Practicke* becomes steadily more attenuated. In *Blundefielde*, the litharge is dissolved in *aqua vitae* to make "mercury exuberated" or "Maydens milke" (*lac virginis*)—a process reminiscent of Ripley's and Guido's vegetable stone.[72] This is transformed into sulphur of nature through repeated grinding with the original litharge, until the powder is fine enough to sublime up into the head of an alembic in the form of "white snowe or Christall stones." The "white Cristalline graynes" are gently melted, and the oil of gold is added slowly, drop by drop.[73] This fermentation process yields the elixir, one ounce of which will multiply 100 ounces of common quicksilver into further elixir, which can then transmute lead into gold.

Blomfild uses essentially the same procedure as the basis for the *Practicke*, while omitting much of the "how-to." He gives a more theorized reading, interpreting processes in terms of the composition of the metals: thus, the process of calcination was "first invented of the philosophers." Its purpose is to purify the metals' unclean mercurial and sulphureous components, by drying out their "evill humidity" and burning away their "evill fetor" to leave only the purest and most incombustible substance behind. The later stages of the process are considerably pared down, reduced to little more than a series

---

70. Ashmole 1492, pt. 9, 154.

71. The *Practicke* and *Regiment of Life* give twenty days, *Blundefielde* twenty-three.

72. Ashmole 1415, fols. 98v-99r: "& this the || Philosophers call [mercury] exuberated, the Philosophers stone, & Maydens milke."

73. Ibid., fol. 99v. The *Practicke*, in Ashmole 1492, pt. 9, 154, also describes the very gradual addition of the oil of gold: "dropping vpon it droppe after droppe."

of substances that can then be assigned letters, in the manner of a Lullian alphabet:

> L signifyeth the mercury exuberate with the menstrue which is called Lac Virgineum also, and this once imbibed into the body agayne and sublymed will arise into the head of the glasse like a ryme of frost or small Christall stones and then it is called sulphur naturae that is signifyed by M.[74]

The letters used to designate these substances, L and M, do not appear elsewhere in the *Practicke*. As Blomfild reminds his readers, he had already described these stages in his earlier offering to the council, the *Alphabet*, "and therfore we will not write them heere agayne but superficially."[75]

Through these hints, we can see how Blomfild manipulated both text and image to generate philosophical authority for his own alchemical program. Inspired by the figures used in pseudo-Lullian treatises like the *Testamentum* and *De secretis naturae*, and engaged in practices that were themselves heavily Lullian in flavor, Blomfild devised his own series of alphabets and wheels to render the various stages of his signature process. From hints in the *Practicke* we can infer that the lost *Alphabet*, which presumably expounded the theoretical principles of Blomfild's approach in greater detail, included a wheel that set out the later stages of the work: H (multiplication), I (spirits), L (mercury exuberated), M (sulphur of nature), N (tincture), and O (oil). These steps were plotted around a mean term, K, signifying the single vessel in which each of these processes took place—a theme that Blomfild would return to in the *Blossoms*. In the *Practicke*, however, he concentrates primarily on the earlier stages of the work, particularly the importance of calcining metals. Accordingly, his wheel plots the seven metals (designated as STVXYZ) around the central mean of "Calcination."

Through such techniques, Blomfild masked his original sequence of practical operations with a Lullian gloss—a strategy eminently suitable to the genre of philosophical treatise that he sought to imitate. This passage also sheds light on the alchemy of the *Blossoms*. To suit the conventions of allegorical verse, Blomfild replaces the technical, pseudo-Lullian vocabulary of alphabets and wheels with a straightforward analogy between planets and metals:

74. Ashmole 1492, pt. 9, 154.
75. Ibid.

The mastery thou gettest not of these planetes seven,
But by a misty meaninge, known onely vnto vs.
Bring them first to hell & afterward to heauen.[76]

For a reader of the *Practicke*, this misty meaning is easily expounded: "hell" must refer to the calcination of the metals, and "heauen" to the solution of their calxes in lunary, the quintessence. *Practicae* like *Blundefielde* and the recipe portion of the *Regiment of Life* provide more detailed instructions, allowing us to plot Blomfild's trajectory further, even in the absence of his theoretical *Alphabet*. For Blomfild's readers on the privy council, who presumably did not have such convenient expositions to hand, the advice of the *Practicke* would have been harder to follow. We must conclude that, for all Blomfild's protestations, the primary aim of his treatise was not to make life easy for a proxy practitioner appointed by the council, but to whet the council's appetite to the point where it would release him to work for them. At the end of the day, the procedure would work best if "wrought by us"— that is to say, if the writer himself were released from prison and allowed to get on with the job.

THE CIRCLE OF WILLIAM BLOMFILD

While the prison writings offer unexpected insight into the evolution of Blomfild's alchemical philosophy, they were not intended to be merely informational. As petitions to the privy council, they were designed to win him an opportunity to plead his case, a respite from close confinement in irons, and grounds for dropping the case against him. These ends are embedded in the structure of the works themselves, particularly the more polished *Practicke*. As we see when we delve deeper, Blomfild's detailed elaboration of pseudo-Lullian alchemy does more than set out his philosophical credentials; it also distinguishes his alchemy from exactly the kind of magic that he was charged with practicing.

While Blomfild's alchemical failures may have contributed to his arrest, the main charge against him was related to conjuring. On 29 July 1546, Blomfild's servant John Morvell deposed that Blomfild intended to draw a magic circle to conjure a spirit, "that shall come & carrye vs bothe Awaye whyther I shall appoynt hym."[77] He alleged that Blomfild planned to carry out the

76. Blomfild, *Blossoms*, 33 (ll. 337–39).
77. SP 1/222, fol. 134r.

practice in the "leads" under the roof of his workplace, to avoid detection through the busy comings and goings in the "chambre" where he usually worked. Since Blomfild sent Morvell back home to fetch a book on weather magic, this building cannot have been his own residence, but may have been the home of a patron or supervisor—perhaps the harsh "overseer" to whom Blomfild alludes in the *Practicke*. Morvell also describes carrying a child to safety after his master noted a possible danger ("the worste ys I shall neades destrowe parte of the howse"). In a timely if startling intervention, demolition was forestalled by the arrival of Sir William Paget, secretary of state, who, says Morvell, "preventyd the matter."[78]

Even if we disregard the more astonishing features of Morvell's testimony, there is no reason to doubt that Blomfild, like many of his contemporaries, was intrigued by the accounts of magic circles and incantations described in contemporary magical texts like Agrippa's *De occulta philosophia*. Had he attempted such an action, as Morvell alleges, it would have breached the recent statute against conjuring, which specifically forbade the invocation of spirits. Coupled with his three earlier failures in alchemical practice, this might account for Blomfild's bad odor among several of Henry's most influential councillors. In *An Incomparable Work*, he begs an unnamed interlocutor to appease his ill-wishers, offering to "gladly . . . prosterate my bodye vnder their feete, to obtayne their good will."[79] In particular, he has offended "my lorde Bishope of Winchester" (the religious conservative Stephen Gardiner) and "my lorde chanceler" (probably Audley's immediate successor, Thomas Wriothesley, appointed in April 1544).[80] He also alludes to unnamed enemies who have spread "manye ivell tallis" about him, and begs his interlocutor to ignore "whatsoeuer any man saye," and "take mee, as yow fynde mee." Possibly he is addressing Paget, who joined the council in 1543 and seems, on Morvell's testimony, to have had a prior connection with Blomfild, perhaps as a patron or customer. As in Jones's case, however, the surviving evidence can take us only so far—and unfortunately not far enough to reconstruct Blomfild's alchemical adventures in full.

What we do know still sheds considerable light on the composition of the *Practicke*. Blomfild never mentions the charge of conjuring in his work,

---

78. Ibid. Interruptions by household members and social callers were an ever-present concern for practitioners of magic. John Dee experienced similar embarrassments, as recounted in Deborah E. Harkness, "Managing an Experimental Household: The Dees of Mortlake and the Practice of Natural Philosophy," *Isis* 88 (1997): 247–62, on 259.

79. Sloane 2170, fol. 59v.

80. Ibid.

but his alchemical exposition impinges on magic in ways that are, given the context, extremely revealing. His recipe begins by directing the reader to take, in the name of Jesus, four gallons of "the sweete Juice of Lunary," from which is distilled four pints of "the purest spirit lively."[81] As we have seen, the herb "lunaria" was indeed a staple of pseudo-Lullian alchemy, appearing in the *Testamentum* and *De secretis naturae* as the cover name for a kind of "vegetable" mercury. Yet rather than proceeding with technical instructions, Blomfild breaks off to explain the peculiarities of alchemical compositional strategies. The alchemical lunary shares its name with a magical herb, but this is not to be taken literally: it is merely the *Deckname* for a solvent that the philosophers have "mistically named" to allude to its secret virtue. Previously, ignorant readers have failed to grasp the allegorical sense of their authorities, mistaking lunary for a magical substance:

> some fooles before our tyme have red in the secret workes of philosophers
> and not vnderstanding the meaning of them they have imagined another
> Lunary which is an hearbe so called that they feigne to have the vertues to
> open lockes and to make them invisible which opinion is very dead is false
> and deceaveth and bringeth the vsers of it into misery.[82]

As seen in the writings of several medieval practitioners, including Hortulanus and John Sawtrey, the use of plant names to disguise chemical ingredients was part and parcel of the act of alchemical composition.[83] In alchemy, literal readings are the province of the uneducated layman rather than the philosopher—indeed, this kind of obfuscation is exactly what authorities like pseudo-Arnald have in mind when they warn that their words are intended to teach wise men and mock fools. Yet such arguments, however flattering to the wisdom of the philosophers, leave open the dangerous possibility that officials unfamiliar with such reading strategies might impute a more sinister meaning to otherwise harmless cover names.

The herb lunaria thus serves as an emblem for the complicated relationship between alchemy and magic, a relationship already explored in late medieval sources, including CCCC 395, one of the pseudo-Lullian

81. Ashmole 1492, pt. 9, 154.

82. Ibid.

83. John Sawtrey, *De occulta philosophia*, in Harley 3542, fol. 71r: "Vegetalis uel Herbalis est quia ex succo .3. herbarum equali proporcione coniunctarum postquam steterint in igne humido per .24. dies emanabit"; Hortulanus, *Rosarium parvum*, in Sloane 1091, fol. 128v: "I shall take iij. herbys whych he fownd plantyd in his rose gardin."

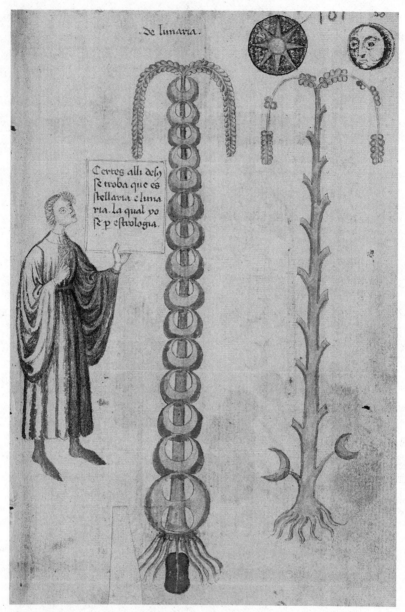

. De lunaria .

Certes alli dels
se troba que es
stellaria e luna
ria. La qual yo
se p estrologia.

FIGURE 6. Illustration of lunaria, in a fifteenth-century manuscript later owned by Giles
Du Wes, Robert Greene of Welby, and probably John Dee. Cambridge, Corpus Christi
College, MS 395, fol. 50r. By permission of the Parker Library, Corpus Christi College,
Cambridge.

collections of Giles Du Wes.[84] Here, an early fifteenth-century copy of John of Rupescissa's *De consideratione* is accompanied by several drawings of lunaria grass, to which Du Wes later appended a short treatise, "On the Virtues of Lunaria" (fig. 6).[85]

The content of the text (copied, says Du Wes, from the "Book of Hermes") describes both alchemical and magical applications for the mysterious herb. If mercury is boiled in the juice of lunaria and hoopoe blood, it makes a red stone that will transmute copper into gold. Both plant and stone have other wonderful properties, too: added to a ring, they will make the wearer invisible or transport him wherever he wishes, while the juice alone, touched to the ears, allows the user to understand the language of birds and animals.[86] Yet it is clear from Du Wes's annotations that, like Blomfild, he identifies "lunaria" as an alchemical cover name rather than a magical object. In one note, he claims that lunaria is the philosophers' prime matter, possessing hidden virtues that enable it to heal all sickness, besides making precious stones, gold, and silver. However, its working "is shown metaphorically under the appearance of a herb."[87] The method for making the stone is described in a "mistical or enigmatic way," and will not be understood by those who do not recognize the nature of the hoopoe bird.[88] For Du Wes, who later interpreted Lactantius's poem on the phoenix as an alchemical allegory, the identification of both lunaria and the hoopoe's blood as alchemical *Decknamen* rather than actual plant and animal products must have seemed obvious—even if this reading fails to capture the intent of the original medieval writer. Furthermore, the disguised process seems to be endorsed by other alchem-

---

84. Outside the pseudo-Lullian corpus, "lunaria" appears in various Latin and Middle English texts, including a fifteenth-century poem, "Her ys an Erbe men call Lunaryie," which Ashmole later published in the *Theatrum*; Harley 2407, fols. 7r-v; *TCB*, 348–49. Harley 2407, which was later owned by Dee, also includes a short Middle English prose text on lunaria that describes how the plant's leaves grow with the waxing moon: "I schal yow tel of an erbe [th]at men cal lunarie," fol. 5v.

85. CCCC 395, fols. 49v-51v. Possibly the scribe included the figures after noticing the significance of "lunaria" (in its alchemical sense, as the quintessence of wine) in the pseudo-Lullian *Tertia distinctio*, which is included later in the manuscript.

86. Ibid., fols. 48v-49r.

87. Giles Du Wes, note to CCCC 395, fol. 48v: "ista herba est illa prima materia de qua philosophi infinita scripserunt propter virtutem eius occultam cum qua sanantur omnes infirmitates et componitur lapides preciosi aurum & argentum . . . & hic metaphorice sub herbe specie docetur."

88. Ibid.: "Nota quod ista prima virtus est compositio lapidis philosophici mistico modo seu enigmatice declarato. Nota quia mirabile & magis quam mirabile intelligenti/ quod non intellige pro vpupa aue."

ical and medical authorities. Concerning the herb's supposed rejuvenating power, which causes white hair to fall out and regrow as black, Du Wes notes that Raymond says the same in *De secretis naturae*, as does Roger Bacon in his book on treating the elderly, *De universali regimine senum et seniorum*.[89]

In his own disquisition on lunaria, Blomfild differentiates between alchemical and magical ingredients, but also between alchemical reading strategies and more straightforward approaches to text. And he demonstrates his own skill at alchemical reading and practice, by revealing the true sense of "our lunary": it is "the pure invisible spirit of wyne whome we also do call aqua*m* ardente*m*, menstruu*m* coelu*m* or quinta essentia."[90] In what amounts to a brief tutorial on alchemical reading, he unpacks each of these names in turn. The spirit of wine is called *aqua ardens* because of its burning virtue, and *menstruum* because it nourishes the stone just as menstrual blood nourishes an infant within the womb. It is also called *coelum*, or "heaven," because it is infused with celestial virtues. Lastly, it is *quinta essentia* "because it is a || thing invisible resting in the Corrumpte body compact of the 4 elemente*s* and it self is the 5th extract from the same."[91] To convince readers of his own probity, Blomfild folds back generations of philosophical accretion to reveal the one true, physical substance at the core of his practice. Mystification aside, it is nothing more than spirit of wine, "howbeit howsoever ye call it, or howsoever ye read of it in olde booke*s*."

Blomfild's maneuvers need to be read in light of the terms of Misselden's license, which warned the merchant to steer clear of necromancy and operate only within the bounds of respectable natural philosophy. By attacking magical practices and inducting his readers into the multilayered mysteries of alchemical composition, Blomfild implicitly distances himself from the necromantic arts, while revealing his practice as nothing more than plain science. This reading requires us to revisit Blomfild's inclusion of a Lullian figure and alphabet in his *Practicke*. More than just gimmicks intended to illustrate his familiarity with up-to-date literature, these may have helped

---

89. Ibid., fol. 49r: "ho*c* tenet raimo*n*du*s* et in sua 5ᵗᵃ. essen*t*ia. affirmat et baco*n* in libro de regi-*mine senium.*" Du Wes refers here to a short tract that often accompanies the better-known *De retardatione accidentium senectutis* in manuscript, edited as Roger Bacon, *De universali regimine senum et seniorum*, ed. Andrew G. Little and Edward Withington, in *Opera hactenus inedita Rogeri Bacon*, fasc. 9 (Oxford: Clarendon Press, 1928), 90–95. See M. Teresa Tavormina, "Roger Bacon: Two Extracts on the Prolongation of Life," in *Sex, Aging, and Death in a Medieval Medical Compendium: Trinity College Cambridge MS R.14.52, Its Texts, Language, and Scribe*, ed. M. Teresa Tavormina (Tempe: Arizona Center for Medieval and Renaissance Studies, 2006), 1:327–72.

90. Ashmole 1492, pt. 9, 152.

91. Ibid., 152–53.

defuse the otherwise troubling associations of circular figures with necromantic practice. Their use is particularly telling in light of Morvell's testimony, which described Blomfild's plan to conjure a spirit by drawing a circle,
in connection with perusal of a book. In the *Practicke*, books and circles are
stripped of their magical connotations, as Blomfild deciphers his alphabetical wheel within the philosophically respectable context of pseudo-Lullian
alchemy.

This is not to say that Blomfild was innocent of the charge of conjuring, but to say that whether or not he actually attempted ritual magic is,
in a sense, beside the point. The *Practicke* offers plausible deniability: the
possibility that an ignorant servant might have mistaken his master's philosophical exercises for evidence of sorcery, or imputed necromantic meaning
to a strange book embellished with wheels, letters, and astronomical symbols. By his own admission, Blomfild has used circles to conjure spirits, but
his "circles" are no more than Lullian wheels, and his "spirits" are the pure
essences drawn out of wine and metallic bodies. The quintessence cannot
confer invisibility on men, but it is (as Blomfild observes twice in the text)
invisible in itself by virtue of its great subtlety. And although the philosophers' lunary cannot open locks like its magical analogue, it does have the
power of opening metallic bodies so as to separate their "forme and first
matter" from the confused substance of their original forms.[92] The *Practicke*
offers Blomfild's readers a reliable practice that is based not on illicit magic,
but on experiment: safely grounded in the plain science of philosophy.

THE PASSING OF THE BOOKS

With the passing of the monasteries, alchemically inclined religious either
abandoned their pursuits, like Ellys, or took up these activities in the secular
sphere, like Blomfild. Legally, alchemy remained a double-edged sword: an
object of suspicion through its dangerous connections with multiplication
and natural magic, but also a potential source of benefit, as grounds for a
license to practice, or even as an opportunity to "get out of jail free." Others besides Jones and Blomfild attempted the latter strategy. In 1551, Robert
Allen, a "prophesyer" held in the Tower, also sought an audience with the
privy council, claiming that despite his lack of Latin "he cowld saye more

---

92. Blomfild seems to tease this usage in the *Blossoms*, where he describes the dreamer attempting to open twelve locks to the Camp of Philosophy. The key that opens all the locks is knowledge
of the correct prime matter, which is called, among other things, lunary (*Blossoms*, 24, ll. 89–91).

concerning astrologie and astronomy than all the lerned men within the universities of Oxford or Cambridge," and furthermore that he had the secret of "the grett alyxor [i.e., elixir]."[93]

Books as well as people could be tainted by association with superstitious practices; books, however, could not plead their cases, and perished in large numbers. The contents of Oxford and Cambridge college libraries were purged during the 1540s and 1550s in the course of Edward VI's reforms. Many manuscripts were hidden or otherwise salvaged by lay and ecclesiastical collectors—a process that gained momentum in the second half of the century as neglect was succeeded by a new appreciation of the value of pre-Reformation books. Monastic books preserved evidence for England's heritage, including the history of the pristine English Church that anteceded Rome.[94] But for collectors of alchemical manuscripts, these books had more than antiquarian value; they also preserved past practices, still ripe for use. The best-known lay collector of the midcentury, John Dee, was assiduous in garnering manuscripts from colleges and former religious houses, including works of science, medicine, and alchemy from the great abbeys of St. Albans and St. Augustine, Canterbury.[95]

Dee had his ecclesiastical counterpart in Matthew Parker (1504–1575), Elizabeth I's archbishop of Canterbury.[96] Much of the content of Parker's priceless collection of manuscripts originated in monasteries and religious

93. Harley 424, art. 7; cited in John Gough Nichols, *Narratives of the Days of the Reformation, Chiefly from the Manuscripts of John Foxe, Martyrologist* (London: Camden Society, 1849), 329. Alchemy might also provide an "alibi" for currency crime: Parsons, *Making Money in Sixteenth-Century France*, 229.

94. Roberts and Watson, *Dee's Library Catalogue*, 14; Sherman, *John Dee*, 118–20. On the dispersal of monastic books during the Reformation, see David N. Bell, "Monastic Libraries: 1400–1557," in *Cambridge History of the Book in Britain*, 3:229–54; James P. Carley, "Monastic Collections and Their Dispersal," in *The Cambridge History of the Book in Britain*, vol. 4, *1557–1695*, ed. John Bernard and D. F. McKenzie, with Maureen Bell (Cambridge: Cambridge University Press, 2002), 339–48.

95. Roberts and Watson, *Dee's Library Catalogue*, 14. These include CCC 125, a collection of texts on magic, alchemy, and pigments from St. Augustine's, discussed at length in Page, *Magic in the Cloister*.

96. David J. Crankshaw and Alexandra Gillespie, "Parker, Matthew (1504–1575)," *ODNB*. On Parker's books, see M. R. James, *The Sources of Archbishop Parker's Collection of MSS at Corpus Christi College, Cambridge* (Cambridge: Cambridge Antiquarian Society, 1899); Bruce Dickins, "The Making of the Parker Library," *Transactions of the Cambridge Bibliographical Society* 6 (1972–76): 19–34; R. I. Page, *Matthew Parker and His Books* (Kalamazoo, MI: Medieval Institute Publications, 1993); Anthony Grafton, "Matthew Parker: The Book as Archive," *History of Humanities* 2 (2017): 15–50. It should be noted that many of Parker's books now rest not in the "Parker Library," but among the British Library's Cotton manuscripts.

foundations; the spoils of the Reformation gathered to provide the "ground-plat" for a new English history. Although best known for its Anglo-Saxon treasures, Parker's collection—part of which is now housed in the library named after him in Corpus Christi College, Cambridge—included a small but high-quality collection of alchemical manuscripts with a distinctly pseudo-Lullian bent. Rather than relics of monastic practice, however, these connect us to the secular collecting practices of Robert Greene and Giles Du Wes, underscoring the significance of these men as conduits for alchemical knowledge in the first half of the sixteenth century. For instance, Corpus Christi College MS 112 includes a *Testamentum* apparently copied during the late fifteenth or early sixteenth century from Greene's important manuscript, CCC 244, including John Kirkeby's colophon.[97] Parker secured an even greater prize in the form of CCCC 395, the repository of Du Wes's reflections on lunaria as well as a precious copy of the *Tertia distinctio*, which after Greene's death may have reached Parker's hands via those of Dee.[98]

The overlapping collections of Dee and Parker offer a bridge between the pre-Reformation mixed economy and the Elizabethan revival of interest in England's alchemical heritage. In this liminal moment, survivors from the earlier period—both monks and books—were still available for consultation. Dee could have read Blomfild's *Blossoms*, but he was also able to consult the author in person. On 16 May 1561, he presented "his most sincere friend Master Blomfild" with a copy of Higden's *Polychronicon* formerly owned by the abbey of St. Augustine, Canterbury, raising the fascinating possibility that the two men at some point discussed alchemical secrets.[99] By this time, however, the landscape of English practice had irrevocably shifted, and a new, Protestant vision of alchemical history had taken hold.

97. Kirkeby colophon on CCCC 112, 358.

98. Although CCCC 395 is not annotated by Dee, several of its contents are listed in Dee's list of alchemical books he read in July 1556; see Roberts and Watson, *Dee's Library Catalogue*, 191–92. Another of his alchemical manuscripts, CCCC 99, containing works of Dastin, Bacon, and Rupescissa, was tentatively linked to Dee by James on the basis of the letter Δ (often used by Dee to stand in for his own name) on fol. 1r; M. R. James, *A Descriptive Catalogue of the Manuscripts in the Library of Corpus Christi College, Cambridge* (Cambridge, 1912), 1:186. On Dee's alchemical reading, see chap. 8, below.

99. Ranulf Higden, *Polychronicon*, Oxford, Queen's College, MS 307 (fifteenth century), fol. 2r: "Joannes Dee 1561. 16 Maij Amico suo Integerrimo Magistro Blomefelde dono dedit." DM161 in Roberts and Watson, *Dee's Library Catalogue*, 182; see also Schuler, "William Blomfild," 80.

CHAPTER SIX

Time and Money

And sso in olde tyme, Emperrours, kyngs, and princes dessieryd this science, more for a roialltye off the thinge, then holie for the dessire off golde, and they dyd maynetayne noble and learned philosophers for that pourpose.[1]

In 1565, English alchemists were shaken by the news that the queen of England had agreed to fund the activities of a Dutch alchemist—had, in fact, set him up in a fully equipped workspace in Somerset House, supported by a generous pension.[2] Although hints survive of earlier royal enterprises, including Richard Carter's practice at Edward IV's manor of Woodstock, and the even scantier evidence for Richard Jones's work on behalf of Henry VIII, Cornelius de Lannoy is the first alchemical philosopher known to have received substantial royal support on this scale. News traveled fast, raising the hopes of other practitioners. Now based in France, the English cosmographer Richard Eden learned of it from his Protestant patron Jean de Ferrières, Vidame de Chartres.[3] He wrote to William Cecil, Elizabeth's

1. Thomas Charnock, *Booke Dedicated vnto the Queenes Maiestie*, Lansdowne 703, fol. 39r.

2. *Calendar of State Papers Domestic: Edward, Mary and Elizabeth, 1547–80* (London: Her Majesty's Stationery Office, 1856), 249–50, 255–57, 275–77; *Calendar of State Papers Domestic: Elizabeth, Addenda, 1566–79* (London: Her Majesty's Stationery Office, 1871), 10–13. Elizabeth's sponsorship of de Lannoy and other alchemists is discussed in detail in James Stuart Campbell, "The Alchemical Patronage of Sir William Cecil, Lord Burghley" (Master's thesis, Victoria University of Wellington, 2009), 78–87; see also Glyn Parry, *The Arch-Conjuror of England: John Dee and Magic at the Courts of Renaissance Europe* (New Haven: Yale University Press, 2012), 74–78; Harkness, *Jewel House*, 170, 172.

3. David Gwyn, "Richard Eden, Cosmographer and Alchemist," *Sixteenth Century Journal* 15 (1984): 13–34, on 30–33. On Eden, see also Campbell, "Alchemical Patronage"; Andrew Hadfield, "Eden, Richard (*c*.1520–1576)," *ODNB*; Christopher Kitching, "Alchemy in the Reign of Edward VI: An Episode in the Careers of Richard Whalley and Richard Eden," *Bulletin of the Institute of Historical Research* 44 (1971): 308–15. Eden's translations, prefaces, and a wealth of bio-bibliographical information are gathered in Edward Arber, "The Life and Labors of Richard Eden, Scholar, and

secretary of state, to congratulate him on retaining "the greate philosopher which my Lord tolde me woorketh for the queenes Maiestie," praying that the queen obtain success in obtaining the stone, "as treuly as I beleve the possibilitie to be trewe and as I iudge her most woorthye so excellent A gyfte of god." Not one to let an opportunity slip, he ended by hinting that his own advice on alchemical matters was available if required: "wherof I wolde write more vnto your Lordship if I knewe howe it wolde be taken. Sed piscator ictus. &c."[4]

Eden ends by wryly quoting an Erasmian adage, *Piscator ictus sapiet* ("But once bitten, twice shy")—an apt reflection on the reverses of his own alchemical career.[5] In 1546 he was appointed as master of Henry VIII's distillery, only to be denied that position by the king's death the following year. In 1550 he undertook alchemical practices on behalf of an important client of the lord protector, but fell foul of the statute against multiplication when his enterprise was betrayed to the local authorities. Bound to desist from practicing alchemy on his native soil, Eden eventually left England in the suite of a foreign patron, de Ferrières, one of the leading lights of French Protestantism and a patron of Paracelsian medicine.[6] For all his congratulations, the news that his sovereign had finally elected to support an alchemist—but a foreigner rather than an Englishman—must have come as a bitter blow.

Eden's quotation referred to his own earlier brush with the law, but it was also prophetic. After the collapse of de Lannoy's enterprise it proved difficult to convince the queen to finance further operations, although Cecil and other administrators and courtiers continued to support promising metallurgical projects, including several related to alchemy.[7] Elizabeth I's patron-

---

Man of Science," in *The First Three English Books on America: [?1511]–1555 A.D.*, ed. Arber (Birmingham: [Printed by Turnbull & Spears, Edinburgh], 1885), xxxvii-xlviii. See also Susanna L. B. De Schepper, "'Foreign' Books for English Readers: Published Translations of Navigation Manuals and Their Audience in the English Renaissance, 1500–1640" (PhD diss., University of Warwick, 2012).

4. Richard Eden to William Cecil, Lord Burghley, 12 October 1565, SP 70/80, fol. 125v.

5. Erasmus, *Adagia* 1.20. *Piscator ictus sapiet* translates literally as "A stung fisherman will be wise"—a reference to the bought experience of a careless fisherman who plunges a hand into his net, only to grasp a sea-scorpion.

6. On the Vidame's interest in alchemy and Paracelsianism, see François Secret, "Réforme et alchimie," *Bulletin de la Société de l'histoire du protestantisme français (1903–2015)*, 124 (1978): 173–86, on 173–76.

7. In a letter of July 1568, Cecil recorded his interest in the claims of a certain Italian, but noted that the queen considered such activities "chargeable without Fruit"; this may well have been related to alchemy, as suggested by Parry, *Arch-Conjuror of England*, 78.

age of de Lannoy nonetheless triggered a flurry of suits from English petitioners who hoped that their own efforts might receive similar attention.

This chapter follows the unfolding of alchemical patronage in the wake of the Reformation.[8] Between the reigns of Edward VI and Elizabeth I, new kinds of patronage and different kinds of practitioner emerged as the locus of alchemical activity switched entirely to secular contexts, including private households and the mint. Mint officials, merchants, artisans, and secular clergy devised alchemical projects, applied for licenses, and sought funding, sometimes with the support (official or unofficial) of senior administrators within the kingdom. As interest spread, demand also increased for copies of medieval alchemical texts, particularly in English translation. This demand enhanced the prestige of fifteenth-century adepts like Ripley and Thomas Norton even as it shaped practitioners' perception of their own past, contributing to their own sense of themselves as inheritors of a long-lived and autonomous tradition that, like the English Church, need no longer be viewed as subservient to continental authorities.

Many of these books survive, including the manuscripts of Richard Eden, Thomas Charnock, and Richard Walton, all of whom petitioned Elizabeth for licenses to practice. Pseudo-Lullian alchemy predominates in these collections, as does the name of Ripley: a philosopher now hailed not only as an accomplished commentator on Raymond, but also as a representative of a distinguished English tradition of alchemical philosophy. While the process of recovering ancient knowledge of alchemy was already underway in lay communities even before the dissolution, it accelerated with the translation of Latin texts into English, the appearance of medieval treatises in print, and the attempts of practitioners to shape these sources into an intelligible "history" of English practice. Even in the new intellectual landscape of Protestant England, alchemical practitioners could not elude their medieval past.

THE ENGLISHING OF ALCHEMY

For sixteenth-century reader-practitioners, reconstructing alchemical history, like reconstructing alchemical practice, depended on securing appropriate books. Against a background of increasing literacy and a rise in the population of urban centers, readers from diverse social and educational

---

8. On scientific patronage in early modern England more generally, see Stephen Pumfrey and Frances Dawbarn, "Science and Patronage in England, 1570–1626: A Preliminary Study," *History of Science* 42 (2004): 137–88; Harkness, *Jewel House*, chap. 4.

backgrounds collected alchemical manuscripts and had copies made, either for their own use or as gifts or commissions for others.[9] The pace of transcriptions picked up as England's merchant class expanded, feeding interest in the profitable metallurgical and medicinal secrets of medieval adepts, and with it an audience for vernacular translations of Latin works.

As the sixteenth century progressed, increasing numbers of alchemical texts became available in print, mainly in the form of collections of medieval treatises of the kind solicited by the Nuremberg printer Petreius.[10] In July 1556, John Dee worked his way through several of the new printed compendia, including Petreius's *De alchimia* (1541) and the first volume of *De Alchimia opuscula complura veterum philosophorum* (1550) from the Frankfurt printer Cyriacus Jacob.[11] While these continental productions offered precious editions of medieval treatises like the *Summa perfectionis* of pseudo-Geber and the anonymous *Scala philosophorum*, they were not enough on their own to supply the needs of budding practitioners, who continued to rely heavily on scribal publication. Despite its tremendous influence on alchemy as actually practiced, the pseudo-Lullian corpus was still poorly represented in print prior to 1561, while the writings of English philosophers like Ripley and Norton remained accessible only in manuscript. Such absences drove an active culture of sharing and copying alchemical books.

Reader-practitioners responded by seeking texts in any available format. During his early investigations in the 1550s and 1560s, Dee supplemented the printed books in his list with an impressive collection of alchemical manuscripts, including several works of Giles Du Wes.[12] Many of the books stored in the stillhouse of Sir Thomas Smith (1513–1577) are described as "written,"

9. R. A. Houston, *Literacy in Early Modern Europe: Culture and Education, 1500–1800* (London: Longman, 1988).

10. Petreius, *De Alchemia*, 374. On these early ventures in alchemical publication, see Kahn, *Alchimie et Paracelsisme*, 100–108.

11. *De Alchimia opuscula complura veterum philosophorum* . . . (Frankfurt am Main: Cyriacus Iacobus, 1550). Dee does not mention reading the *Rosarium philosophorum*, the lengthy florilegium that comprises the whole of the second volume of Jacob's edition. His list also includes an edition of *De secretis naturae* (he does not state which one); Petrus Bonus, *Pretiosa margarita novella de thesauro, ac pretiosissimo philosophorum lapide*, ed. Giovanni Lacinio (Venice: Aldo Manuzio, 1546); and Philipp Ulstad's *Coelum philosophorum*, a popular work on philosophical distillation in the tradition of John of Rupescissa, first published as *Coelum philosophorum seu de secretis naturae* (Fribourg [Strasbourg]: [Johann Grüninger], 1525), and frequently reprinted.

12. Early examples include the fifteenth-century manuscripts Sloane 2128 and 2325, both in Dee's hands by 1557 and inscribed with his name and the year. The contents of the list coincide with a number of items in Du Wes's former manuscripts, including CCCC 395 (such as *Averrois super hermetem* and the *Dialogus inter Hilardum et spiritum*), Harley 3528 (the *Semita semitae*, a text attributed to Arnald of Villanova with the incipit "Venerande pater"), and Trinity O.8.25 (Du Wes's *Epistola scientiam enucleans*).

including at least three manuscripts of Ripley and a "written booke of Norton in blew velvet," in addition to a copy of the pseudo-Lullian *Experimenta*, a gift from Eden, his former student.[13] Print could also morph into script. Eden copied items from Guglielmo Gratarolo's 1561 edition into his own compendium alongside a plethora of unpublished texts.[14] Cecil's papers include a complete English translation of the 1545 edition of *Alchemiae Gebri Arabis*, a reprint of the 1541 volume previously studied by Dee—most likely a gift from a petitioner hoping to appeal to the minister's known alchemical interests.[15]

The mixing of print and manuscript sources reflects an expanding market for alchemical material in which readers sought to consult a range of philosophical models and practical instructions, whether for personal use or with a view to exchange or presentation as gifts. It also reflects the strong medical interest in distilled remedies during the early decades of the sixteenth century, exemplified by the best-selling status of Hieronymus Brunschwig's vernacular collection of distilled remedies, the *Liber de arte distillandi de simplicibus* (1500), which in 1527 became the first book of alchemical techniques to be printed in English.[16] This trend led publishers to prioritize medico-alchemical works like Rupescissa's *De consideratione* and the first

---

13. "An Inventarie," Queens' College MS 49, fol. 117v. The list includes "Ripley written," "Collectanea Riplei written," and "A written booke of Ripley which was Jo[hn] Busshops." On Smith's books, see Richard Simpson, "Sir Thomas Smith's Stillhouse at Hill Hall: Books, Practice, Antiquity, and Innovation," in *The Intellectual Culture of the English Country House, 1500–1700*, ed. Matthew Dimmock, Andrew Hadfield, and Margaret Healy (Manchester: Manchester University Press, 2015), 101–16.

14. Guglielmo Gratarolo, ed., *Verae alchemiae artisque metallicae, citra aenigmata, doctrina, certusque modus, scriptis tum novis tum veteribus nunc primum & fideliter maiori ex parte editis, comprehensus* (Basel: Heinrich Petri and Peter Perna, 1561). Eden's book, Trinity R.14.56 (discussed below), includes several extracts from this edition, particularly at fols. 78r-79v—e.g., "Opus ex Mercurio solo. Ex libro Raymundi qui dicitur Summaria lapidis consideratio. Ex magno libro Guilhelmi Grataroli. fol. 162" (fol. 78r).

15. Hatfield House, Cecil Papers 271/1. This is an English translation of Chrysogonus Polydorus, ed., *Alchemiae Gebri Arabis* (Bern: Mathias Apiarius, 1545), itself a reprint of Chrysogonus Polydorus, ed., *In hoc volumine de alchimia continentur* (Nuremberg: Iohannes Petreius, 1541).

16. Hieronymus Brunschwig, *Liber de arte distillandi de simplicibus* (Strassburg: J. Grüninger, 1500); translated as *The vertuose boke of distyllacyon of the waters of all maner of herbes* (London: Laurence Andrewe, 1527). On Brunschwig, see Alisha Rankin, "How to Cure the Golden Vein: Medical Remedies as *Wissenschaft* in Early Modern Germany," in *Ways of Making and Knowing: The Material Culture of Empirical Knowledge*, ed. Pamela H. Smith, Amy R. W. Mayers, and Harold J. Cook (Ann Arbor: University of Michigan Press, 2014), 113–37; Tillmann Taape, "Distilling Reliable Remedies: Hieronymus Brunschwig's 'Liber de arte distillandi' (1500) between Alchemical Learning and Craft Practice," *Ambix* 61 (2014): 236–56; Taape, "Hieronymus Brunschwig and the Making of Vernacular Knowledge in Early German Print" (PhD diss., Pembroke College, University of Cambridge, 2017). Conrad Gessner's *De remediis secretis*, published under the name of Euonymus, was another important source of distilled remedies: *Thesavrvs Evonymi Philiatri De*

two medical books of *De secretis naturae*, which appeared in print considerably earlier than the later chapters on transmutation. Such partial editions did little to satisfy the craving of chrysopoeians for Raymond's elusive *Tertia distinctio*, but they could still save time for those who did not wish to copy out the entire text. When the London mercer Robert Freelove transcribed a fifteenth-century manuscript of *De secretis naturae* in 1536, he copied only the third and fourth books, recognizing that the first two were already available in print:

> The First and Second Distinctions are held to be books of Raymund Lull, which are called *On the Testament and Practica, etc.* However, you should not look for them in this volume as they are found in another book on the art of medicine by one Johannes Matheus de Gradi (so called), the *Consilia of Johannes Matheus de Gradi, etc.*, printed in Venice in 1521 AD on 4 August. And since I bought this book, in which I discovered the aforesaid First and Second Distinctions, which are called the *Book Concerning the Quintessence*, as noted above, for the sake of brevity I therefore saved myself the effort of recopying these two Distinctions.[17]

Freelove's engagement with Lullian alchemy offers a link between the humanist world of Du Wes and Greene, characterized by its attention to early manuscripts and the correction of Latin texts, and the vernacular, mercantile context within which alchemy would flourish in sixteenth-century England. A member of the Worshipful Company of Mercers, by December 1550 Freelove also held the position of clerk to the Company of Merchant Adventurers, and may have found his mercantile connections useful in sourcing alchemical books from abroad.[18] Having once acquired this mate-

---

remediis Secretis . . . (Zurich: Andreas Gesner and Rudolf Wyssenbach, 1552); translated as *The newe iewell of health wherein is contayned the most excellent secretes of phisicke and philosophie. . .* , trans. George Baker (London: Henrie Denham, 1576).

17. Sloane 3604, fol. 64v: "Distinccio Prima et Secunda habentur inter libros Raymundi Lullij que dicuntur De Testamento & Practica cum ceteris/ Attamen non queras in hoc voluminem/ quia habentur in alio libro de Arte medica cuisdam Johannis Mathei de gradi (sic appellato). Consilium Johannis Mathei de gradi &c Venetys inpresso Anno domini .1521. die .4. augusti et quia hunc librum emi. in quo inveni predictam Primam et secundam Distinccionem que dicitur liber de quinta essencia vt predicum est, laborem igitur harum duarum distinccionum rescribendarum breuitatis causa omisi." On Gradi's edition, see p. 145, note 29, above.

18. Ashmole 1478, pt. 2, fol. 96r. The Merchant Adventurers' primary business in the early sixteenth century was with the cloth trade, and the company was consequently dominated by mercers. On London's mercantile activities during this period, see Kenneth R. Andrews, *Trade, Plunder and Settlement: Maritime Enterprise and the Genesis of the British Empire, 1480–1630* (Cam-

rial, he converted it into a fresh source of intellectual and economic capital in the form of transcriptions and English translations, crafting his own philosophical persona through new presentations of medieval sources.

In April 1536, a year after the death of Du Wes, but two years before Greene's self-professed retirement from practice, Freelove acquired one of England's most important pseudo-Lullian compendia, CCC 244, apparently directly from Greene. He transcribed most of its contents, including the elaborate diagrams and even Kirkeby's marginalia, as well as the partial copy of *De secretis naturae* noted above, into a handsome compilation, now Sloane 3406. He also left his mark on his medieval exemplar. In CCC 244, at the point where Greene supplied missing text to the *Testamentum* in an unsigned marginal note, Freelove set his pen to the parchment to provide information of his own, adding the attribution "Sir Robert Greene W" in his most elegant hand. Beneath, he signed his own name in transliterated Greek characters: "φελοβε" (F[r]elove) (fig. 7).[19]

Freelove's intervention suggests that he valued the manuscript not just as a useful exemplar, but as an "association copy" related to a practitioner whom he knew either directly through personal acquaintance, or indirectly through books. He also took Greene as his model in the art of alchemical reading and writing. Like Greene he furnished his transcriptions with extensive colophons, providing not only his name and the date, but also the time of day and, in some cases, the prevailing astrological conditions.[20] And, like Greene, he sought to show off his smattering of Greek. On 13 May 1536, he signed his copy of the *Codicillus* by awkwardly transliterating his name and profession, "Sir Robertom F[r]elove Mercer," into Greek characters.[21] While Greene adopted a Greek word, *chloros* (green) to render his surname, Freelove adopted a Greek name, Eleutherius (Ελευθέριος, "Liberator"), to pun on the first syllable of his own—and then lived up to the sobriquet by strewing it liberally across the pages of Sloane 3406.[22]

Although the manuscript is unfinished and bears no explicit statement of Freelove's patronage aspirations, the care lavished on Sloane 3406 suggests

bridge: Cambridge University Press, 1984); Robert Brenner, *Merchants and Revolution: Commercial Change, Political Conflict, and London's Overseas Traders, 1550–1653* (London: Verso, 2003); Stephen Alford, *London's Triumph: Merchant Adventurers and the Tudor City* (London: Allen Lane, 2017).

19. CCC 244, fol. 37r.

20. For instance, Sloane 3604, fol. 14v.

21. Ibid., fol. 106r. Since Freelove never uses the title "Sir" elsewhere, this may suggest further imitation of Greene's style.

22. See, for instance, ibid., fols. 39r, 63v, 141r.

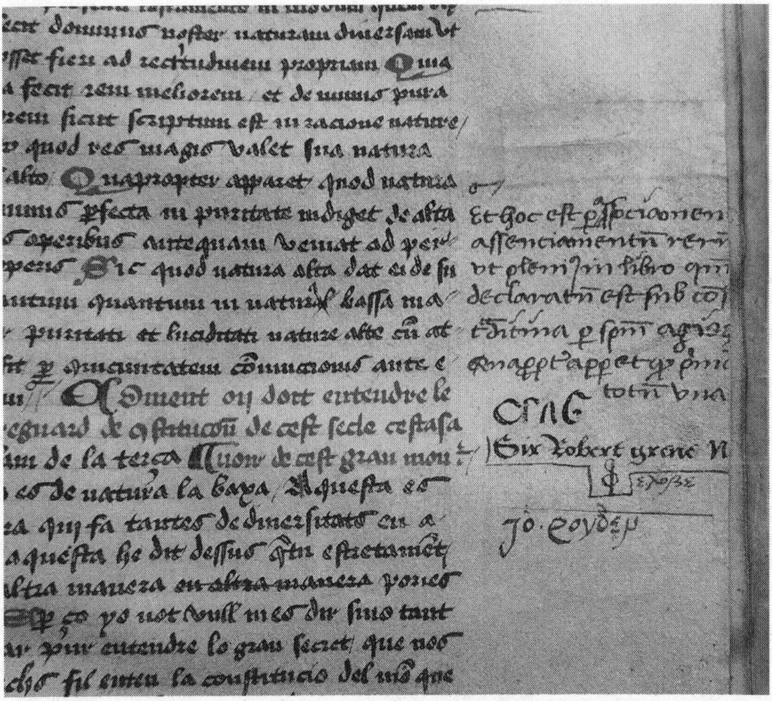

FIGURE 7. Marginal notes added by Robert Greene of Welby and Robert Freelove to the pseudo-Lullian *Testamentum*. Oxford, Corpus Christi College Library, MS 244, fol. 37r. By permission of the President and Fellows of Corpus Christi College, Oxford.

that it was conceived of as a presentation volume—perhaps an indication that the City mercer hoped to attract the attention of Henry VIII himself. For instance, Freelove altered the order of texts in CCC 244 so that his own collection would open with Raymond's *Compendium animae transmutationis metallorum*, a treatise on the making of precious stones in which Raymond addresses his patron, King Robert. The incipit of this work grandly adorns the first page of Freelove's manuscript: "Fulgeat regis diadema Roberti" ("May the diadem of [King] Robert shine"). Besides being well suited to a royal audience, this opening seems custom-made for Freelove, who shared his Christian name with Raymond's royal interlocutor, and the initial letter of his surname with the first letter of the incipit. To emphasize both points, Freelove tucked his own name, "Robertus Eleutherius Freelove," into the coils of the magnificent letter *F*, in both Latin and Greek script (fig. 8). Finally, he worked a miniature portrait of Henry VIII into the historiated *I* that opens the *Compendium*'s *Practica* (fig. 9).

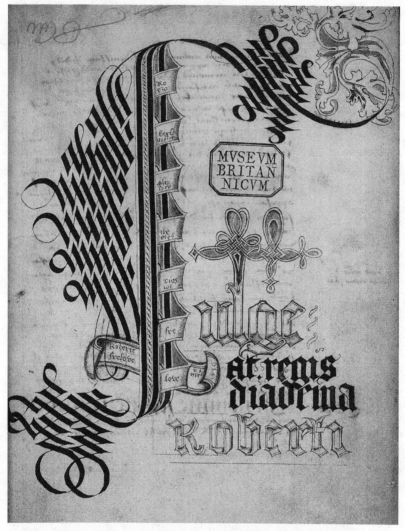

FIGURE 8. Robert Freelove, title page of the pseudo-Lullian *Compendium animae transmutationis metallorum*. © The British Library Board, MS Sloane 3604, fol. 3r.

Whether or not Freelove approached prospective patrons on his own behalf, he was well aware of the value of a book of this kind to practitioners still unable to obtain pseudo-Lullian texts in print—Sloane 3406 later sold for the considerable sum of £20 in English money.[23] Even in humbler produc-

<hr />

23. Sloane 3604, fol. 290v: "Hec Liber valet vigenti libras legalis monete anglie."

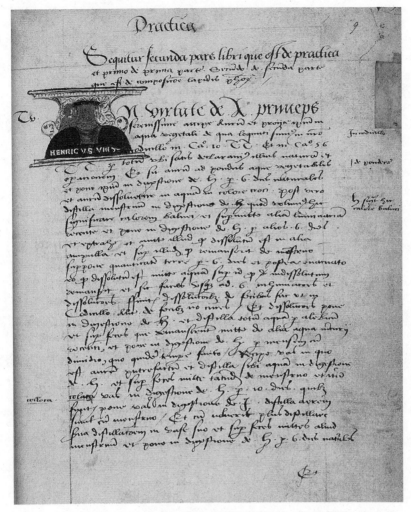

FIGURE 9. Robert Freelove's portrait of Henry VIII in a historiated initial *I*. Pseudo-Lull, "Practica," *Compendium animae transmutationis metallorum.* © The British Library Board, MS Sloane 3604, fol. 9r.

tions, the care with which copyists like Du Wes, Greene, and Freelove signed their transcriptions testifies to the importance they placed on the process of recording earlier sources. Such elaborate colophons were a period style, allowing scribes to write themselves into the history of their text: an act that had particular meaning in the context of transmitting alchemical writings, by associating the writer's contribution with that of past philosophers.

This care extended to the work for which Freelove is now primarily remembered: translating medieval texts. Freelove is one of the earliest identifiable translators of English alchemica, and one whose translation career spanned the pre- and post-Reformation period—a fact reflected in his colophons. His first known production is the translation of a French treatise attributed to Jean de Meung on the "blessyd vegetable stone," completed on 3 May 1522, sometime between three and four o'clock, and dedicated to "[th]e maker & forme of all thyng*is* & to [th]e worship of his blessyd mother Mary."[24] Invocations of the Virgin are conspicuously absent from his translations of Latin treatises made during the reign of Edward VI, the *Privitie or Secret of Avicen*, completed on 16 December 1550, and the pseudo-Baconian *Radix mundi*, finished on 16 February 1550/1, which reflect Protestant sentiment by offering praise to God alone.[25]

The 1550s and 1560s would emerge as a golden age of vernacular translation, as English versions of medieval classics became available in increasing numbers. One of the period's greatest feats was the Englishing of the *Testamentum* in 1558 by Walter Atherton (occupation unknown), whose contribution was recorded by Richard Walton, a haberdasher living in the Old Change immediately to the east of St. Paul's Cathedral. Like Greene and Freelove before him, Walton folded his own role into the history of the text, providing a genealogy of its dissemination: "Thus endyth thys bok wrytten in latten by mathew lond' the 17 of dece*m*ber an*n*o 1472/ & translatyd into ynglyshe by Wa[lter] atherton the 29 of august an*n*o 1558/ & copyed by rychard waultown haburdassher of london an*n*o do*m*ini 1563 the 16 of marche."[26]

This translation appeared just as Raymond's authority was coming to be rivaled—even overhauled—by that of his English exponent, Ripley. Over the next few decades, the core texts of the "Ripley Corpus" came to dominate both vernacular compendia and English patronage suits, as Ripleian works that had previously been known only in Latin became available in new English translations. The most important of these was the *Medulla alchimiae*,

24. Beinecke Library, MS Mellon 33, fol. 59v. This copy is not in Freelove's own hand. Jean de Meung, one of the poets of the famous twelfth-century French romance the *Roman de la rose*, later became established as an alchemical authority on the basis of his brief treatment of alchemy in the poem; see Newman, *Promethean Ambitions*, 77–82.

25. Pseudo-Avicenna in Ashmole 1478, pt. 2, fol. 96r, in Freelove's hand. The pseudo-Bacon is preserved in Mellon 33, fol. 97r, and in Bodleian Library, MS Digby 133, fol. 36r; neither copy in Freelove's own hand. An adjacent translation in Mellon 33, a treatise attributed to Khālid ibn Yazīd (fols. 96v-110v), may also be by Freelove; although unsigned, the colophon is similar in style to those of his other translations: "traunslatyd owt of latyn in to englyshe an[n]o do*m*ini 1542" (fol. 110v).

26. Ashmole 1479, fol. 140v.

first translated by the Protestant divine David Whitehead in 1552, and Englished again later in the century by John Higgins, vicar of Winchester.[27] The *Concordantia* was probably translated not long after.[28]

Walton's handwritten compendium, now Ashmole 1479, provides a treasure trove of these early translations, including the Ripleian *Marrow, Concordance, Philorcium,* and *Cantilena,* several of which he also dated—the dates, between 1 February 1561/2 and 20 October 1565, providing the terminus ante quem for their appearance during a moment of unprecedented energy and expansion in English alchemical literacy.[29] These translations presented anglophone readers with a body of "new" content that was quickly shared and copied; often, as we shall see, into whole compendia of Ripleian and pseudo-Lullian works. Together with Ripley's famous *Compound* and *Epistle to Edward IV,* as well as two Middle English treatises commonly ascribed to him (the *Pupilla alchemiae* and *Accurtations of Raymond*), these works formed the basis of the canon's alchemical reputation during the 1560s.

The Englishing of the corpus served practical as well as antiquarian ends, coinciding with a historical moment in which princes and other investors were already eyeing alchemical expertise as the key to unlocking mineral wealth through improved mining and metallurgical techniques.[30] In England, the desire to exploit native mineral resources (including those associated with former monastic lands), coupled with the still-parlous state of the coinage—a problem that would not be resolved until well into the reign of Elizabeth I—heightened interest in alchemy among prospective patrons, as

27. Sloane 1842, fol. 78. On the *Medulla*'s translation history, see CRC, on 130–31. The *Medulla* is CRC 16.

28. Bale seems to refer to an English version of Ripley's *Concordantia* in the *Catalogus* of 1557. The *Index* records two incipits for the work. The first, "Quia Raimundus dicit, loquens de fermento lapidis," is the Latin original, known to Bale by 1548; John Bale, *Index Britanniae Scriptorum: John Bale's Index of British and Other Writers,* ed. Reginald Lane Poole and Mary Bateson (1902; repr., Cambridge: D. S. Brewer, 1990), 85. The second, "Asserit Raymundus de fermento loquens," seems to be his rendering of the English translation back into Latin; Bale, *Catalogus,* 623.

29. Ashmole 1479: Walton transcribed the "marie of allchimy" on 1 February 1561/2 (fols. 35r-42v) and the "concordance bytwyne guido & raymonde lully ij phylosophers" by May 1563 (fols. 43r-44r). The volume also includes the start of an English translation of the *Cantilena* (fol. 228v; possible leaf missing); and the "phylorsium of [th]e alchymists" (fols. 229r-35r).

30. On the development of English mining operations during the second half of the sixteenth century, see Eric H. Ash, *Power, Knowledge, and Expertise in Elizabethan England* (Baltimore: Johns Hopkins University Press, 2004), chap. 1; M. B. Donald, *Elizabethan Monopolies: The History of the Company of Mineral and Battery Works from 1565 to 1604* (Edinburgh: Oliver & Boyd, 1961). On mining culture in early Stuart England, see Cesare Pastorino, "Weighing Experience: Francis Bacon, the Inventions of the Mechanical Arts, and the Emergence of Modern Experiment" (PhD diss., University of Indiana Bloomington, 2011).

well as merchants and craftsmen, and fed demand for alchemical books. In the event, Ripley's reputation fared rather better than his sixteenth-century readers. The promise and pitfalls of alchemical patronage would be jointly realized during the 1550s, as practitioners became embroiled in a series of scandals connected to senior figures in the royal mint, including members of the circle of Edward Seymour, Duke of Somerset (ca. 1500–1552), who, as Lord Protector, was effective ruler of England.

BLOMFILD'S RIVALS

In the wake of the Reformation, the popular impression of alchemical misbehavior as the province of monks, canons, and friars changed—a transition already apparent in Blomfild's 1557 *Blossoms*, which contrasts past philosophers (mostly religious) with the contemporary exploits of frauds and fools (mostly mint officials). The result is a distinctively midcentury vision of the traditional polarity between "alchemists" and "philosophers." Whereas earlier satires, like those of Chaucer and Ripley, involved fictional, anonymous characters, Blomfild describes the activities of documented practitioners active during the reigns of Henry VIII and Edward VI, many of whom were involved in scandals arising from bad monetary policy or outright currency crime.

In the *Blossoms*, Philosophy's garden recreates the timeless space of medieval romance narrative, in which the dreamer meets those adepts in person whom in reality Blomfild met only in the pages of his books. Among them are several whose work had by this time become associated with English practice: Albertus Magnus, Roger Bacon, and Raymond Lull, as well as Ripley, "the chanon of Bridleington so profound."[31] However, a topical note is introduced into the dreamlike landscape by the appearance of Blomfild's rival practitioners, whom he relegates to the unprofitable wasteland beyond Philosophy's gates:

---

31. Blomfild, *Blossoms*, 24–25 (ll. 106–10). Blomfild also includes an unnamed "monke," although the identity of this practitioner is unclear. Albertus Magnus appears as a protagonist in several Middle English alchemical writings, probably reflecting the enormous success of the pseudo-Albertine *Semita recta*; see Peter Grund, "Albertus Magnus and the Queen of the Elves: A 15th-Century English Verse Dialogue on Alchemy," *Anglia: Zeitschrift für englische Philologie* 122 (2004): 640–62; Grund, "'ffor to make Azure as Albert biddes': Medieval English Alchemical Writings in the Pseudo-Albertan Tradition," *Ambix* 53 (2006): 21–42. Albert is named as an English adept in Thomas Charnock's *Booke*; Lansdowne 703, fols. 16r-18r, 51v.

These were Broke the preste & yorke in cotes gay,
Which robbed king henry of a million of gold;
Martin pery, mayre, & thomas De Lahaye,
Saying that the king they greatly enrich would.
They wispered in his eare, & this tale him told:
"Wee will worke for your highnes the *Elixer vite*,
A princely worke called *opus regale*."

Then brought they in the vicar of Maldon
With his lyon greene, that most royall secret,
Richard record & little Master Edon
(Their mettals by corosiue[s] to calcinate & fret);
Hugh oldcastle & Sir Robert greene with them mett,
Rosting & broileinge all thinges out of kinde,
Like [Foolosophers] left off with losse in the end.[32]

Whatever the relationship between Blomfild's condemnation and the reality of practice, his choice of targets reflects the extent to which alchemical practitioners and their projects had become visible in English public life by the 1550s. Most of them held offices in the mint, and several were associated with unpopular economic policies—particularly Robert Brock (also spelled Brooke and Broke), comptroller of one of the Tower mints, who carried much of the public blame for Henry VIII's catastrophic debasement of the coinage.[33] Blomfild was not alone in lampooning him. Early in Edward VI's reign, Hugh Latimer, the former Bishop of Worcester, assailed worldly prelates who garnered royal offices rather than tending to their flocks. His target in the polemical "Sermon on the Plough" was clearly Brock. "I would fain know," Latimer demanded, "who controlleth the devil at home in his parish, while he controlleth the mint? . . . I cannot tell you; but the saying is, that since priests have been minters, money hath been worse than it was before."[34]

The management of English coin did not improve under Edward VI.

---

32. Blomfild, *Blossoms*, 25 (ll. 127–40).

33. The Tower of London housed two mints; Brock was comptroller of Tower I. C. E. Challis, "Mint Officials and Moneyers of the Tudor Period," *British Numismatic Journal* 45 (1975): 51–76, on 57.

34. Hugh Latimer, "A Sermon of the Reverend Father Master Hugh Latimer, Preached in the Shrouds at Paul's Church in London, on the Eighteenth Day of January, Anno 1548," in *Sermons*

Blomfild's colorful York with "coates gay" is Sir John York (d. 1569), sheriff of London and Middlesex and a prominent client of John Dudley, Earl of Warwick (soon to be Duke of Northumberland). By 1550 York was one of the most influential men in London, and the effective manager of the Tower and Southwark mints. His later reputation for financial misdemeanor stems primarily from an incident in 1551 when he lost £4000 of Edward's silver bullion through unlucky speculation on the Antwerp exchange, although he was subsequently pardoned.[35] Such transgressions apparently lived on in public memory, to judge by his inclusion in the *Blossoms*. According to Blomfild, York was not the only culprit: three more of his targets, Martin Pirry (d. 1552), John Maire (fl. 1548–1561), and Robert Recorde (ca. 1512–1558), served in various mints during the reign of Edward VI, and all attracted scandal for one reason or another.[36] Pirry, or Pery, was appointed comptroller of the Dublin mint despite having previously fled to France on suspicion of clipping, while the priest John Maire, or Mayre, an assay-master at the ecclesiastical mint at Durham House, was accused by one suspected conjuror of using magic to locate buried treasure—an activity expressly banned by the Witchcraft Act.[37]

While Blomfild was hardly a disinterested witness, his poem does hint at a more general perception of the mint as a hotbed of multiplication and currency crime, reflecting popular dissatisfaction with the state of English coin. The rulers who oversaw these economic miscalculations did not escape blame. When Edward Seymour, Duke of Somerset and former Lord Protector, who had governed England in the name of Edward VI until his fall, was executed in 1552, one of the complaints made against him was that he

(New York: E.P. Dutton, 1906); Project Canterbury, http://anglicanhistory.org (accessed 7 July 2018). Challis considers the contemporary view that Brock was responsible for the debasement to be unfounded; Challis, *Tudor Coinage*, 87n146.

35. J. G. Elzinga, "York, Sir John (d. 1569)," *ODNB*.

36. Blomfild actually names Richard Recorde, the brother of the mathematician Robert Recorde (ca. 1510–1558) who became mayor of Tenby in 1559; Edward Kaplan, "Robert Recorde (c. 1510–1558): Studies in the Life and Works of a Tudor Scientist" (PhD diss., New York University, 1960), 1. However, Robert's position in the mint, and the fact that he was later imprisoned for debt, make it plausible that he is the alchemist intended. Robert evidently was interested in alchemy, since Bale lists several of his alchemical books, including a copy of Norton's *De transmutatione metallorum* (possibly an alternative title for the *Ordinal*), in Bale, *Index Britanniae Scriptorum*, 179.

37. Challis, "Mint Officials and Moneyers," 65 and 63, respectively. On Pery, see Schuler, "Three Renaissance Poems," 47. On Maire, see the deposition of William Wycherley, who accused Maire of conjuring while himself under investigation for the same offense; Nichols, *Narratives of the Days of the Reformation*, 330, 332–34.

"commaunded multiplication and Alcumistry to be practised, to abuse the kinges coyne."[38] It is telling that one of Somerset's major clients, Richard Whalley, did in fact employ an alchemist, although this came to light only after Seymour had gone to the scaffold. His accomplice was the same "little Master Edon" named in Blomfild's roll call of shame: the humanist cosmographer whose example illustrates both the pitfalls of practice, and the role played by books and communities of readers in establishing alchemy as an English enterprise.

THE ENTERPRISE OF RICHARD EDEN

Eden is known today for his correspondence with the great ministers of Elizabeth I's reign, William Cecil and Thomas Smith, and better still for his influential translations of works on cosmography and navigation, which contributed to England's store of seafaring knowledge just at the moment when the kingdom began to establish itself as a maritime power.[39] Paradoxically, this place in the history of science arose from his failure to derive a steady income from his primary interests in alchemy and medicinal distillation. Throughout his career he attempted to develop these interests by working the patronage system: cultivating personal and familial connections, and seeking access to diverse sites of chemical practice, including the court, the mint, and a gentleman's household.

Eden seems to have acquired his enthusiasm for scientific knowledge while a student at Queens' College, Cambridge, under the influence of his friend and tutor, the brilliant mathematician, classical scholar, and professor of civil law Thomas Smith.[40] Through Smith, Eden gained the entrée to a humanist circle that included John Cheke, the Regius Professor of Greek, as well as Cheke's promising students Roger Ascham and William Cecil. This circle also proved a congenial environment for cultivating alchemical tastes. Writing to Cecil in 1563, Eden later recalled that Cheke was partly responsible for instilling "the divine sparke of knowleage that is in your Honour"—

38. "K. Edvvard. 6. The Troubles and Death of the Duke of Somerset," in John Foxe, *The Unabridged Acts and Monuments Online (1570 edition)* (HRI Online Publications, Sheffield, 2011), 1587; http//www.johnfoxe.org (accessed 29 January 2018).

39. On Eden, see note 3, above.

40. Ian W. Archer, "Smith, Sir Thomas (1513–1577)," *ODNB*; Mary Dewar, *Sir Thomas Smith, a Tudor Intellectual in Office* (London: Athlone Press, 1964).

meaning an appreciation for alchemical secrets.[41] Both Smith and Cecil would retain this appreciation throughout their careers, allowing opportunities for practitioners like Eden to press their own expertise in chemical matters.

Smith's influence may have helped Eden secure his first job in the Treasury, where, as Eden himself later recalled, "he remained for two years until the King's death; who, when dying, did not forget him, but assigned to him the office of the distillery."[42] Henry VIII placed a high value on distilled remedies, to judge from the list of fifty-four waters distilled for his use between May and July 1539, presumably by Eden's predecessor, Thomas Seex.[43] That Eden was appointed to replace Seex in Henry's household implies that he had already acquired a reputation for skill in chemical operations by the time of his appointment in December 1547. When Somerset bestowed the office elsewhere, Eden was forced to seek other ways of exploiting that expertise.

In early 1550, following advice from Sir John York, he determined to try his fortune with the Earl of Warwick, a noted patron of natural philosophy whose political star was on the rise. While waiting at Lion Quay for a boat to Greenwich, Eden fell in with a gentleman he had previously met at a dinner party. Richard Whalley was not an aristocrat, but a gentleman of respectable birth who had profited from the acquisition of former monastic lands under Henry VIII. Whalley offered Eden a seat in his boat, and the two men beguiled the voyage by discussing a chemical book that, Eden explains, he had brought as a gift for Warwick:

41. British Library, MS Lansdowne 101, fol. 19v. On the alchemical component of this letter, see Campbell, "Alchemical Patronage," 25; Parry, *Arch-Conjuror of England*, 76. On its place in English translation studies, see Neil Rhodes, Gordon Kendal, and Louise Wilson, eds., *English Renaissance Translation Theory* (London: Modern Human Research Association, 2013), 305–7.

42. Arber, "Life and Labors," xlv. Eden refers to himself in the third person in this text, a curriculum vitae addressed to Elizabeth I.

43. Royal MS 7 C XVI, fol. 19r: one list of fifty-four waters, titled "These be the names of the waters which where stilled this yere from the begynyng of Maye vntyll the xvth daye of July in the xxxith yere of the raign of our soueragne lorde King Henry the viij [i.e., 1539]," and a second list of twenty-four, "Old watters of the last yeres remayne." Seex was granted an annuity in December 1546 for his services in distilling waters for the king: "Henry VIII: December 1546, 26–31," *Letters and Papers, Foreign and Domestic, Henry VIII*, vol. 21, pt. 2, *September 1546–January 1547*, ed. James Gairdner and R. H. Brodie (London: Her Majesty's Stationery Office, 1910), 313–48; *British History Online*, http://www.british-history.ac.uk/report.aspx?compid=80889 (accessed 21 November 2008).

Therewith I shewed hym a boke which I hade then abowte me, towchinge
thes matters, wrytten with myne owne hande & gathered owte of sundrye
Auctours, declaringe forther to hym that, at the request of Syr John Yorke
I entended to present that boke to my Lorde of Warwike, nowe Duke of
Northumberlande. Thus we passed the tyme redinge & reasoninge untill we
came to Grenewich, where we parted.[44]

For both men, the trip was a fishing expedition. As Eden had determined at
their earlier dinner meeting, Whalley was interested in mining enterprises,
but also intrigued by the possibility of alchemy and curious about the extent
of Eden's own skill. For Whalley's part, he knew Eden's reputation in metallurgy and medicinal distillation—an impression presumably confirmed while
discussing his manuscript compilation. Eden and Whalley later traveled back
to London, this time accompanied by Warwick's client, York. The men dined
together, and Whalley took Eden aside into the garden to offer him a position and laboratory space in his own household. Eden gladly agreed. Two
months later, he and his wife moved into Whalley's house, site of the former
abbey of Welbeck, ready to start work on practices "concernynge metalles &
Quinta Essentia," based on the distillation of wine.[45]

Although Whalley's offer of a place in his household with a salary of
£20 per year must have looked particularly appealing to a recently married
chemist of a scholarly disposition, it turned out to be a political miscalcula-
tion. Whalley was chamberlain, councillor, and kinsman by marriage to the
Duke of Somerset, who had been arrested in October 1549, shortly before
Whalley's first encounter with Eden. As prominent members of the ducal
retinue, both Whalley and Cecil were briefly imprisoned in the Tower after
Somerset's fall; after their release in January 1550, both began to hastily forge
ties with the now-dominant Warwick. Now responsible for the expenses of
Somerset's household, and with his own position uncertain, it was around
this time that Whalley began to think seriously about alchemy.[46]

Given the legal risks, we might wonder what prompted this hardheaded
administrator to give house room to an alchemist at such a politically dangerous moment. Possibly Whalley's interest arose from a desire to develop the

44. National Archives, SP 46/2, fols. 164–67; see Kitching, "Alchemy," 312.

45. Kitching, "Alchemy," 312.

46. Eden later reported that Whalley offered him the position when "at liberty"—that is to
say, following his release after his first imprisonment on 25 January 1550. Eden, cited in Kitching,
"Alchemy," 312.

substantial property he had acquired through purchasing former monastic lands, which included several mines.[47] Eden later reported that Whalley had quizzed him on his mining knowledge during their first meeting, and commissioned him to translate Biringuccio's work on mining and metallurgy, *Pirotechnia*, from Italian into English.[48] However, Eden explicitly denied expertise in working with "grose mynes" (metallic ores), and in fact "never sawe the places where mynes ar engendred." Rather, his skill lay in "workes of greater subtilite"—namely, "the philosopher's stone, *Aurum potabile*, and *Quinta Essentia*."[49] On this basis, then, Eden was employed to pursue both transmutation and medicinal remedies, including the quintessence and potable gold that he initially pitched to Whalley. His new patron threw himself into this new enterprise by purchasing "many bokes of Alchemye," as well as glasses and other equipment that, as Eden later deposed, "yett remayne in hys howse."

Over the next two years, Whalley's enthusiasm for alchemy dwindled with his own fortunes. In February 1551, he was imprisoned in the Fleet on suspicion of attempting to reinstate Somerset as Lord Protector. In October that year he was rearrested and spent a spell in the Tower in the run-up to Somerset's trial and eventual execution. Eden continued to serve his patron during this trying time, sometimes visiting him in the Tower to discuss business, and defending his lack of results when Whalley demanded "what practyses I had had all that tyme."[50] Shortly after Whalley's release, Eden seems to have betrayed his alchemical activities to William Bolles of Osberton (1495–1583), auditor for the Court of Augmentations in Nottinghamshire. Whalley was arrested for a third time in October 1552, although the charges of illegal multiplying were eventually dropped. After making his deposition, Eden himself was released, bound to the sum of £200 to refrain from practicing alchemy again.

After Somerset's fall, Eden attempted to retrieve his position with the new regime by deploying his linguistic skills in translating and editing scientific and cosmographical works, although, understandably given his recent

47. Whalley started to purchase former monastic land in 1536, when he acquired the estates of Welbeck Abbey. In 1545 he was appointed receiver for the court of augmentations for Yorkshire, purchasing Worksop Priory in the same year. Alan Bryson, "Whalley, Richard (1498/9–1583)," *ODNB*.

48. Kitching, "Alchemy," 314; Biringuccio, *De la pirotechnia*.

49. Kitching, "Alchemy," 311.

50. Ibid., 314.

embarrassments, he refrained from publishing on alchemical topics.[51] Eden's service to England's mercantile and maritime interests did not banish public memory of the Whalley affair, which led Lawrence Humphrey, Master of Magdalen College, Oxford, to describe him in 1558 as "cosmographer and alchemist."[52] In 1557, as we have seen, Blomfild ranked him among the "Foolosophers" who had brought the art into disrepute. Yet when Eden's readings and practices are set alongside those of Blomfild and other contemporary practitioners, they do not look so very different. All were concerned with the same, multistranded tradition of pseudo-Lullian alchemy, an approach that offered both chrysopoetic and medicinal benefits, and was elicited using alchemical reading techniques from their shared medieval sources.

THE ENGLISH ALCHEMICAL COMPENDIUM

Eden's experience offers a revealing account of the mechanics of alchemical patronage in post-Reformation England, but it is also a story about the role played by books. As both a gift for presentation to Warwick, and a pretext for philosophical improvisation on the voyage to Greenwich, Eden's 1550 compilation of "sundrye Auctours" served as a vital prop and tool for engaging the interest of powerful men. He continued to employ this strategy during his travels in France, sending a copy of the pseudo-Lullian *Experimenta* to his former tutor, Smith, a keen distiller in his own right, who set about reconstructing some of the book's contents.[53] He also compiled texts for his own use. While he left no alchemical treatise of his own, one collec-

51. Eden sought Northumberland's favor with a translation of Sebastian Münster, *A treatyse of the newe India with other new founde landes and islandes, aswell eastwarde as westwarde, as they are knowen and found in these oure dayes...* , trans. Richard Eden (London: S. Mierdman for Edward Sutton, [1553]), dedicated to the duke. He may also have served as Cecil's secretary, as suggested by Arber, "Life and Labors," xxxviii: "Around this date [i.e., 1552] Eden was, I believe, acting as private secretary to Sir W. Cecil. I have, however, lost the reference to the authority for this." His most important publication, the *Decades*, is primarily a translation of the first three decades of Pietro Martire d'Anghiera, *De orbe novo decades* ([Alcalá]: [Arnaldi Guillelmi], [1516]), and Gonzalo Fernández Oviedo, *Historia general y natural de las Indias* (Seville, 1530–55); Richard Eden, *The Decades of the newe worlde or west India, Conteyning the nauigations and conquestes of the Spanyardes, with the particular description of the moste ryche and large landes and Ilandes lately founde in the west Ocean perteynyng to the inheritaunce of the kinges of Spayne* (London: Richard Jugge, 1555).

52. Lawrence Humphrey, *Interpretatio linguarum: seu de ratione conuertendi & explicandi autores tam sacros quam profanos, libri tres* (Basel: Hieronymus Froben, 1559), sig. L4: "Joannes [sic] Eden, Cosmographus et Alchumista." The tag was picked up by Bale, *Catalogus*, sig. 3; cited in Arber, "Life and Labors," xl.

53. Queens' College MS 49, fol. 117v: "Experimenta Rai: Lull: ex dono Ric: Eden written." Eden maintained his correspondence with Smith, who wrote to him on 9 March 1572/3; National

tion of Latin and English authorities survives that records Eden's fascination with sericonian alchemy.[54]

Although not previously connected to Eden, Trinity R.14.56 is written primarily in his hand and bears his name, "Richardus Edenus," on the fly-leaf.[55] The book also carries his dense annotations throughout, several of which are signed with his abbreviated name, "Ed." Although the original manuscript has lost some fifty folios and been considerably reorganized since its original production, enough remains to offer insight into not just Eden's alchemy, but also the community of readers whose books he read.[56] Crammed with recipes and annotations, cross-referenced against one another and against other books, its content confirms the strongly pseudo-Lullian tenor of Eden's practice. Heavily annotated extracts from the *Testamentum* and *Epistola accurtationis* point to his efforts to construe these enigmatic texts—in particular, the nature of the *Testamentum*'s "G. vegetable," as well as mysterious cognates like the "menstruum resoluble" and "Green Lion."[57] Through such reflections, Eden reveals the practical preoccupations that underwrote his reading of medieval sources, including English alchemical poems, which he approached with the same rigor and seriousness as he did Latin prose.[58]

Eden's book underscores what we already know from other contemporary manuscripts and patronage suits: that both the mineral and the vege-

---

Archives, SP 70/146, 60. On Smith's distillation practice, see Simpson, "Sir Thomas Smith's Stillhouse."

54. The contents of this manuscript, Trinity R.14.56, are described in detail in Timmermann, "Alchemy in Cambridge," 450–59.

55. Trinity R.14.56, flyleaf. Elizabethans often cultivated different hands for different occasions, as may also be seen in the case of John Dee. Although Eden's book is written in a less formal secretary hand than his correspondence with Cecil, most of the individual letterforms are identical. In particular, the signature and annotations match those of Eden's annotated copy of Martire's *Decades* (Johns Hopkins University Library), and Eden's signatures in Bodleian Library, MS Savile 18, fols. 37v and 171r. Savile 18 is a book of fourteenth- and fifteenth-century transcriptions that Eden co-owned with his fellow queensman Edward Gascoyn, and that points to Eden's early scientific interests. It includes Roger Bacon's writings on perspective and the multiplication of species.

56. On the physical construction of the manuscript, see Timmermann, *Verse and Transmutation*, 144n6.

57. See, for instance, Trinity R.14.56, fols. 12v, 129r-v.

58. As Anke Timmermann notes, the poems in this collection were studied as intensively for their practical meaning as any of the prose treatises in the collection, possibly even more so. She examines the note-taking strategies in this manuscript (although without connecting it to Eden) in detail in Timmermann, *Verse and Transmutation*, chap. 5, particularly in relation to the verse contents.

table strands of the pseudo-Lullian tradition continued to play a major role in English alchemy as actually practiced well into the second half of the century. Whether posterity branded them as fools or philosophers, virtually all of the English practitioners who petitioned Elizabeth I invoked the vegetable stone in their suits, and most did so on the basis of Ripley's authority. Thus, in spite of Blomfild's poetical indictment of Eden as a charlatan, the two men were probably engaged in broadly similar activities. Both regarded Raymond and Ripley as authorities, and each used the promise of the quintessence as a means of soliciting patronage. For instance, we know from Eden's deposition that he was working on the pseudo-Lullian process for the vegetable stone extracted from the *Epistola*—or, as he termed it, "the worke of Raymundus cauled Accurtatio."[59] One of his arguments with Whalley in fact arose over the difficulty of securing an adequate supply of red wine, the prime ingredient in this work.

Blomfild's own devotion to pseudo-Lullian alchemy has already been noted, and was still in evidence around 1574, when he wrote the *Regiment of Life* in an attempt to secure Elizabeth's favor.[60] The work charmingly describes how "our heaven" (the quintessence) may be beautified by the stars (metallic bodies): specifically by saturating a pound of powdered lead in spirit of wine for eight days, drawing off the quintessence, and distilling it to produce an oil that "hath the taste of suger."[61] Blomfild hails this sweet, leaden compound as "very medicinable for diuers Infirmiteis of mannys body."[62]

Although we lack Blomfild's own manuscript notes, those of Eden, his contemporary rival, reveal how strategies of alchemical reading continued to underpin practical attempts to reproduce the vegetable stone. For instance, the title of one process shows Eden reading the *Epistola* against the *Testamentum*: "The Accurtation of Raymond concerning tartar and the wine of the philosophers, and G. vegetable, and also the philosopher's salt of tartar, which is the mercury of the *Testamentum*."[63] Even as he set down the procedure, Eden sought to reconcile it with other pseudo-Lullian writings—in this

---

59. Kitching, "Alchemy," 313–14.

60. *Regiment of Life*, fol. 15v. Many of the recipes employ the "heuynly quyntacens" (as at fol. 10v), and are intended for medicinal purposes, following the categories of disease set out in *De secretis naturae*.

61. Ibid., fols. 11v-12v. This section draws directly from John of Rupescissa's *De consideratione*.

62. Ibid., fol. 12r.

63. Trinity R.14.56, fol. 129r: "Accurtatio Ray[mundi]. De tartaro et vino ph*ilosophorum*, atq*ue* de .G. vegetabili, simulq*ue* de sale Tartari ph*ilosoph*i, q*uod* est M*er*curius Testamentarius."

case, using the *Epistola*'s black tartar (the famous "black blacker than black") to gloss the Magister's mysterious "G."

Eden would also have received a heavy hint about the value of tartar from another of his major authorities, Ripley's *Medulla*. Although the compendium includes only a few of Ripley's attributed works (extracts from the *Accurtations of Raymond* and the Englished *Concordance*), hardly a page goes by without a reference to the canon's writings, suggesting that Eden worked with a collection of Ripleian texts close to hand, and that he used them as keys to interpreting his pseudo-Lullian sources. One note (signed "Ed.") compares a passage from *De secretis naturae* on the dissolution of pearls with a similar process in the *Medulla*, while others show Eden grappling with Ripleian terms like "sericon" and "Adrop."[64]

Sometimes these allow us to trace his attempts to reconcile texts with his own practical experience. One procedure starts by iterating the names of Adrop: "*Saturnus* in Latin; that is, the Stone, or Antimony." Eden remarks that he has found Adrop referred to this way in "a certain ancient book," but his own experiments failed to secure a firm interpretation of its meaning.[65] The recipe directs the practitioner to take four pounds of Adrop, ground into a fine powder, and place it in an earthen pan with four gallons of strong distilled vinegar, stirring with a stick until it dissolves into a clear, crystalline water. Eden realized that red lead could not be easily dissolved by this method, noting, "therefore it is not minium of lead."[66] Yet Adrop did not obviously correspond to antimony, either, since the recipe states that the resulting liquid is clear. "Therefore it isn't antimony," Eden mused in the margin, "because that would always be more red or purple."[67] What, then, was sericon?

In construing such terms, Eden did not have to rely solely on his own resources. His manuscript shows how an alchemical book functioned in company, and in use—not just a resource for private study, but a microcosm of interactions between different readers and collectors. Other hands point

64. Ibid., fol. 40v.

65. Ibid., fol. 41r: "Adrop. Latinae Saturnus, id est. Lapis; Vel Antimonium . . . Sic in quodam antiquo libro."

66. Ibid., fol. 41r: "Take iiij. libri of Adrop \.scilicet. combustum/ and grynde it into fyne pouder \ergo non minium plumbe./ Then do it in an erthen pan, and put therto iiij galons of stronge vineger distilled. And stere them and labour them well togyther with A staffe. And so do iij or iiij tymes in the day. Then lett it stonde and cleare as cristall." Eden added a note to the final comment, recognizing that Ripley had described a similar, crystalline water in the *Medulla*: "Medulla into A water ponderous."

67. Ibid.: "Ergo non est Antimonium, quia semper esset magis rubeum vel purpureum."

to the intervention of either Eden's friends and contemporaries, or later owners who continued where Eden left off.[68] Marginal and interlinear notes show that Eden was also consulting other manuscript collections compiled by peers with mint or mining connections. These include the books and annotations of "Baptista," the Venetian chemist Giovanni Battista Agnello, who, like Eden, later benefited from the support of de Ferrières, Vidame de Chartres.[69] Eden also refers to the pseudo-Lullian books belonging to one "Mr Bolles"—almost certainly the same man who betrayed Whalley's and Eden's practice to the court.[70]

William Bolles has not previously been identified as an alchemist, yet he was well positioned to acquire and hone skill at metallurgy. He was a teller at the Tower mint between March 1549 and December 1550, exactly the period when Eden was seeking a position there through the offices of York, and immediately prior to Whalley's offer of employment.[71] Bolles was also a beneficiary of the dissolution. Like Whalley he was responsible for assessing and distributing their spoils, having been appointed by Cromwell in April 1536 as auditor of the Court of Augmentations for Derby, Nottinghamshire, and Cheshire. He was granted the former Augustine priory of Felley in Northamptonshire, and in 1541 he also purchased part of the estate of Osberton, near Worksop, from his fellow commissioner Robert Dighton.[72] At the

68. Perhaps the most surprising feature of the book is that it seems to have remained in, or returned to, the hands of the Whalley family. It was eventually presented to Trinity College, Cambridge, by Richard Whalley's grandson Thomas, vice-master of the college. One of the book's annotators gives his initials as "TW"—presumably the vice-master himself, or else his uncle Thomas, Richard's eldest son, who helped Eden pay for supplies of red wine and who was also (as Eden later informed Cecil) his witness to "a secreate practise"; Richard Eden to William Cecil, 1 August 1562, Lansdowne 101, fol. 19v; Kitching, "Alchemy," 313. On the various generations of the Whalley family, see Robert Thoroton, *The Antiquities of Nottinghamshire*, ed. John Throsby, 2nd ed. (Nottingham: G. Burbage, 1790), 1:250. On the donation of books by Thomas Whalley, vice-master of Trinity, see Timmermann, *Verse and Transmutation*, 144; Timmermann, "Alchemy in Cambridge," 352.

69. On Agnello, see Campbell, "Alchemical Patronage," 121–24; Harkness, *Jewel House*, 174–78.

70. At the end of a process for dissolving mercury in "common water," someone—probably Eden himself—has struck through his own initaled note, adding, "vid[e]. in fine Ray*mundus*. Bolles."("See at the end of Raymundus. Bolles"), fol. 66r. Eden also cross-referenced processes against multiple books, such as a recipe for an elixir drawn from oil of mercury, found in one of Bolles's books as well as in another of his own: "Vid[e]. lib[er]. Bolles. fol. 67. et in fine nos*ter* Ray. de Elix. ex oleo M*ercur*ij" ("See Bolles's Book, fol. 67, and at the end of our [copy of] Raymond on the elixir drawn from oil of mercury"), fol. 24v. Another note refers to Bolles's *Liber niger*: "Mr Bolles in fine lib. nig." (fol. 69r).

71. Challis, "Mint Officials and Moneyers," 56.

72. On Felley, see Patent Rolls, 30 Henry VIII, pt. 6, m.19, 1 Sept. See J. Charles Cox, "The Religious Pension Roll of Derbyshire, temp. Edward VI," *Journal of the Derbyshire Archaeological*

same time, Dighton sold another part of the Osberton estate to Whalley, and in this way the two men became not merely colleagues, but neighbors and future litigants.[73]

Bolles's proximity to both Eden and Whalley helps to solve one of the puzzles of their story: how he first became aware of their illicit practice. It now appears that Bolles was himself interested in alchemy, if not as a practitioner then certainly as a reader. A manuscript inscribed "William Bolles, possessor," now Glasgow University Library, MS Ferguson 102, provides evidence of his pseudo-Lullian interests.[74] This book is also written by the same scribe who added several treatises toward the end of Eden's manuscript, providing a physical link between the two books.[75]

Whatever the fate of their personal relationship in the wake of the Whalley affair, their books show that Eden and Bolles shared a common interest in the alchemy of Raymond and Ripley. Yet they also differ in significant ways. Written mostly in Latin, Eden's compendium is made up primarily of extracts rather than complete works, while the nature and density of its annotations suggest that he viewed it as a working copy and a platform for further exegetical efforts. Bolles's book, on the other hand, provides an early example of a type of document that would become increasingly common throughout the later sixteenth and seventeenth centuries: the English alchemical compendium, a collection dominated by works in translation as well as those originally composed in English. Such compendia, although they might furnish content for commonplace books and working copies like Eden's, were clearly also intended to supply complete, reference copies of texts. As such, they served a vital role in disseminating alchemical treatises

---

and Natural History Society 28 (1906): 10–43, on 15–16. In October 1552, a commission appointed to investigate the payment of annuities discovered that he had purchased the annuity of a former religious for twenty nobles; ibid., 19–20. On the disposition of Osberton to Whalley and Bolles, see Robert Thoroton, "Osberton," in *Thoroton's History of Nottinghamshire: Volume 3, Republished With Large Additions By John Throsby*, ed. John Throsby (Nottingham, 1796), 401–2; *British History Online*, http://www.british-history.ac.uk/thoroton-notts/vol3/pp401–402 (accessed 14 April 2018).

73. Eden's original deposition, as well as a record of the circumstances under which his practice first came to light, survives because Bolles revived the matter while fending off a lawsuit brought by Whalley in 1556; National Archives, SP 46/8, fol. 168r; Kitching, "Alchemy," 310–11.

74. MS Ferguson 102, fol. 3v.

75. Trinity R.14.56, fols. 80v-83r. The shared hand suggests either that Bolles contributed to both books himself (if the hand is his), or that Ferguson 102 was written by another, unknown scribe, one of whose other books later came into Bolles's possession, and who also contributed several pieces to Trinity R.14.56. It is not certain, however, that texts by the "Bolles scribe" were added to Trinity R.14.56 while it was still in Eden's possession, since these pages are not annotated by him.

outside humanist and scholarly circles, bringing translated texts into the hands of a wide variety of potential practitioners, some of whom repurposed their content with a view to securing royal permission to practice. The manuscripts of alchemical readers like Eden, Bolles, Dee, and Walton reveal the outlines of a corpus that shaped not only the content of English alchemy, but also how its history was perceived.

THE ENGLISH ALCHEMICAL PATRONAGE SUIT

The scandals of the 1550s can hardly have promoted the cause of English alchemy during subsequent decades. Nor were Englishmen necessarily well equipped to compete with practitioners from abroad. During the 1560s, English mining enterprises still relied heavily on "strangers" from overseas, particularly German engineers who possessed the metallurgical expertise that their English counterparts lacked.[76] In the field of alchemy, too, English adepts lagged reputationally behind their Italian and German brethren. The solution—to demonstrate technical expertise through practical means—was further stymied by the old statute against multipliers, which forced practitioners to confess (at least in their correspondence with the queen) that they had not fully attempted the stone. The law thus placed them at a considerable disadvantage compared to foreign alchemists who could point to experience gained abroad. Under the circumstances, it is not surprising that most of the chemists associated with funded Elizabethan projects hailed from overseas, including the first and last alchemical "philosopher" to receive serious Crown sponsorship: Cornelius de Lannoy, a Dutch alchemist educated in Cracow.[77]

De Lannoy floated a variety of medicinal and chrysopoetic benefits in his initial approach to the queen, some of which are detailed in his treatise dedicated to her, *De conficiendo divino elixire sive lapide philosophico* (On Making the Divine Elixir or Philosophers' Stone). Once he grasped that Elizabeth's primary interest was in gold and silver bullion, he responded by proposing an ambitious gold-making schedule, including likely year-on-year yields.[78]

76. Ash, *Power, Knowledge, and Expertise*, chap. 1. On the role played by immigrants, or "strangers," in Elizabethan science, see Harkness, *Jewel House*.

77. De Lannoy introduces his credentials as "philosopher et Iatromathematicae Doctor almae Craconienses academiae" in his letter to Elizabeth I of 9 February 1965; National Archives, SP 12/36, fol. 25r.

78. Cornelius Alnetanus [de Lannoy], *De conficiendo divino elixire sive lapide philosophico* (14 July 1565); printed as "Libellus Elizabetae Reginae Angliae dicatus, tractat de conficiendis duobus olcis pro Elixire diuino ad transmutandum metalla imperfecta," in *Secreta secretorum Ray-*

In the event, this enterprise played out along similar lines to those of the "entrepreneurial alchemists" active in the German lands, whose activities have been charted by Tara Nummedal.[79] An early atmosphere of optimism soon gave way to complaints over inadequate materials and apparatus on the part of the alchemist, and concern over slow progress on the part of the prince, rapidly eroding trust. Less than a year after de Lannoy started work, Cecil came to suspect that the alchemist was planning to fly England with another prospective patron, Princess Cecilia of Sweden, and acted to protect the queen's investment. By July 1566, de Lannoy and his practice had been relocated to the Tower. By February 1567, Elizabeth's experiment with chrysopoeia was effectively over, although the Tower retained de Lannoy's person until at least 1572, after which no more is heard of him.[80]

Even before the patron-client relationship had soured, the regime's apparent willingness to engage with alchemical projects prompted English practitioners from diverse social and educational backgrounds to proffer their own services. Edward Cradock, appointed in 1565 as Lady Margaret Professor of Divinity at the University of Oxford, wrote works in both English and Latin on the philosophers' stone, complete with a lengthy dedication to the queen in Latin verse.[81] Thomas Charnock wrote an even lengthier treatise over the winter of 1565, volunteering to be confined within the Tower as surety for his pledge to manufacture the philosophers' stone.[82] He later recorded his dissatisfaction at learning that his proposal had been set aside while Cecil investigated de Lannoy's claims.[83] The haberdasher Richard Walton probably requested a license in the same year. The 1570s saw a fresh wave of peti-

---

mundi Lulli et hermetis philosophorum in libros tres divisa (Cologne: Goswin Cholinus, 1592), 143–55. The treatise had been translated into English by 1605, when it was copied by Thomas Robson: Ashmole 1418, fols. 43r-47v; another copy is in Sloane 3654, 4r-6v.

79. Nummedal, *Alchemy and Authority*, esp. chaps. 3–4.

80. Barbara de Lannoy petitioned Cecil (now Lord Burghley) for her husband's release in February 1571/2, as identified by Campbell, "Alchemical Patronage," 86; Longleat House, The Dudley Papers, MS DUI, fol. 209r.

81. For Cradock, see Mordechai Feingold, "The Occult Tradition in the English Universities of the Renaissance: A Reassessment," in *Occult and Scientific Mentalities in the Renaissance*, ed. Brian Vickers (Cambridge: Cambridge University Press, 1984), 73–94, on 86; Schuler, *Alchemical Poetry*, 3–48. The latter includes an edition of Cradock's "Treatise Touching the Philosopher's Stone." Cradock's works are preserved in Bodleian Library, MSS Ashmole 1445, pt. 6, and Rawlinson poet 182. Cradock also presented Cecil with the text of a Greek oration in MS Lansdowne 19, fol. 57r (art. 25).

82. Lansdowne 703, fols. 9r-v, 10v-11r, 52r.

83. Trinity O.8.32, fol.44r: "In [th]e yere off our Lorde god .1566. I dyd dedicate a booke off philosophie to Quene Elizabeth and delyveryd him to hir cheiffe secrettorie named secretorye Sicyll: But be cawse the Quene and hir counsell had set some a worke in somerset place in \London/ before I came and had wrought there by the space off one year therefore my booke

tions, including approaches from familiar figures like Dee, Eden, and Blom-fild, as well as suits from Humfrey Lock, an engineer hoping to be recalled from Moscow; Francis Thynne, an antiquarian scholar imprisoned for debt; and Samuel Norton, a Somerset gentleman anxious for preferment.[84]

Given that European authorities led the way in alchemical theory, style, and practice during the first half of the sixteenth century, we might expect to see continental innovations dominate in these English patronage suits, even outside humanist circles. As we have seen, Agrippa was hailed in England as a serious adept from the first appearance of *De occulta philosophia*. "Voarchadumia," a term of art devised by Giovanni Agostino Pantheo to distinguish his own cabalistically infused philosophy from the "alchemy" illegal in his native Venice, was adopted by Dee, but it also had wider appeal, particularly after de Lannoy used it in his own suit.[85] Eden awarded himself the cognomen "Voarchadumus" in his own alchemical compendium, Walton and Thynne adopted the term in their petitions, and even Ripley's *Medulla alchimiae* was retitled *Medulla Warchadumia* in Higgins's translation, suggesting an attempt to avoid the negative associations of "alchemy."[86]

For practitioners without a sound reading knowledge of Latin, or who lacked access to the latest books from the Continent, another source of material for patronage suits lay more readily to hand: in English alchemical compendia, packed with the wisdom of their own medieval antecedents. These books also provided sources of historical information for readers like

---

was layd asyde ffor a tyme: and was put in the Quenes librarie: and in this book I dyd wrete that uppon payne off loseinge off my hede that I wolde do the thinge that all this realme showlde not do agayne: quoth thomas charnocke."

84. On Lock, see Grund, *Misticall Wordes*. On Thynne, see Campbell, "Alchemical Patronage," 45–51; David Carlson, "The Writings and Manuscript Collection of the Elizabethan Alchemist, Antiquary, and Herald, Francis Thynne," *Huntington Library Quarterly* 52 (1989): 203–72; Louis A. Knafla, "Thynne, Francis (1545?–1608)," *ODNB*. Norton is discussed in chap. 7, below.

85. Giovanni Agostino Pantheo, *Voarchadumia contra Alchimiam: Ars distincta ab Archimia, et Sophia* (Venice: Giovanni Tacuino, 1530), fols. 8r-9v. On Dee's use of Pantheo in his *Monas hieroglyphica*, see Deborah E. Harkness, *John Dee's Conversations with Angels: Cabala, Alchemy, and the End of Nature* (Cambridge: Cambridge University Press, 1999), 88–89; Hilde Norrgrén, "Interpretation and the Hieroglyphic Monad: John Dee's Reading of Pantheus's Voarchadumia," *Ambix* 52 (2005): 217–45. De Lannoy refers to the science as "Boarchadamia" (*sic*) in his correspondence with Elizabeth: SP 12/36, fol. 25r; see Campbell, "Alchemical Patronage," 79; Parry, *Arch-Conjuror of England*, 75.

86. Eden places the term after his own name in his manuscript compendium, although it was later deleted: Trinity R.14.56, fol. 1r: "Richardus Edenus Voa[rcha]d[umus]." On Walton's use, see Sloane 3654, fol. 14v; on Thynne's, see "A discourse vpon the Lorde Burghleyghe his Creste," Trinity R.14.14, fol. 69r. On the *Medulla Warchadumia*, see Sloane 1842, fol. 78.

Charnock, Walton, Thynne, and Samuel Norton, whose petitions are all distinguished by attempts to situate their own practice in relation to that of earlier English adepts—and, in some cases, to the patronage of earlier English kings. Just as alchemical reading helped to shape their practical aspirations, so it dictated the form and content of their petitions. Thus, despite Elizabeth's and Cecil's immediate interest in strengthening Crown finances, their petitioners persisted in lauding the medical contributions of alchemy alongside its transmutational value: an approach that can be traced directly to the influence of Raymond and Ripley, and their advocacy of distinct animal, vegetable, and mineral stones.

The approach is exemplified by Walton's efforts to sculpt his own reading matter and practical program into a petition suitable for presentation to the queen. Although his original letter has not survived, a copy is preserved in a seventeenth-century manuscript, Sloane 3654, where a scribal error in the title (misspelling the author's name as "Walker") means that it has previously escaped notice.[87] Walton probably wrote it sometime after October 1565, the month in which he finished transcribing Norton's *Ordinal* into Ashmole 1479—a work he refers to several times in his petition, and which seems to have been fresh in his mind.[88] Given the timing, we can speculate that Walton was responding to news of Elizabeth's known and generous sponsorship of de Lannoy. He seeks more tangible forms of support than a license alone, claiming that "the keeping of my house is so costlie that I shall not be able to goe thorough with the Charges therof."[89] He has lost over £600 in the "time of trouble": perhaps a reference to the difficulties experienced by devout Protestants during Mary's reign. Walton and his wife, Izabell, certainly had their fair share of grief and hardship. Walton's notes on a spare leaf of Ashmole 1479 show that nine of his fourteen children born between 1548 and 1566 did not survive infancy, possibly victims of the great plague epidemic that devastated the city in 1563.[90]

---

87. Sloane 3654, fols. 14v-17r. Walton is unmistakably identified in the colophon, as "your humble Subiect Richard waltone haberdasher dwellinge in the ould Change by Paules Church." This manuscript also contains the English translation of de Lannoy's treatise to Elizabeth I, *De conficiendo* (fols. 4r-6v), suggesting that the compiler had access to royal petitions. I have not identified any copy of Walton's work in the state papers.

88. Ashmole 1479, fol. 300r; Sloane 3654, fols. 15v-16v.

89. Sloane 3654, fol. 16r.

90. Ashmole 1479, fol. 222v: "The yeare and [th]e daye of [th]e byrthe of all [th]e chullderne of Rychard Waultowne alias Walton & Izabell hys wyffe whyche 2 were maryed [th]e 14 daye of auguste in anno domini 1547." Walton states that he was age forty-eight at the time of drawing up the list. On episodes of plague in Elizabethan London, see Paul Slack, *The Impact of Plague in Tudor and Stuart England* (Oxford: Clarendon Press, 1990).

Walton's *Letter* is the epitome of a vernacular patronage suit of this period, and a model of what we can by now characterize as a distinctively "English" approach to alchemical practice. Not only is it written in English, but it is composed on the basis of the same vernacular sources, including translated texts, that Walton had previously assembled in his own alchemical compendium. These comprise primarily English authorities like Bacon, Ripley, and Norton, although Raymond also occupies a place of honor in both the *Letter* and Ashmole 1479. The suit is also modeled on an English exemplar: Ripley's appeal to the bishop as set down in the *Marrow*, portions of which Walton silently paraphrases. He even incorporates Ripley's estimate of the quantity of gold required for the vegetable work, calculating that the canon's "lesse pound" of gold would equate to seventy-two angels in coin. He subsequently uses this figure as the basis for costing his own proposal, concluding that the entire work will cost £244, of which he requests a £100 subsidy from the queen.[91]

The result is a suit firmly grounded in the alchemy of Raymond and the Ripley Corpus, particularly in the contrary operations of the two "fires" that provide the technical core of Ripley's *Medulla*. In Ashmole 1479, in addition to transcribing and annotating texts, Walton attempts to set down concrete procedures for the two menstrua, "The true makyng of [th]e fyre of nature, [th]e w*hich* ys a quyntessence" and "[Th]e p*erfecte* makyng of [th]e fyre agaynste nature after Raymond," as well as an "exposytion" of Ripley's four fires extracted from the *Marrow*.[92] These passages inform the practice set down in his *Letter* to Elizabeth, where, in a passage larded with pseudo-Lullian terms (the natural fire, the Green Lion) he advertises the value of the quintessence as both "the elixar of life and [th]e Elixer of mettallis."[93]

Whatever he may have lacked in formal education, Walton accepted that mastery of *Decknamen* was an essential part of acquiring this philosophical craft. While annotating Ripley's *Compound* (another source for the *Letter*), he observes that "every syence hathe hys *proper* tearmes"—a quotation from Norton's *Ordinal*.[94] Where Ripley warns that the stone has an infinity of names, Walton points the moral: "therfore of e*very* tearme learne [th]e meanyng of them or ellse you shallt be no ph*ilosoph*ors but a broyler."[95] Yet he does not romanticize the art, and is not coy about suggesting amendments where appropriate. While Ripley recommended using an athanor

91. Sloane 3654, fol. 17r. On his source in the Medulla, see p. 112, above.

92. Ashmole 1479, fols. 33r-34r, 52v, and 218v-19r, respectively.

93. Sloane 3654, fol. 17r.

94. Ashmole 1479, fol. 8v; *Ordinal*, 55 (l. 1730).

95. Ashmole 1479, fol. 8v; Walton's note on Ripley, "Solution," stanza 21.

to conserve heat, Walton dismisses this technology in favor of more recent innovations: "other furnyssys inventyd sence more profytable then ever was [th]e antynor."[96] Like a philosopher, he also read his books one against another to resolve doubts. When Ripley's process for the philosophical mercury seemed too arduous, Walton reminded himself to "loke in norton" for a comparison.[97] Studying the *Philorcium*, he was surprised to see that the recipe for a vegetable and mineral water differed from the *Medulla*'s famous process for the compound water, noting that "thys composition ys nothyng lycke yt in [th]e marye [i.e., *Marrow*]."[98] But he also read texts against his own practice, striking out common salt from his recipe for the fire against nature, on the grounds that "salte wyll hurte your work."[99]

Walton's engagement with his fourteenth- and fifteenth-century sources shows us a would-be English alchemical philosopher at work: acquiring and studying texts, testing their contents, and reflecting on their interactions. Such activities cannot have been carried out in isolation, but probably took place within a community of readers whose contours we can now only sketch. For instance, by February 1561/2 Walton had secured what must have been a very recent English translation of the prefatory verses to the *Medulla*, "put in myttor by wylliam bolisse"—almost certainly the Mr. Bolles who, as we have already seen, was keenly interested in the same Lullian and Ripleian sources as Walton.[100]

Landowners like Bolles, however, were able to draw an income from their estates. Walton struggled with the difficulty of reconciling his expensive and time-consuming practice with the demands of his own trade, "for he that doth this can do nothing ells."[101] Patronage was therefore essential. He argues as much to the queen:

> I finde that in the Aunctient time, that none were admitted to the studdy of the 7 liberall sciences, but such as were noble or rich merchantes sonnes such as their freindes were able to leaue vnto them of their wordly goods sufficient to their liuing that they might giue their Mindes wholly to their

---

96. Ibid., fol. 14r; note on Ripley, "Putrefaction," stanza 7.

97. Ibid., fol. 8v; note on Ripley, "Solution," stanza 22.

98. Ibid., fol. 313v.

99. Ibid., fol. 52v.

100. Ashmole 1479, fol. 32r. Ashmole later included Bolles's translation, although without attributing it to him, in the *Theatrum*; *TCB*, 389–92.

101. Sloane 3654, fol. 17r.

studdy, and to the practize of those things by which by their learning they had the speculation.[102]

Possibly he has Raymond's advice from his own Englished copy of the *Testamentum* in mind—that alchemy can be pursued only by those endowed with money, wisdom, and books. Yet Walton avoids referring to the monastic life that had formerly provided religious practitioners with the solitude required for both practice and speculation. Instead, he chooses to historicize Raymond's aphorism by grounding alchemy in an ancient, pagan, and secular past. If God permitted such profitable "speculations" even for heathens, he suggests, it must be apter still for Christians to apprehend the holy science of "Coarchadumia or Alchimia." A license and royal investment would allow him to follow in the footsteps of those ancients, with the added advantage of Reformed religion on his side. Through such maneuvers, executed with a view to acquiring patronage, English alchemy acquired its history.

ALCHEMICAL CHRONICLING

The first "history" of English alchemy is written into a patronage suit: the work of the alchemist and unlicensed medical practitioner Thomas Charnock. Born in Faversham in Kent, Charnock later moved to Somerset near his wife's hometown of Stockland Bristol. He is now best known for his celebrated *Breviary of Natural Philosophy*, finished on 1 January 1557/8: a poem that differs from Blomfild's *Blossoms* in its autobiographical and discursive style, which takes Norton's *Ordinal* as its model rather than Ripley's more doctrinally focused *Compound*.[103] Charnock's preference for anecdote and racy narrative also characterizes his *Booke Dedicated vnto the Queenes Maiestie*, written over the winter of 1565/6: a work that probably never made its way to Elizabeth, but was instead filed among Cecil's papers.[104]

A self-described "unlettered scholar," Charnock understood the necessity of book learning as the key to both practical success and the fashion-

102. Ibid., fol. 14v.

103. Thomas Charnock, "Breviary of Natural Philosophy," in *TCB*, 291–303.

104. Lansdowne 703. On the discovery and contents of this manuscript, see Alan Pritchard, "Thomas Charnock's Book Dedicated to Queen Elizabeth," *Ambix* 26 (1979): 56–73. On Charnock, see also F. Sherwood Taylor, "Thomas Charnock," *Ambix* 2 (1946): 148–76; Robert M. Schuler, "Charnock, Thomas (1524/6–1581)," *ODNB*. Charnock himself gives three different dates for the composition of the *Booke*: 25 November 1565 for the prefatory epistle to the queen; 5 December 1565 for the commencement of the "Confabulation"; and 1566 for the actual sending of the manuscript to Elizabeth, according to the note referenced in note 83, above.

ing of a philosophical persona. Hampered by the availability of texts as well as his own restricted Latinity, his solution was to compile a history primarily from vernacular, alchemical sources, producing an idiosyncratic narrative that owes more to the tradition of medieval chronicles than to humanist histories.[105] Indeed, Charnock seems to have valued alchemical treatises and poems in part for the light they shed on the activities of past philosophers—a history, he notes, "which I coulde neuer rede in no yngelyshe chronicles."[106]

The *Booke* presents his findings in the form of a "Confabulation" between the alchemist and a learned visitor from Oxford, whom Charnock regales with stories of past alchemists, as well as details of the "enterpryse" he hopes to carry out on the queen's behalf. The format allows Charnock to present himself in the mold of well-read humanist philosophers like Dee or Eden, rather than a compromised alchemist of the type lampooned by Ripley, Norton, and Chaucer, whose satirical verses he cites at length. To that end, he has the Oxfordman compliment him on the outfitting of his study with globes, maps, and other scientific instruments, as well as "a fayre liberary off bookes & well augmentid sends [i.e., since] I was here last."[107]

For a provincial practitioner, securing such a library was no sinecure. In his treatise, Charnock claims that the only books he ever obtained from London were those he inherited at the age of twelve from his uncle, also called Thomas Charnock, a Dominican doctor of divinity based in Blackfriars, Ludgate.[108] Since then, his hopes of visiting London to purchase further alchemical titles have been thwarted by press of business. Yet Charnock's own books suggest that he was resourceful in acquiring material. Several items from his "fayre liberary" survive, allowing us to read his petition, like those of Eden and Walton, in light of his own collecting and annotating practices.

Between 1562 and 1579 Charnock annotated several fifteenth- and early sixteenth-century compendia, now preserved in Trinity O.2.16 (pt. 1) and

---

105. On the structure and narrative style of English chronicles, see Chris Given-Wilson, *Chronicles: The Writing of History in Medieval England* (Hambledon: Hambledon Continuum, 2004). On the early modern decline of the format, see D. R. Woolf, *Reading History in Early Modern England* (Cambridge: Cambridge University Press, 2000), chap. 1.

106. Lansdowne 703, fol. 21v.

107. Ibid., fol. 7v.

108. Ibid., fol. 25r. *Alumni Oxonienses* records "Thomas Charnoke" as "Dominican; B.D. 15 June, 1528, D.D. 8 April, 1530." "Chaffey-Chivers," in *Alumni Oxonienses, 1500–1714*, ed. Joseph Foster (Oxford: Parker and Co., 1891), 255–73; *British History Online*, http://www.british-history.ac.uk/alumni-oxon/1500-1714/pp255-273 (accessed 1 February 2018).

O.8.32 (fig. 10).[109] These show that although he regarded himself as a man "gyven to great soliterines," Charnock was by no means isolated when it came to obtaining alchemical material.[110] One booklet previously belonged to John Mayre, probably the mint-assayer shamed by Blomfild, who signed and dated it in 1560, just two years before it came into Charnock's possession.[111] Another is copied from Du Wes's compilation in Trinity O.8.24, including the Ripleian *Cantilena* and other illustrated texts that Charnock embellished with his own notes.[112] Several recipes are in the hand of John Coch, vicar of Stockland, whom Charnock mentions in the *Booke* as a collaborator on medical matters.[113] A compendium of pseudo-Lullian texts once belonged to Richard Atkins, an otherwise obscure figure who also owned a fifteenth-century astrological compendium now in the Ashmole collection.[114]

The nature of his sources and the scattershot results of his collecting helped to shape Charnock's view of alchemical history. The *Booke* relates the adventures of English adepts, most of whom were members of religious communities. Charnock reconstructs this history primarily from earlier writings, including the satirical passage on false alchemists in Ripley's *Compound*, and Norton's story of Thomas Daulton (a monk abducted for his alchemical knowledge) from the *Ordinal*, which he treats as factual accounts. Yet his ability to provide a complete history is limited by the difficulty of obtaining suitable texts, and he laments the lack of good copies of "manye engelyshe philosophers" whose proficiency he is unable to verify, "because I have not their bookes hole and perfet, whereby I ame not able to make my avctoritie good." He lists the books that he hopes to obtain in order to make "a greatter collection," such as a complete copy of the work of Merlin—not, he points out, the false prophet of Wales or Scotland, but Merlin the alchemical philosopher, of whose book "I could neuer gett but a vj. or viiij.

109. Ashmole later made copies of Charnock's notes in these manuscripts; see chap. 9, below.
110. Trinity O.8.32, fol. 102r.
111. Trinity O.2.16, pt. 1, fol. 82r.
112. Ibid., fols. 25r-31v.
113. Ibid., fol. 79r: "for makyng of Cerys [i.e., ceruse] per *Master* Coche." The hand resembles that of Coch's medical compendium in Harley 1887. Charnock mentions him in Lansdowne 703, fol. 47r, as "that learned man Mr Jhon' Coche vycker off stockelande and parson of coussenton, with whome I dyd confer in phisicke."
114. Trinity O.8.32, fol. 1r: "Richard Atkins his book." Cf. Ashmole 391, fol. 1: "Richard Atkins his boocke."

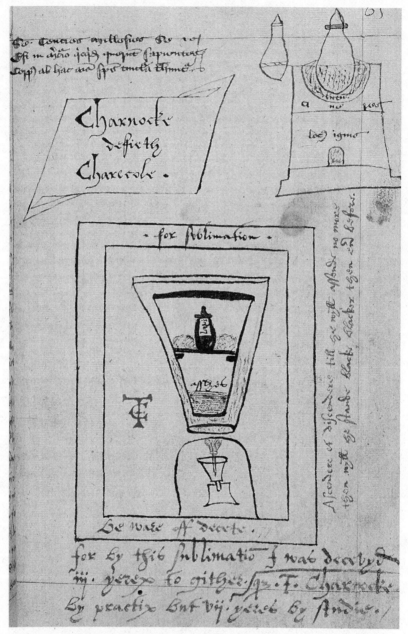

FIGURE 10. "Charnocke defieth Charcole." Thomas Charnock's notes and drawings of furnaces added in the waste spaces of a fifteenth-century manuscript. Cambridge, Trinity College Library, MS O.2.16, fol. 65r. By permission of the Master and Fellows of Trinity College, Cambridge.

leves."[115] Like his rival poet, Blomfild, Charnock was concerned to distance his own activities from any hint of magic or superstition, instead reserving magical practices for Catholic monks and friars like Bacon, whose history (strongly influenced by later legends of Bacon as a notorious conjuror) he unfolds in some detail.[116]

Charnock's attempt to distinguish between the multiple identities of Merlin—the prophet and magician of British legendary history, but also the author of a distinguished set of Latin alchemical verses—illustrates the problems faced by practitioners who relied on books for both their knowledge of alchemical practice and their understanding of its past. For readers who treated alchemical texts as legitimate sources of historical information, the pseudonymous attributions of many core treatises resulted in an inadvertently skewed chronology. Such pseudepigrapha generated an alternative past, analogous to the legendary history of Geoffrey of Monmouth and his medieval successors, or the productions of ecclesiastical historians like Matthew Parker who scoured British antiquity for evidence of an ancient, pristine English Church. Within this largely fictive "alchemical history," the existence of verses attributed to Merlin implied that alchemy had been practiced in the British Isles long before the twelfth-century translation of Arabic texts into Latin, while the prefaces of pseudo-Lullian and Ripleian texts suggested regular collaboration between philosophers and historical English kings.

Not all interventions were accidental. For all his desire for authority, Charnock had no compunction in inventing or elaborating episodes in the interest of furthering his own patronage aspirations. His suit offers Elizabeth several examples of royal sponsorship of alchemists, from Edward IV in England, who adopted Norton as his philosopher, to Emperor Charles V, who treasured the alchemical expertise of Cornelius Agrippa more for "the roialltie off the thing" than for "the yerely value off golde."[117] Such collabora-

<hr/>

115. Lansdowne 703, fol. 24v-25r. Charnock is referring to the twelfth-century tradition of two Merlins: the Welsh Merlinus Ambrosius and the Scottish Merlinus Silvestris or Caledonius, both derived from Geoffrey of Monmouth's version of the Myrddin myth in the *Historia regum Britanniae* and *Vita Merlini*. On the development of the Merlin legend, see J. S. P. Tatlock. "Geoffrey of Monmouth's Vita Merlini," *Speculum* 18 (1943): 265–87. On the alchemical Merlin, see Didier Kahn, "Littérature et alchimie au Moyen Age: De quelques textes alchimiques attribués à Arthur et Merlin," *Micrologus* 3 (1995): 227–62.

116. Lansdowne 703, fols. 16r-18r. The passage on Bacon is evidently influenced by contemporary works like *The Famous Historie of Fryer Bacon*, an anonymous, midcentury treatise that initially circulated in manuscript, and that presented Bacon as a master conjuror who (as in Charnock's version) eventually repented of his dealings with spirits. See George Molland, "Bacon, Roger (c.1214–1292?)," *ODNB*.

117. Lansdowne 703, fol. 38v.

tions culminate in the person of Henry VII, a philosopher-king who (Charnock assures Elizabeth) had the stone in his own right:

> I meane the Queenes most excellent graces gravndefather Henry the vij.
> who was a great and a wyse philosopher, and had this stone a .v. yeres before
> he dyed off a iove bowerne [i.e., a Jew born], as I have learned owt off my
> vncles bookes off this same science . . . who as he writes was the kyngs con-
> fessor [th]e last yere off his maiesties raigne.[118]

While many patronage suits gesture toward alchemy's philosophical lineage, none is as explicit as Charnock's in forging historical links between philosophers and kings, nor as overt in presenting the writer as an integral part of that history. Through the alleged association between his uncle and Henry VII, Charnock wrote his family into English alchemical history: flattering Elizabeth through the compliment to her grandfather, while setting up an obvious analogy with himself and the queen whose philosopher he hoped to become. Dee or Eden might have winced at such crude methods, but for a practitioner like Charnock, lacking courtly connections or Latin scholarship, concocting alchemical histories may have offered the only plausible route to preferment: a link to the past made concrete through his inheritance of books—and, by implication, skill.

PROVIDENTIAL HISTORIES OF ALCHEMY

If Charnock gives us a "chronicling" kind of alchemical history, largely grounded in the activities of English monks, then another history of a more "providential" type can be elicited from the petitions of Walton and Blomfild, practitioners of a puritanical bent who sought to legitimize their science with reference to scriptural authority rather than royal precedent. While Charnock scoured his sources for connections between alchemists and kings, Walton came to a different conclusion: that no English prince had ever succeeded in acquiring the stone, because God granted it only to poor and humble men.

As one of the overriding themes of both academic theology and popular preaching in later sixteenth- and seventeenth-century England, providentialism inflected all aspects of English cultural life, including the history of England itself. Alexandra Walsham has pointed to the providential view of

118. Ibid., fol. 22r.

history captured in Holinshed's *Chronicles* and other Elizabethan histories and martyrologies, in which both political events and individual misfortunes were read as evidence for God's will and direct intervention in creation.[119] This presentation of history, exemplified by Foxe's *Book of Martyrs*, served Protestant ends by presenting the Reformation as divinely ordained, but it also suggested a special role for alchemical knowledge. While the philosophers' stone was a wholly material product, the difficulty of its manufacture implied the hand of providence at work: a variation of the long-standing *donum dei* trope that was particularly appropriate for Protestant alchemists.

In his own day, and in Protestant England, Walton seemed to envisage just such a providential order, whereby God extends the true knowledge of alchemy to humble subjects rather than to kings directly:

> For I haue not redd since the conquest yt any manner of kinge had it. Although that raymond lully was in kinge Edward the 3des time, and Ripley in king Edward the 4th time, and Norton in [th]e latter end of king Edward the 4th & kinge Richards time, and king henerie the seaunths time, yet none of all these taught itt to any of these kings. . . . Thus it pleaseth god to bestow his great grace vpon pure and simple men such as semeth the outcasts of all men . . . that they should cast themselues downe att the feete of ther princes and commonwelth.[120]

Under these conditions, Walton's own lowly status counts in his favor by making him a likely candidate for grace. It also appears from this passage that Walton was unaware of (or deliberately chose to ignore) the burgeoning legends that linked past adepts like Raymond and Ripley to English kings.[121] On the other hand, he readily concluded from the existence of alchemical works attributed to Mary the Prophetess that "our lord God himselfe of his great mercye did giue [the secret] to Moyses, ffor I haue found that Mary

119. Alexandra Walsham, "Providentialism," *The Oxford Handbook of Holinshed's Chronicles*, ed. Felicity Heal, Ian W. Archer, and Paulina Kewes (Oxford: Oxford University Press, 2012), 427–42; Walsham, *Providence in Early Modern England* (Oxford: Oxford University Press, 2001). See also Nicholas Popper, *Walter Ralegh's "History of the World" and the Historical Culture of the Late Renaissance* (Chicago: University of Chicago Press, 2012).

120. Sloane 3654, fol. 14v.

121. Since Ashmole 1479 does not include the *Epistle to Edward IV*, Walton may have been unaware of the tradition connecting Ripley to Edward. On the other hand, the colophon in his English copy of the *Testamentum* states quite clearly that the work is sent into the custody of "Kyng edwarde of Woodstok," suggesting that he was either unusually skeptical with regard to the attribution, or else chose to ignore it: Ashmole 1479, fol. 215r.

the sister of Moyses made a booke of the same science."[122] Such connections, mined from his reading, allowed Walton to frame himself and his work not just in relation to earlier English philosophers like Ripley and Norton, but also to the overmastering authority of God's word.

Walton may have felt that claiming alchemy as a gift of God helped to mitigate his presumption in approaching the queen directly, but it also reveals how English writers adjusted their conception of alchemy in line with Protestant doctrine. Walton notes that God repeatedly singles out poor men to carry out his tasks, and punishes them when they falter: for Christ "will not haue [tha]t any man shall hide his candle vnder a Bushell."[123] Those to whom God grants alchemical secrets cannot, accordingly, bury their talent like the unprofitable servant, as condemned in Matthew 25—a standard trope of craft petitions since Theophilus's *De diversis artibus*, but one that Walton here adjusts to demonstrate his commitment to the doctrine of salvation by grace rather than by works.[124] According to the doctrine, satisfactory performance alone cannot guarantee salvation, but men should nonetheless rejoice in doing well ("not that we meritt in doein[g]e good"). Their comfort rests in God, who bestows great gifts even upon his unprofitable servants—"nott for our deserts but for his owne sake," in and through Christ's mercy.[125] Modesty aside, Walton's identification of himself with the profitable servant implies that he viewed himself as a recipient of God-given knowledge, and hence as one of the fortunate elect.

That the elect philosopher had a role to play within both alchemical history and sacred history appears from other Protestant alchemical tracts. Thus, for Blomfild, religious and alchemical doctrine intertwine. Even the ability to comprehend the invisible potential of matter is proposed as an indicator of elect status, since those who lack grace will fail to grasp the secrets of nature, and hence fail in practice also. Grace is thus correlated to exegetical ability, as he hints in the *Blossoms*:

> From god it commeth, & god maketh it sencible
> To some preelect; to other doth it denay.[126]

122. Sloane 3654, fol. 14v.

123. Ibid., fol. 15r. The proverb is taken from Matthew 5:15 (Bishop's Bible): "Neyther do men lyght a candell, and put it vnder a busshell: but on a candelsticke, and it geueth lyght vnto all that are in the house"; cf. Mark 4:21; Luke 8:16.

124. Sloane 3654, fol. 15r.

125. Ibid., fol. 15r-v.

126. Blomfild, *Blossoms*, 23 (ll. 61–62). On Blomfild's Puritan inclinations, see Schuler, "William Blomfild," 82–85.

Blomfild deploys the "unprofitable servant" trope in three of his prose works—*An Incomparable Work,* the *Practicke,* and the *Regiment of Life*—as justification for approaching Henry VIII and Elizabeth I, but also in support of his claims to both practical and spiritual prowess.[127] It is particularly marked in the *Regiment* (the only one of his treatises composed under a securely Protestant regime) where his offer to share the secrets of medicinal alchemy rests on his claim to have "receyuyd [th]e talente not at man*n*es hand, but only of god, the wich is gyuer of al con*n*yng."[128] This claim of direct revelation of alchemical knowledge has its parallel in Blomfild's account of his religious practice: specifically, "soch godly excercises that I had in hand for a tyme."[129]

It is important to note that, while acknowledging the divine origin of alchemical "cunning," none of our petitioners treats the actual practice of alchemy as involving anything other than material processes. The religious dimension of their work relates explicitly to the superior level of insight required to construe philosophical texts, and the care and skill needed to reconstruct their contents successfully. These practitioners do not claim that the stone itself possesses any kind of supernatural power; in fact, Blomfild's most religiously imbued work, the *Regiment,* does not even describe the stone, but offers a set of medicinal recipes based on the quintessence.

Of course, physical alchemy might still serve religious ends in other ways—most urgently, by preserving the life of the Protestant queen. Elizabeth's health was a matter of universal interest to her subjects, particularly during the 1570s when the strongest claim to the English throne rested with a Catholic claimant, Mary, Queen of Scots. Such considerations could only enhance the value of alchemical medicines, ranging from the universal stone, which Charnock hoped might serve "for the presarvinge off her graces roiall lyffe, in health,"[130] to specific remedies like potable gold, which Blomfild viewed as particularly suitable for a queen, since "Special things of price to

127. Sloane 2170, fol. 59v: "If I showlde not vtter this knowledge, that god hathe gyven mee, I showld be condempned with the vnfrutefull servante, that \hid/ his lord*es* talent. in the grownde"; Ashmole 1492 (152): "one for cause of discharge of my Conscience least I should be reproved of Chryste with the vnprofitable servant to hide my Lord*es* talent vnder the earth"; *Regiment of Life,* fol. 2v: "And bycause I shuld not be iudged wit*h* the vnprofitable seru*a*nt to hide my lordes talent in [th]e erth. I wyl now vncou*er* [th]e secret hid tresure."

128. *Regiment of Life,* fol. 2v.

129. Ibid., fol. 1v. Blomfild seems to have been engaged in "exercises" or "prophesyings" based on the free interpretation of scripture, a practice of Puritan ministers that was later condemned by Elizabeth. See Schuler, "William Blomfild," 83–84.

130. Lansdowne 703, fol. 39r.

special princely parsonages belong."[131] Seeking to persuade Elizabeth, Charnock framed the stone as a treasure equivalent to three shiploads of gold, jewels, and spices brought back from the Americas and East Indies: its value lying not so much in its gold-making power, "but that it is the greattest cordiall in [th]e worlde."[132]

In allowing petitioners to address the queen directly, license applications offered an opportunity for devout Protestants to voice such concerns—and, for those who felt that Elizabeth was too moderate in her reform of the church, to offer their own opinion on religious matters. Blomfild takes full advantage of this opportunity in the *Regiment*, the most confident of his works, and the one most strongly inflected by his Puritan radicalism. His preface, addressed to Elizabeth, installs her within a firmly providential history, in which God has "by sondry wayes, as by rebellion, clansed & purged you, [tha]t you shuld bryng forth more frute."[133] Yet his praise is tempered by an unmistakable hint that God's favor may also be retracted. He exhorts the queen to persecute Catholics, reminding her of the fate of the biblical king Ahab who spared the life of the Assyrian king after defeating him in battle, only for the Lord to strike him down in his enemy's place.[134] Written around 1574, after the Catholic-orchestrated Ridolfi Plot, it is difficult not to see this passage as an allusion to Elizabeth's perceived clemency toward Mary, whose execution was already an object for the hotter sort of Protestant.[135] Although Blomfild hoped that his petition, accompanied by the gift of medicinal recipes, would help him to regain his parish and resume his "godly exercises," such imprecations were unlikely to further his cause with the more conservative queen—although they do preserve a taste of the forthright qualities that doubtless contributed to his rejection by his former parishioners.

AN ASPIRATIONAL SCIENCE?

In August 1572, Richard Eden came up against the hard edge of the wars of religion when he and his Huguenot patron narrowly escaped with their

131. *Regiment of Life*, fol. 10v.

132. Lansdowne 703, fols. 4v-5r.

133. Ibid., fol. 7r.

134. Ibid., fol. 8v; referring to Kings 3 20:28–43 (Bishops Bible), particularly 20:42: "Thus sayth the Lorde: Because thou hast let go out of thy hande a man that is in my curse, thy lyfe shall go for his lyfe, and thy people for his people."

135. Pius V's bull *Regnans in excelsis* excommunicated Elizabeth I in 1570, heightening anti-Catholic sentiment in England and increasing concern over Mary Stuart's strong claim to the English throne. Carol Z. Wiener, "The Beleaguered Isle: A Study of Elizabethan and Early Jacobean Anti-Catholicism," *Past & Present* 51 (1971): 27–62.

lives during the St. Bartholomew's Day massacre. "Losse of goods, and ᵔ ger of lyfe," he later noted, "hath dryuen me home agayne into my nat countrey."[136] Returning to England, he was irritated to find that foreigne still appeared to be enjoying favors that he had been denied throughout hiᵓ own career. In September 1573, he made a final bid for royal support in his alchemical endeavors, by requesting a license to pursue the manufacture of Paracelsian remedies:

> Who can at this present time, in England, compound the admirable medica-
> ments of Paracelsus from metals and minerals (which are symbolized by an
> alchymical method), without immediately incurring from ignorant calum-
> niators the infamy and perils of practising alchemy, which is prohibited by
> the laws. To obviate this evil, a royal licencse is needed. Neither do I doubt
> that since it is permitted to Brocardus and many other foreigners freely to
> practise [the art]; that the same will, with more justice, be granted to me by
> the royal authority.[137]

Eden's comment reflects the ongoing perception among English practitioners that strangers were preferred in chemical and medical matters—
ironically, a complaint voiced in the context of his own bid to pursue
Paracelsian practices imported from the Continent.[138] The aura of foreign
competition sharpened the emphasis on England's own alchemical tradition
that is apparent in so many English patronage suits. Yet concern over rivals
from abroad does not explain away the genuine practical interest in pseudo-
Lullian and Ripleian alchemy in the second half of the sixteenth century,
which English practitioners continued to explore alongside the more recent,
continental innovations of Agrippa, Pantheo, and Paracelsus. This success
rested in part on the ability of sericonian alchemy to reflect or absorb such

---

136. SP 92/32, translated by Arber, "Life and Labors," xlvi.

137. Richard Eden, dedication to Sir William Winter, in Jean Taisner, *A Very Necessarie and Profitable Booke Concerning Navigation*, trans. Eden (London: Richard Jugge, [1575]); cited in Arber, "Life and Labors," xlvi.

138. Campbell suggests that his particular target may have been the German physician and mining specialist Burchard Kranich (ca. 1515–1578), the same "Dr Burcot" who was later credited with preserving the queen from smallpox in 1562. Campbell, "Alchemical Patronage," 124–25. On Burchard, see also M. B. Donald, "Burchard Kranich (c. 1515–1578), Miner and Queen's Physician, Cornish Mining Stamps, Antimony, and Frobisher's Gold," *Annals of Science* 6 (1950): 308–22; Donald, "A Further Note on Burchard Kranich," *Annals of Science* 7 (1951): 107–8. Another possible candidate, the Italian Protestant Jacopo Brocardo (anglicized as James Brocard), who wrote on Christian Cabala, was not in England until around 1580: Antonio Rotondò, "Brocardo, Jacopo," *Dizionario biografico degli italiani* 14 (1972), http://www.treccani.it/enciclopedia/iacopo-brocardo_(Dizionario-Biografico)/ (accessed 1 May 2018).

adaptations: the blending of medical distillation and metallic ingredients serving as the basis for further experimentation, including increased attention to the use of tartar, antimony, and other readings of "sericon." As a practical approach, this was one that could also be attempted by practitioners at all levels of the social hierarchy, thanks to the accessibility of core texts in English, and the relative availability and low cost of most of its ingredients.

One of the most remarkable aspects of English alchemy in this period is that, even in the absence of courtly connections, practitioners felt empowered by their possession of alchemical knowledge to approach Elizabeth directly. They recognized that their social status and educational background were not sufficient to compel the queen's attention; nor, in the shadow of the statute, could they claim to have physically completed work on the stone. Rather, they sought to convince readers that their humble outward appearance concealed privileged access to philosophical secrets, whether acquired through expertise in natural philosophy, ingenuity in alchemical reading, or the special grace of God. In contradistinction to their foreign rivals, they pressed the Englishness of alchemy itself: a tradition of knowledge retrieved from the hints and fragments encountered in their reading, and woven into whole cloth in their own ingenious petitions. As practitioners turned increasingly toward Paracelsian chemical medicine in the last quarter of the century, they did so within this frame: at once medicinal and chrysopoetic, medieval and Reformed, and persistently sericonian.

PART THREE

The Legacy of Medieval Alchemy in Early Modern England

CHAPTER SEVEN

Recovery and Revision

At the last better waighinge my Master Ripley his words, I larned to stand
vpright, wher I was wont to fall, for he it is whose only hand hath Rowled
away the stombinge stone wherat men vsially fell.[1]

At St. John's College, Cambridge, on 20 July 1577, the Somerset alchemist
Samuel Norton (1548–1621) dedicated his *Key of Alchemie* to Elizabeth I. In
doing so, he joined the growing number of Englishmen who sought to legitimize and fund their alchemical practice by obtaining royal patronage. Like
other English practitioners, Norton's conventional professions of expertise
were supported by invoking the authority of his predecessors, the great
fifteenth-century adepts George Ripley and Thomas Norton. Yet Samuel also
claimed to have access to a source of knowledge unavailable to his peers. In
addition to Ripley's well-known writings, Samuel's *Key* presented findings
discovered in an old commonplace book, "thought to bee the hand writing
of Mr George Rypley Chanon."[2]

Norton's rediscovery of Ripley's *Bosome Book*, a compendium crammed
with texts, recipes, and verses, and liberally dotted with the initials "G.R.,"
underscores Ripley's recognized importance in the 1570s, while ushering in a
new and more intense period of engagement with the English canon's works.
In this context of heightened interest in alchemical history, the appearance
of the *Book* was a thrilling find, both as a valuable antiquity and as a repository of practical advice and secrets accumulated by a famous English philosopher who was presumed to have made the stone. These factors gave the
contents of the *Book* tremendous prestige as objects of exchange, and copies
quickly proliferated. In a process that has so far gone unnoticed, since it took

---

1. Samuel Norton, *Key of Alchemy*, Ashmole 1421, fol. 173r.
2. Sloane 2175, fol. 148r.

place almost entirely through scribal copies rather than print, the influence of the *Book* rippled outward across Europe.

With the *Book*, the themes of alchemical history, practical exegesis, and patronage fuse into a single stream. This relic of England's past prompted attempts at experimental reconstruction, as practitioners sought to replicate Ripley's experiments and legendary results—effects that Norton also hoped would prove worthy of royal interest. His find was the catalyst for a new wave of translations and transcriptions both in England and abroad, which, extending well into the seventeenth century, left a lasting mark on Ripley's alchemical reception in print and manuscript. At the same time, Norton's enthusiastic use of Ripley illustrates how powerfully English authority might be put to work in the service of both state and subject, as a repository of tried-and-tested experimental information that could be replicated and adapted for the benefit of queen and country.

For historians, Ripley's lost book offers a way into the alchemy of the fifteenth century: in particular, insights into his influential but obscure masterpiece, the *Compound of Alchemy*, and the life of the man himself. But there is a twist in this tale of alchemical recovery. The original *Book* does not survive, except in copies and translations made by alchemical enthusasists of the Elizabethan and early Stuart age, none of which are identical, and all of which can be assumed to include both omissions and interpolated material. This means that our own reconstruction of Ripley's experimental practice is mediated through the experience and expectations of his early modern readers: those for whom Ripley was already a revered authority. Reading the *Book* thus requires us to look at it through their eyes, and to assess its value and impact in terms of their priorities. For these readers, the *Book* provided a window into alchemical history, while also offering new opportunities to participate in that history—by reconstructing Ripley's own experiments and, in the process, sculpting a new tradition of English alchemy.

ASSIMILATING MEDIEVAL ALCHEMY

After the failure of Cornelius de Lannoy's alchemical enterprise, the 1570s saw a fresh wave of suits from alchemical practitioners addressed to Elizabeth I and William Cecil, the latter ennobled in 1571 as Baron Burghley. Some of these may have been prompted by evidence of courtly interest in chemical medicine and unusual metallurgical projects. From 1571 to 1576, the Society for the New Art financed an iron-to-copper transmutation process devised by the alchemist William Medley, with Cecil and Sir Thomas Smith as major investors. The Crown also invested in Martin Frobisher's return voyage to North America, after a black ore brought back in 1576 was pronounced gold-bearing.[3] Cecil's known curiosity about transmutation and alchemical remedies encouraged new approaches, sometimes precipitated by personal crises, or prompted by the discovery of intriguing new documents.

These petitioners did not necessarily seek to preside over major chemical enterprises. For instance, the alchemist and antiquary Francis Thynne desired relief from habitual financial dire straits, but his longer-term goal was to secure a position commensurate with his scholarly aspirations. His presentation volume addressed to Burghley, which includes several discourses on coats of arms (including Cecil's own) as well as an elaborate alchemical poem, therefore tends to emphasize his antiquarian interests over his practical knowledge of alchemy. This gift may have proved its worth in March 1576 when, imprisoned for debt in the White Lion prison in Southwark, Thynne petitioned Cecil once more for assistance.[4] He was released around May 1576; years later, in 1602, he succeeded in winning the post of Lancaster Herald.[5]

Another English alchemist, Humfrey Lock, turned to alchemy when seeking to return to his native land from an unpalatable posting abroad. Peter Grund has identified Lock as an English craftsman, possibly an engineer or builder, who in 1567 entered the service of Czar Ivan Vassilivitch (Ivan the Terrible).[6] In letters to Cecil and the Earl of Leicester, Lock complained of his treatment by members of the English merchant community in Moscow, particularly the ambassador, and begged permission to return. His petition, probably written around 1572, was accompanied by a treatise that he hoped might intrigue his patron to the point of recalling him:

> For when I compiled it, I ment to haue sent it into Ingland as a present & mediator to help me home out of Russia, wherfore I made it the more darke that I might the sonner be sente for home for to doe it myselfe.[7]

Writing under very different circumstances, Thynne and Lock nonetheless turned to a common reservoir of English authority when composing their respective treatises. Thynne's poetical "metalls *Metamorphosis*," com-

---

3. On these ill-fated projects, see Campbell, "Alchemical Patronage"; Harkness, *Jewel House*.
4. Francis Thynne to William Cecil, Lord Burghley, 13 March 1576, British Library, MS Lansdowne 21/57.
5. Campbell, "Alchemical Patronage," 47–48; Knafla, "Thynne, Francis (1545?–1608)."
6. On Lock, see Grund, *"Misticall Wordes and Names Infinite."*
7. Ibid., 11.

posed for Burghley in 1573, demonstrates a patriotic regard for such English adepts as "Th'englishe freer olde *Bacon*, and the good Britishe *Riplye*," as well as Thomas Norton, whose *Ordinal* he extensively cites.[8] His collecting habits also suggest connections to an earlier generation of English practitioners. On 18 October 1573, Thynne copied the Middle English poem "Merlyne and Morien" from an original owned by one Thomas Peter, which he claims to have transcribed with Peter's help—raising the fascinating possibility that this was the same man who wrought havoc at the priory of Little Leighs some forty years earlier.[9] Even during his period of imprisonment, Thynne managed to secure one of Giles Du Wes's manuscripts: Trinity O.8.24, previously gifted to Greene. In late 1574 he copied most of its contents, including Du Wes's marginal illustrations.[10]

Even farther afield, in Moscow, Lock collated excerpts from medieval sources, which he grouped into chapters on the animal, vegetable, and mineral stones as the basis for his own lengthy treatise. Like Richard Eden's book intended for the Earl of Warwick, Lock's *Treatise* is essentially compiled from earlier sources—among them, English translations of Guido de Montanor's *De arte chymica* and Ripley's *Medulla* and *Concordantia*, as well as the Middle English *Mirror of Lights*, a text based on the *Semita recta* pseudonymously attributed to Albertus Magnus.[11] Unlike Eden, however, Lock did more than cut and paste from earlier authorities; he also edited and rearranged them in ways that offer insight into his own practical interpretation of their contents. For instance, although his chapter on the vegetable stone quotes heavily from the *Medulla*, he omits Ripley's vital reference to sericon,

---

8. Francis Thynne, "Another Discourse vpon the Philosophers ARMES," Trinity R.14.14, fol. 138r. Thynne's transcriptions of alchemical treatises, preserved in Warminster, Longleat House, MS 178, include both Ripley's *Compound* (fols. 58r-86r), dated 5 April 1578, and an *Ordinall of Alchemy* dated 3 June 1574 (fols. 10v-48r), which Thynne annotated carefully.

9. Longleat House, MS 178, fol. 105v: "Copyed out of the originall the 18 of october 1573 by me Francis Thynne whiche originall I had of Mr Tho: Peter—written withe thayde of the same Thomas Peter but I thinke this worke is imperfecte because as yt semeth theire lacketh some verses to furnyshe the ryme \but/ not withstandinge I haue followed the copye. Laus deo in Eternum." On Peter, see chap. 4, above.

10. British Library, MS Add. 11388; discussed in Rampling, "Alchemy of George Ripley," chap. 5. According to Thynne's dates, these tracts were copied as follows: *De vetula*, 20 September 1574; *De pomo*, 28 September 1574; Lactantius, 1 November 1574; Claudianus, 3 November 1574; and the *Cantilena*, 18 November 1574. The texts have subsequently been bound in a different order.

11. On the composition of the text, see Grund, "*Misticall Wordes and Names Infinite.*"

the "body calcined to red," instead substituting tartar—a substance that Ripley alluded to earlier in his text, but in a different context.[12]

The cases of Thynne and Lock illustrate the ongoing role played by the scribal circulation of alchemical texts during the 1570s, and the variety of responses to earlier authorities among English reader-practitioners; for instance, by composing "new" treatises on the basis of old, invoking lineages of English adepts, and deploying them for personal advancement. It is no coincidence that so much of our evidence for late medieval and Henrician alchemy survives in copies from this golden age of transcription. The most striking products of this era include the so-called "Ripley Scrolls": a series of emblematic rolls, deliberately archaic in appearance, that incorporate English alchemical verses and related imagery, copied from a late fifteenth-century exemplar.[13] It is against this background, characterized by copying, translating, editing, and questing for patronage, that Samuel Norton's own petition took shape.

The son of the Somerset gentleman Sir George Norton, Samuel was still a young man when he completed the *Key* for Elizabeth I at St. John's College, Cambridge, in 1577.[14] He had already been a student of alchemy for some years, and his interest would continue long after his succession in 1584 to the family estate in Abbots Leigh near Bristol, and his appointment as justice of the peace and sheriff of Somerset.[15] During the 1580s he dedicated another, shorter work on Hermetic and Paracelsian themes to Lord Burghley, while citations and interpretations first set down in the *Key* recur in his *Libri tres tabulorum arboris philosophicalis* (Three Books of Tables of the Philosophical Tree), completed in 1599—by which time he was no longer a student, but fifty-one years old and a respected local magistrate.[16]

---

12. Lock, *Treatise*, ed. Grund, in *Misticall Wordes and Names Infinite*, 224 (ll. 25–28); discussed below, pp. 337–38.

13. Although the earliest extant roll is Bodleian Library, Bodley Rolls 1 (late fifteenth century), most copies date from the late sixteenth and seventeenth centuries. I discuss these in detail in Rampling, *Hidden Stone* (forthcoming). For an overview, see R. Ian McCallum, "Alchemical Scrolls Associated with George Ripley," in *Mystical Metal of Gold*, ed. Linden, 161–88; on the Scroll verses, Timmermann, *Verse and Transmutation*, chap. 4 and 294–303.

14. Getty Research Institute, MS 18, vol. 10, pt. 2 (hereafter Getty 18/10), 13: "[F]rom St Johnes in Cantabrige The 20 of July 1577." On the provenance of this copy, see note 32, below.

15. Scott Mandelbrote, "Norton, Samuel (1548–1621)," *ODNB*; Campbell, "Alchemical Patronage," 71–74; Rampling, "Transmuting Sericon," 29–31.

16. Samuel Norton, *Summarie Collections of true natural Magick grounded vpon principles diuine: and from the writings of Hermes Trimegistus and others the learned Auncients: conteining the true Philosophie and Physick drawen into commune places*, Cambridge University Library, MS KK.1.3, pt. 3, fols. 32r-52r. This previously neglected treatise was first discussed by Campbell, "Alchemical

Although a particular devotee of George Ripley, Samuel was also proud of his own alchemical heritage as a descendant of Ripley's contemporary, Thomas Norton. Samuel's choice of alchemical cognomen, "Rinville," seems a conscious homage to his "great-grandfather, Thomas Rinvile Norton."[17] Descent from one of the "aristocrats" of English alchemy, coupled with his own learning, underpins Samuel's declaration, proudly appended to the *Libri tres*, that its author is "Samuel Rinville, alias Norton, bearing arms in philosophizing."[18]

Like Thynne and Lock, Norton sought to join the ranks of the *adepti* through close textual and practical engagement with an existing tradition. His earliest work, the *Key*, a book in seven chapters with a dedicatory letter, verses, and preamble, provides a sustained commentary on the works of Ripley and his pseudo-Lullian sources. Although it is uncertain whether the *Key* ever came to the attention of the queen, it represents a serious attempt to attract support for a wide-ranging alchemical program. Norton's dedication and preamble employ many of the conventions observed in such appeals, as he invokes earlier relationships between rulers and alchemists. Besides the traditional pairings of Aristotle and Alexander, Morienus and Khālid, and Raymond and King Robert of Seville, Norton also offers an English model, citing the favor in which his forebear Thomas Norton was held by King Edward IV.[19] Such exemplars laid the ground for his own program for preparing transmutational and medicinal elixirs. The earnestness of Norton's proposal may be gauged from the detailed drawings of apparatus and furnaces with which he closes the treatise, in addition to an itemized estimate of the cost of his project. At just £63 6s. 4d., the work represented a considerable improvement on the thousands invested in de Lannoy's costly enterprise.[20]

Norton was also concerned with more recent themes in English alchemy—most importantly, rising interest in Paracelsian chemical medicine, which drew on alchemical and medical knowledge in a way already apparent in medieval writings on distillation. Although Paracelsus proposed a different theoretical basis for his own medicinal "quintessences," their

---

Patronage," 71–74. Sloane 3667 contains an incomplete copy of the *Libri tres*, including a series of handsome "tabulae," or tree diagrams (fols. 24r–89v; further "tabulae" at fols. 11r, 12r, 15r). A complete copy is in Ashmole 1478, pt. 6, fols. 42r–96v, dated Bristol, 20 May 1599 (fol. 104v).

17. Sloane 3667, fol. 71r: "[P]roavus meus Thomas Rinuile Norton."

18. Ashmole 1478, pt. 6, fol. 104v: "Samuele Rinuillo alias Nortono armigero philosophante."

19. Getty 18/10, 15.

20. Ibid., 156–68.

preparation still owed much to the tradition exemplified by John of Rupescissa and pseudo-Lull, ensuring the continued relevance of these authorities even as their reputations became increasingly intertwined.[21] The arrival of Paracelsianism also affected how alchemists related their own history, shifting to accommodate the Swiss physician's iconoclastic approach within a more traditional lineage of medical and hermetic knowledge.

Norton's suit offers an early example of the reception of Paracelsianism in England, suggesting an attempt to rehabilitate the controversial physician by positioning him alongside authorities who already enjoyed a reputation for alchemical medicine, such as Lull and Ripley.[22] In the *Key*, Paracelsus is thus revealed to be a pseudo-Lullian at heart: for the "aqua vitae" that he used in preparing his sublimate was, Norton claims, "our Aquavitae, & not of a vine, but our true Quintessence & [mercury] vegetable."[23] By equating Paracelsian *aqua vitae* with the "vegetable mercury" described by Raymond and Ripley, Norton confidently reads the new remedy back into an earlier medieval tradition.

This concordance follows a far more influential synthesis of Paracelsian and traditional medicine published just a few years earlier: the *Idea medicinae philosophicae* (Ideal of Philosophical Medicine, 1571) of Peder Sørensen (Petrus Severinus, 1542–1602), physician to successive kings of Denmark.[24] While it is unclear whether Norton himself knew this treatise, Severinus's

21. Paracelsus's medical writings became available in print during the 1550s and 1560s. On their early reception in England, see Paul H. Kocher, "Paracelsan Medicine in England: The First Thirty Years (ca. 1570–1600)," *Journal of the History of Medicine* 2 (1947): 451–80; Allen G. Debus, *The English Paracelsians* (London: Oldbourne Press, 1965); Webster, "Alchemical and Paracelsian Medicine."

22. An attempt to insert Paracelsus into a distinguished genealogy of English adepts, including Bacon, Ripley, and Thomas Norton, was later made by the lawyer and member of Parliament Richard Bostocke (ca. 1530–1605): R[ichard] B[ostocke], *The difference betwene the auncient phisicke, first taught by the godly forefathers, consisting in vnitie peace and concord: and the latter phisicke proceeding from idolaters, ethnickes, and heathen: as Gallen, and such other consisting in dualitie, discorde, and contrarietie* ... (London: [G. Robinson] for Robert Walley, 1585), 139. On Bostocke, see Webster, "Alchemical and Paracelsian Medicine," 313 and 329–30; David Harley, "Rychard Bostok of Tandridge, Surrey (c. 1530–1605), M.P., Paracelsian Propagandist and Friend of John Dee," *Ambix* 47 (2000): 29–36. Samuel Norton may have been influenced by Bostocke, whose sharp critcism of Galen he follows in his own "Summarie Collections."

23. Getty 18/10, 130.

24. On Severinus, see Jole Shackelford, *A Philosophical Path for Paracelsian Medicine: The Ideas, Intellectual Context, and Influence of Petrus Severinus (1540/2–1602)* (Copenhagen: Museum Tusculanum Press, 2004). Other Paracelsian physicians would later succeed in winning court positions, including Joseph Du Chesne (ca. 1544–1609), physician-in-ordinary to Henri IV of France; see Kahn, *Alchimie et Paracelsianisme.*

success as both syncretist and successful client might have offered an impressive precedent for an ambitious young alchemist.

In his own attempt at synthesis, Norton promotes the advantages of the new medicine while remaining (at least superficially) respectful of the authority of Galen and Hippocrates: a compromise achieved by plotting an alchemical *via media* between these opposing medical positions. In practice, he observes, there are "sundrie kinds of phisicke between the Galenists & the Chimicall phisitians."[25] While Galenists tend to overprescribe, chemical physicians also risk harming their patients by the "desperate giving of [mercu]ries, & hurtfull mineralls"—a hazardous practice, given that "verie few know how rightfullie to prepare them."[26] For instance, Norton dismisses mercurial sublimates and precipitations as "devils," more effective in surgery than internal medicine. We might note that this is exactly the kind of toxic preparation, made using the fire against nature, that Raymond and Ripley recommended for transmutation while forbidding its use in actual physick—a distinction that Norton, a supporter of separate mineral and vegetable stones, here seems to endorse.

Between these dangerous extremes, Norton proposes a third way, achieved by extracting oils and quintessences from metallic bodies. This method allows "mettalls such as are of knowne vertues" to be safely administered "if they be brought into potable liquors."[27] This *via media* is solidly based on the quintessential alchemy of *De secretis naturae* and Ripley's sericonian practice, a strategy that reflects his high regard for Ripley, the "Raymond of the English," while also preserving the reputation of Paracelsus.[28]

Yet Norton's interests go beyond merely reconciling texts. The *Key* was intended to secure royal support for his own practical endeavors, in which the medicinal vegetable stone ranked more highly than the chrysopoetic mineral work. To that end, Norton sought to recruit the polarized authorities of Galen and Paracelsus in support of the same quintessential medicines as those purveyed in the heyday of Raymond and Ripley—an English tradition, surely calculated to appeal to the tastes of an English queen, and which Norton hoped Elizabeth might invite him to continue. As we shall see, Norton's efforts would indeed alter the course of the history of English alchemy,

25. Getty 18/10, 129.

26. Ibid., 129–30. Norton's comments anticipate later criticism of "bad" (as opposed to "good") Paracelsians; see, for instance, George Clowes's attacks on Paracelsian physicians, discussed in Harkness, *Jewel House*, chap. 2.

27. Getty 18/10, 130.

28. Ibid., 127: "*Anglorum Raimundus*, I mean George Ripley."

although not quite in the manner he envisaged—bringing him success not as an adept in his own right, but as the conduit for a new surge of interest in Ripley's sericonian practice.

RIPLEY'S *BOOK* AND NORTON'S *KEY*

The *Key* was never printed, and in light of Samuel's genealogical claims has been mined primarily as a source of information on Thomas Norton.[29] Yet although Samuel's own compositions never attained the success of his ancestor's *Ordinal*, they still circulated among the knowledgeable. For instance, the *Key* fell into the hands of Thomas Robson, the great copyist of English alchemica, in 1613.[30] Robson copied it immediately, and in 1617 produced a second, almost complete transcription, noting that "yf it were in some mens hands I thinke that they would bring some great matter to perfection by it."[31] A slightly more complete version of the text survives in another copy, later integrated into the collection of the eighteenth-century ship's doctor and alchemist Sigismund Bacstrom.[32]

Like Lock's *Treatise*, Norton's *Key* relies heavily on the structure and contents of Ripley's *Medulla*. In the organization and synthesis of its material, however, the *Key* provides a far more comprehensive treatment, offering guidance on the manufacture of a variety of clearly delineated products. Besides the familiar animal, vegetable, and mineral stones, these include a "mixed" and a "transparent" stone. Norton was evidently struck by Ripley's account of the compound water in the *Medulla*, for he adapts it as the basis for his own chapter on the mixed stone, remarking on its combined mineral and vegetable character. Only the transparent stone has no obvious equivalent in the *Medulla*, Norton drawing this method (used for making pearls

---

29. See, for instance, Reidy, "Introduction," in Norton, *Ordinal*, xxxviii–xlii.

30. Ashmole 1424, pt. 2, 19: "The which booke . . . hapned into my hands in the yeare 1613."

31. Ibid. The complete copy is in Ashmole 1421, fols. 165v–220v ("By Me Thomas Robson 1613," fol. 217v). The second is in Ashmole 1424, pt. 2, 49–90 ("Finnis 1616: february 28," 102). On Robson, see chap. 9, below.

32. Getty 18/10, pt. 2. This is probably the exemplar for a transcription by William Alexander Ayton (1817–1909) in Wellcome Library, MS 1027, since Ayton made other transcriptions from Bacstrom's library: S. A. J. Moorat, *Catalogue of Western Manuscripts on Medicine and Science in the Wellcome Historical Medical Library* (London: Wellcome Institute for the History of Medicine, 1962–73), 2:58. Bacstrom himself transcribed Norton's chapter on the animal stone in Getty 18, vol. 11, fols. 5r–11r. On Bacstrom and his manuscript collection, see Ron Charles Hogart, ed., *Alchemy, a Comprehensive Bibliography of the Manly P. Hall Collection of Books and Manuscripts: Including Related Material on Rosicrucianism and the Writings of Jacob Böhme; Introduction by Manly P. Hall* (Los Angeles: The Philosophical Research Society, 1986), 226–34.

and gems) in part from the *Accurtations of Raymond*, which he also attributes to Ripley. To these five stones he adds chapters on the specific problems of fermentation and multiplication, and one on the elixir of life—the latter including a long passage from *De secretis naturae*.

The *Key* is therefore dominated by the same doctrines and techniques that originate in pseudo-Lullian writings and are exemplified by Ripley's work. For Norton, Ripley even surpasses Raymond in the sheer number of his practices. For instance, while Raymond provides only one recipe for the mixed stone (in the *Epistola*, which Norton cites extensively), Ripley gives three. Norton sums up his reflections on the relationship between the two adepts:

> No marvaile therefore, if [Ripley's] schollers finde so much profit in Raimond; Considering that their M*aste*r was so conversant *w*ith Raimonds works, & was so great an expositor of Raimond, that almost hee might deserve the proverbe of Raimonds Ape; and yet in this hee sure exceeded Raimond, so that looke what soever hee fetched out of him, hee proved it to the vttermost.[33]

Norton's final comment is telling. He sees Ripley's contribution as lying not just in reiterating Raymond's sayings, but in testing them in practice. In the *Key*, the most thorough contemporary exposition of Ripleian alchemy, Norton perhaps deliberately set out to do for Ripley as Ripley had done for Raymond. Yet Norton recognized that the obscurity of Ripley's writing made it inaccessible to many, with a deleterious effect on the canon's reputation. In the preamble to the *Key*, we may even detect a slight defensiveness as Ripley is introduced:

> [O]ur Noble Riplie whome I cannot sufficientlie extoll; Although some there be that mightilie Inveigh against him, whome I will referre over to their owne errors: Yet Riplie [is] not to blame; but such as mistake Riplie, & vnderstand him not. I take God to witness; I never yet found false Conclusion in Riplie, but that the proofe fell iustlie with his speeche.[34]

It is Norton's determination to make *practical* sense of Ripley's alchemy that truly sets the *Key* apart from other contemporary patronage suits, including the bricolage of Lock's *Treatise*. Not content to work from a single

33. Getty 18/10, 113.
34. Ibid., 19–20.

text, or even a single author, Norton evaluates Ripley's writings in relation to other authorities: Raymond, Geber, and the more recent master, Paracelsus. Above all, he examines Ripley's use of pseudo-Lullian texts and processes, paying particular attention to the canon's main areas of expertise: the vegetable stone, the animal stone, and the art of fermentation.[35]

In casting himself in the role of Ripley's expositor, Norton had reason to consider himself unusually well qualified for the task. While other alchemists might evoke Ripley, quote from his works, or even, in the case of Charnock, claim to have been taught by one of Ripley's own disciples, Norton offered a more tangible link with the past: an old Latin commonplace book, written, he claimed, by Ripley himself.[36] Rather than a handed-down transcription, distorted and defaced by careless scribes, this book represented Ripley's personal notes and jottings: "his bosome booke or the booke he daylie used."[37]

Norton describes the find and its value at the start of the *Key*, in his dedication to the queen:

> Although it fortuned mee in manner vnloked for, to hitt vpon the secret bosome booke of Riple, wherby the true grounds are discovered, Of which havinge by profe found so many to be true, and little doubtinge of the accomplishment of the rest; I thought it but a point of dutie to reveall and vppen the Secrets heereof vnto your Highnes.[38]

Norton gives no hint as to how he stumbled upon this relic from England's true golden age. As we have seen, early alchemical manuscripts were prized by late sixteenth-century practitioners, and widely copied. Yet Norton's account, coupled with a lack of earlier manuscript references, suggests that this particular find had not been widely disseminated before the 1570s. Awareness of both its practical importance and its antiquarian value lent additional urgency to the task of translation, which, Norton explains, he struggled to complete in spite of illness:

> Being about Candlemas last in great danger by sicknes; at which time, there was not any one thing, [tha]t more greeved mee to thincke on; then [tha]t I

---

35. Getty 18/10, 21: "Alonely Riplie hath the price of the vegetable stone"; Ashmole 1421, fol. 173v: "Concerninge the anymall stone, Ripley toucheth it above al other writters" (text omitted from Getty 18/10); Getty 18/10, 20: "[S]o doth hee above all the writers of the world, open the secrets of handling the ferment."

36. Charnock claims that one of his masters was taught by Ripley's "boy"; *TCB*, 301.

37. Sloane 2175, fol. 148r.

38. Getty 18/10, 7.

Could not be a meanes for revivinge againe of that, which had so longe laien dead; In which I had found such great likeliehoode, sure tokens, and troth of practick, which forced mee even sicke as I was, to labor right ernestlie to finish the translatione of Riplie.[39]

In February 1573/4, in Bristol, Norton completed the translation. A little over three years later, now in Cambridge, he finished work on his own *Key*, in which he attempted to interpret Ripley's known writings (particularly the *Medulla*) in light of the alchemy of the *Bosome Book*. By the close of the decade, excerpts from Ripley's lost book had escaped the confines of Norton's study and found homes in the libraries of other alchemical aficionados. Within another decade, they reached the imperial capital of Prague.[40] A few more years, and a poem extracted from the *Book* became one of the first English alchemical texts in print.

Happily, Norton's original translation seems to have survived. Written in a heavy, untidy hand with execrable and inconsistent spelling, Sloane 3667 is nonetheless a plausible candidate for Norton's personal copy. It is bound with several other manuscripts of the late sixteenth and early seventeenth centuries, including a copy (in another hand) of Norton's *Libri tres*.[41] The manuscript containing the *Book* opens with a prayer, written in Norton's hometown, "At Bristowe anno 1573," in which the supplicant appeals for divine support in his alchemical endeavors.[42] The prayer is followed by three works cited in the *Key*: the *Medulla* (in Whitehead's English translation), the *Accurtations*, and a short *practica* by a lesser-known alchemist, "Master Ive."[43] The *Bosome Book* then follows:

39. Ibid.

40. On the dissemination of Ripley's writings, see Rampling, "Dee and the Alchemists"; and chap. 8, below.

41. Identifying Norton's hand is hampered by the fact that he seems to have employed a scribe, as the conclusion of the *Key*'s dedication suggests: "I most humblie desire your Hignes, to accept in good part, the green fruict of this my Monthes travaile . . . although it bee not soe exactlie produced, fframed, & penned by mee & my writer, as I would wish" (Getty 18/10, 13). The writer may have been employed to produce the royal copy only; Francis Thynne also seems to have employed a professional for the presentation copy of his *Discourses* to Burghley; Rampling, "Alchemy of George Ripley," chap. 5.

42. Sloane 3667, fol. 91v: "[Y]t I may aske suche thynges as may be to the honnor & glory of thy most holye nam[e] . . . to be instrocted in [th]e knowledge of naturall thynges. . . . Geue me tharfore grace good lord I humblye beseche thee, to knowe theis thynges, and dulye & effectuoslye withowt impediment."

43. The *Medulla* translation in Sloane 3667 is attributed to "Mr Davye Whithede clarke anno 1552" (fol. 104v), and the *Accurtations* to "Raymonde Lully, compackt & Gathered to gether (as yt

The Copye of a old Booke, which is thowght to be [th]e hand writtyng of Mr gorge Rippyle Channon, translated owt of Latten bye samwell norton Esquyer the vth [i.e., 5th] of feberary Anno domini 1573[/4].[44]

To Norton's contemporaries, what follows must have seemed like the textual embodiment of the philosophers' stone. The *Book* is packed with material, including four substantial prose commentaries, two poems, a diagrammatic "tree" of the work, and hundreds of recipes, the latter varying in length from a few lines to several folio pages. Ripley's primary authorities, Raymond and Guido, are heavily cited throughout. In addition to the distinctively Ripleian character of the texts, many are annotated with the canon's name or initials, while several epigrams also incorporate his name. The large amount of content made transcription a tedious task, as indicated by a revealing personal note toward the end of one Latin copy: "Here [th]e writer oute of [thi]s booke beganne to bee wearye, & scipped over as many receipts as remayne here ensueing."[45] Even with these omissions, the manuscript in question, Harley 2411, remains the most complete copy of the original, Latin *Bosome Book*, for no trace remains of Norton's exemplar. While the *Book* lives on in translation and transcription, the source has disappeared as mysteriously as it came.

The loss of the original *Book* raises immediate difficulties for modern readers. The most obvious question is that of authenticity. Was this really a book compiled by Ripley, or simply an alchemical collection with a Ripleian flavor? For skeptics, the book's redundant emphasis on Ripley's name might smack of wish fulfillment or deliberate forgery, an attempt on the part of either Norton or another to generate additional interest in the manuscript and the commentaries it inspired. Given the prevalence of alchemical pseudepigrapha during this period, the appearance of the *Book* and its role in Norton's bid for alchemical patronage seem at first sight highly suspicious. Yet Norton's deeply researched proposal and cross-referenced notes speak of genuine personal conviction concerning both Ripley's authorship and the place of the *Book* in relation to his other works. To argue convincingly for its authenticity, we must examine the *Book* in light of both its bibliographical remains and the alchemical principles discernible from its contents.

---

is thowght) by gorge Rypleye Channon" (fol. 112v). Next comes "A notable worke of Sericon writen *per* master Ive" (fol. 115r).

44. Sloane 3667, fol. 124r.

45. Harley 2411, fol. 85v.

THE *BOSOME BOOK* IN MANUSCRIPT

It is no simple task to survey surviving copies, for the *Book* exists in several versions. The Latin copy in Harley 2411, dating from the early seventeenth century, offers the most complete transcription and therefore provides the model for this discussion. The compendium is here titled "Various Collected Experiments of George Ripley, with some Expositions on Hermes, Aristotle, Guido, and Raymond," followed by a note in English: "I haue often herde this book to bee named Ryplayes Bosom Book."[46] In addition to this Latin copy, Norton's English translation survives in Sloane 3667, which was later amended in a secretary hand. The same seventeenth-century annotator was also responsible for transcribing Sloane 2175, a copy of Sloane 3667 that incorporates both the scribe's earlier emendations and Norton's own extensive marginalia.[47] Shorter extracts from Norton's translation can be identified in other collections dating from about 1579 onward.[48]

That Norton was not the only enthusiast to attempt the translation of the *Bosome Book* appears from a previously unidentified copy, Ashmole 766 (pt. 5). Unusually, the title does not refer to Ripley at all, and also occludes the collection's strong medicinal bent: "A certen booke of alkimy written by an vnknowne Author for the makinge of gould."[49] This translation, made twenty years after Norton's, was completed on 24 July 1593 by one Roger Howes, "for Mr Gawyn Smithe gentleman."[50] Howes was an experienced translator of alchemica, having also Englished copies of works by Petrus Bonus of Ferrara for Smith in 1590.[51] Gawin Smith, whom Howes styles "gentleman Master of her ma*ies*ties Engines," was a prominent engineer in the service of Elizabeth I, who later received letters patent from James I that enabled him to style himself "cheife Enginer of England."[52] A friend of John Dee, he

46. Harley 2411, fol.1r-v: "Georgij Riplaij Experimenta varia, Collectanea, Expositionesq*ue* aliquot in Hermetem, Aristotelem, Guidonem & Raymondu*m* &c."

47. Sloane 2175, fols. 148r-72r. In addition to the marginalia, this copy refers to the foliation and various interpolations in Sloane 3667, as at fol. 153r: "That which followeth was pinned to the Copy in a paper."

48. Recorded in CRC 3. The earliest datable witness is Potter's transcription of an extract, the *Whole Work*, in 1579–80.

49. Ashmole 766, pt. 5, fol. 1r.

50. Ibid.

51. *Introductio in divinam chemiae artem integra* (completed 16 October) and *Pretiosa margarita novella* (30 November), in Sloane 3682. Howes comments that the translation was "the work of 60 days or thereabout" (fol. 285r).

52. Sloane 3682, fol. 1*r; Gawin Smith, "The true Coppie of a peticion deliuerid to [th]e L[ord] mayor & Aldermen of [th]e Cittie of London by Gawin Smith," British Library, MS Cot-

petitioned the queen on Dee's behalf in July 1590.[53] The indefatigable Thomas Robson copied Howes's translation in 1606, and thereafter included extracts in several other of his manuscript compilations, referring to it as the "Uncutt Book."[54]

The structure of the *Bosome Book* also varies between surviving copies. Harley 2411 has more of the appearance of a notebook than Norton's translation in Sloane 3667, for the material is presented in no clear order (table 7.1). For instance, it starts with a collection of recipes, the first a procedure in nine stages titled "My process at Estergate."[55] Norton relegates this process to the final pages of his own translation, along with various other recipes.[56] In its place, he opens with Ripley's *Concordance*, a work also included in Harley 2411, but much later in the compendium. It appears that Norton selectively rearranged the *Book*'s material, privileging theoretical texts over practical recipes in imitation of the more conventional format of a medieval "philosophical" treatise, namely a *theorica* followed by a *practica*. The ordering of Howes's alternative translation is rather closer to that of Harley 2411, suggesting that the latter indeed preserves something of the *Book*'s original format—a conclusion that accords with the fact that Howes was hired to translate rather than to edit the text, whereas Norton already had a royal dedication in mind as he commenced his own English version.[57]

As we saw in chapter 3, much of the content of Norton's and Howes's exemplar connects closely to other Ripleian works, supporting the case for its fifteenth-century provenance. Among the more prominent items are the "Notable Rules Taken from the Book of Guido de Montanor," a set of forty-five aphorisms also quoted in the *Medulla* and *Concordantia*, some of which are embellished in copies of the *Book* with the suggestive note "haec G.R."[58] The *Book* also includes two recipes with the incipit "Put the body which is most weighty into a stillatory, and draw out its sweat"—a pseudo-Lullian

ton Titus B.V, fol. 273r. For other manuscripts referring to Smith, see Harkness, *Jewel House*, 286n56.

53. Smith had visited Dee for several days in Bremen in October 1589. Edward Fenton, ed., *The Diaries of John Dee* (Charlbury: Day Books, 1998), 249, 240.

54. Ashmole 1418, pt. 2, fols. 1r-47v.

55. "Processus meus apud Estergate," Harley 2411, fol. 1r. Estergate, or Eastergate, is a village in West Sussex.

56. "Mr George Ryples Prosedynges at estergate," Sloane 3667, fols. 171v-72r.

57. Howes has nevertheless switched the order of the first two components, so that his copy begins with the "Exposition on Aristotle and Hermes," followed by the "Estergate" process. This order is reversed in Harley 2411.

58. See pp. 114–15, above. The first seventeen of these are also found in the fifteenth-century manuscript Trinity O.8.9, although the set is truncated owing to the loss of several folios.

TABLE 7.1. Major components of George Ripley's *Bosome Book* (compiled ca. 1470), in the order found in MS Harley 2411

Short Title	Description of Content	CRC no. (if applicable)
Processus meus apud Estergate	Recipe in nine steps	
Exposition of Aristotle and Hermes	Commentary on the *Secretum secretorum* and *Emerald Tablet*	
Tree diagram	Diagrammatic scheme of the work, beginning with "Adrop"	
Recipe Adrop	Extended sericonian recipe	Basis for *Practise by Experience*: CRC 26
Attinkar pro mercurio sic fit	Collection of chrysopoetic recipes, citing Guido de Montanor	
Maria dicit	Commentary on a saying of Maria the Prophetess	
Separatio elementorum	Series of linked recipes for isolating the four elements, based on distillation of sericon	Basis for *Whole Work*: CRC 35
Concordantia Guidonis et Raymundi	Concordance of Guido's and Raymond's advice on fermentation and the alchemical body	CRC 10
Notable Rules Taken from Guido	45 aphorisms extracted from works of Guido	CRC 22
Nota de ignibus nostris	Commentary on the pseudo-Lullian natural fire and fire against nature	CRC 15
Compendium totius artis	Commentary on sericonian alchemy; Ripley's rejection of erroneous experiments conducted between 1450 and 1470	
Practical Compendium	Short treatise citing Guido	
Vision	Allegorical poem	CRC 32
Somnium	Allegorical poem	CRC 28
Res naturales sunt hae septem	Table of elements, non-naturals, alchemical "fires," etc.	
Pone corpus quod ponderosius	Short recipe illustrated by labeled diagram of a furnace	Component of *Viaticum*: CRC 31

aphorism that is cited verbatim in the *Medulla*, where Ripley uses it to hint at the preparation of his own vegetable stone.[59] Such examples may be supplemented by dozens more throughout the *Bosome Book*, underlining both the fifteenth-century character of the collection and its many points of intersection with the Ripley Corpus, especially the *Medulla*. The evidence strongly suggests that Samuel Norton had indeed stumbled upon a fifteenth-century commonplace book, either compiled by Ripley himself, or copied from one that was.

THE ALCHEMY OF THE *BOSOME BOOK*

The Ripleian character of the *Bosome Book* is cemented by its theoretical and practical content. Although the collection contains a wide variety of recipes and commentaries, both are dominated by sericonian alchemy. The *Book*'s alchemical content frequently overlaps with that of Ripley's *Medulla*, particularly his famous processes for the vegetable stone and compound water. These similarities may be explained by the fact that the works have two major authorities in common: Raymond and Guido. References to Ripley's two favorite philosophers abound in the *Book*'s more substantial texts, notably in four commentaries based on alchemical sententiae. These include two expositions on passages by single, named authors (Mary the Prophetess and Raymond) and two concordances between different authorities (Guido and Raymond, and Aristotle and Hermes). Each of these reveals further connections with Ripley's oeuvre, as we may see by examining three of these in turn.

De ignibus nostris

Other than the *Concordantia*, the most conspicuously Ripleian of the *Book*'s commentaries is a text titled "A note on our fires [*Nota de ignibus nostris*], without knowledge of which the mastery is not completed, by G.R. canon, the expositor."[60] The subject of the commentary is a short passage attributed to Lull, "Here lie contrary operations," describing Raymond's two contrary

---

59. Harley 2411, fols. 72r, 74r: "Pone corpus quod ponderosius est in distillatorio, et trahe sudorem eius"; cited in *Medulla*, fol. 2r. Variant versions of another of the *Book*'s processes, "Recipe Kibrith," recur both in the *Medulla* (fol. 4v) and in Trinity O.8.9 (fol. 34v).

60. Harley 2411, fols. 57r-61v: "Nota de ignibus nostris sine quorum notitiae magisterium non perficitur per G:R canonicum exponentem." Untitled in Sloane 3667.

fires—the natural fire and the fire against nature.[61] This passage appears verbatim in the *Scala philosophorum* and later in the *Medulla*, where it provides the basis for Ripley's own exposition of the fires.

In fact, *De ignibus nostris* elaborates the doctrine in even greater detail than the *Compound* and *Medulla*. Here, the author explains why mineral corrosives are useful in the lesser alchemical work, but not in the greater:

> Such fire (i.e. the fire against nature) in our worke is not of the vertue and operacion of our mercury, but it is the fire of nature (i.e. of our aqua ardens oylie) which is pure naturall. And therefore Guido sayeth, it is the greatest medicine for mans body that can bee, and that it healeth all infirmities of mans body aboue all the potions of Hippocrates and Galenes. . . . But concerning the fire against nature Raymond sayeth that all Alchymick gold is made of Corosiues, And therefore it doeth frette and destroy nature wherefore it ought not to come in medicines to be ministred for mans body.[62]

Already familiar from Ripley's exposition in the *Medulla*, the distinction between a minerally derived fire against nature and the wine-based natural fire is here made explicit. The sayings of Raymond and Guido cited in this passage also appear in the *Medulla*, again to emphasize the distinction between transmuting and medicinal waters.[63] In sum, it would be hard to conceive of a work more compatible with Ripley's alchemical philosophy than *De ignibus nostris*, either in choice of subject matter or selection of authorities. Yet although the text seems to be related to the *Medulla*, the exact nature of the relationship is less clear. We should bear in mind that Ripley was not the only fifteenth-century alchemist interested in Raymond's exposition of the contrary fires. For instance, an exposition of exactly the same Lullian passage is included, admittedly in far less detail, in the fifteenth-century Sloane 3747.[64]

---

61. The text in Harley 2411 ("Hic iacent contrariae operationes, quia sicut Ignis contra naturam resoluit Spiritum corporis fixi in aquam nubis, et corpus spiritus volatilis constringitur in terram congelatam") may be compared to that used in the *Medulla*; see above, pp. 84–85.

62. Sloane 2175, fol. 156r. I here use the scribe's "improved" version of Norton's translation.

63. *Medulla*, fol. 5r: "Quoniam omne aurum alkimicum vt Raymundus asserit fit ex corrosiuis. Ideoque aurum sic factum non ingreditur medicinas"; "et tunc habet potestatem conuertendi omnia corpora in aurum purum et sanandi omnes infirmitates supra omnes potaciones Ypocratis et Galieni."

64. Sloane 3747, fols. 34v-35r.

Exposition of Aristotle and Hermes

The ambiguous relationship between the *Bosome Book* and the Ripley Corpus is exemplified by another commentary, in which Ripley has apparently imported dicta from his favorite authorities into a preexisting text. The result is a substantial treatise entitled "An Exposition by George Ripley of the Sayings of Aristotle and Hermes, Mutually Reconciling Them."[65] Here the commentator seeks to reconcile two short yet influential texts: the *Emerald Tablet* of Hermes and the famous lemma from the pseudo-Aristotelian *Secretum secretorum*: "Take the anymale, vegytable, & minerall stone, which is not a stone, nether hyt hath yt the nature of a stone."[66] In the course of the exposition he has frequent recourse to Lull and Guido, citing, for instance, Guido's saying that the imperfect metallic body, once cleansed, is a thousand times better than common gold or silver (also quoted in both the *Accurtations* and the *Concordantia*). For a reader like Norton, already familiar with Ripley's works, the wealth of interconnected references coupled with Ripley's known predilection for reconciling awkward authorities doubtless suggested the canon's own hand in this commentary. He admits as much in the *Key*, where he cites "Ripley vpon the Concordance of the words of Hermes and Aristotle" in his chapter on the vegetable stone.[67]

Yet the original "Exposition" is not by Ripley at all. A shorter version, included in the fifteenth-century compendium Sloane 3744, is attributed to "Ricardus de Salopia" (Richard of Shropshire), an alchemist probably of the late fourteenth or early fifteenth century.[68] Richard's surviving recipes focus on sal ammoniac, including a recipe for the "human stone" made by distilling human excrement gathered from the sewer.[69] Given the focus of his practice, it is no surprise that Richard's commentary on the *Secretum* reveals par-

---

65. Harley 2411, fol. 7r: "Expositio G: Riplay dicto*rum* Aristotelis, & Hermetis concordando ea adinvicem."

66. Sloane 3667, fol. 149r.

67. Ashmole 1421, fol. 181r.

68. This dating is suggested on the basis of Richard's cited authorities (which do not yet include any references to Raymond Lull), and the fifteenth-century dating of several manuscripts in which his works appear (see note 69, below).

69. "Text*us* Ari*stoteles*," Sloane 3744 (fifteenth century), fols. 54r-60r. Richard's attributed works include "De proporc*ione* eleme*n*to*rum* dicit Ri*card*us de Salopia," Harley 3703 (fifteenth century), fols. 74v-95r; "Elixer de lapide huma*n*o per Ric. de Salopia," Sloane 3744, fols. 9v-11r; DWS 222 and 223. *De proporcione* reveals a similar preoccupation with the *Secretum secretorum*, reiterating pseudo-Aristotle's advice, "cu*m* hui*us* aq*ua* ex ae*re* et aerem ex igne et igne*m* ex *terra* tu*nc* ha*b*ebis plene arte*m*" (Harley 3703, fol. 74v).

ticular interest in the animal stone: a vegetative principle present in the composition of man's body.[70]

The version of the "Exposition" found in the *Bosome Book* therefore constitutes a later and additional layer of commentary on Richard of Shropshire's original concordance, where the additions represent Ripley's "working up" of the text, similar to his commentary on Lull in the *Medulla*. In the process, the underlying alchemy has been transformed. References to Raymond and Guido, absent from Richard's version, have been added to yield a familiar reading: namely, that "this stone springs in the first place from wine, and in the second from Adrop."[71] The resulting commentary presents us with Ripley's own, sericonian elaboration of Richard's earlier treatise—an act of textual cannibalization and appropriation in the service of the canon's own practical program.

Maria dicit

Ripley's knack for adding a sericonian twist to earlier authorities is further illustrated by "Maria says" (*Maria dicit*), his commentary on a passage attributed to the ancient authority Mary the Jewess, also known as Mary the Prophetess. The passage describes a process using a fixed metallic body: "From it make your water, like running water, from two *Zaybeths* divinely crafted, and afterwards kill it upon the fixed body which is the heart of Saturn."[72] This text is itself adapted from an influential fifteenth-century dialogue, in which "Maria" uses the term "zaibeth" or "zaybeth" as a *Deckname* for mercury.[73] As such, the enigmatic passage is ripe for a sericonian reading.

The first step is to gloss Maria's terms. For Ripley, Maria's fixed body must be the imperfect metal, Adrop, "of which is made that which is called by

---

70. Sloane 3744, fol. 55v: "Aristoteles in libro de secretis secretorum cum dicit Operatio huius uegetativae est in compositione corporis humani."

71. Harley 2411, fol. 20r: "Hic autem lapis oritur primario ex vino, secundario ex Adrop, cuius similitudinem scilicet oleagineam induit, qui ideo propter maculam originalem multis depurationibus indigebit."

72. Ibid., fol. 40r: "Radix scoliae nostrae est corpus indole . . . fac ex eo aquam tuam, sicut aquam currentem, ex duabus Zaybeth divinitus elaboratam, et post interfice eam super corpus fixum quod est de corde Saturni."

73. "Mariae Prophetissae Practica," in *Artis auriferae, quam chemiam vocant*, 3 vols. (Basel: Conrad Waldkirch, 1593): 1:319–24, on 321–21; recently edited from manuscript sources as "Alumen de Hispania," in Timmermann, *Verse and Transmutation*, 305–11 (discussion on 44–45). The gloss "de corde Saturni" is absent from both editions.

the masters sericon." Her "water" is the menstruum drawn out of this body, which has a dual nature, containing both vinegar and the mercury of the body itself.[74] This double menstruum, simultaneously vegetable and mineral, corresponds to Maria's two mysterious "zaybeths." It is used to dissolve the fixed body, or "heart of Saturn"—a term that caused the commentator some initial difficulty:

> I studied and mused with my selfe a long tyme before I coulde vnderstand what that was, that is called the harte of Saturne. A long time I tooke it for gold, which was not verie licklie . . . soe saith Guido: purses are not to be opened for making of greate expences, which in this our arte is not req[u]ired.[75]

Once again, Guido's prohibition of expensive materials allows gold to be ruled out as a major ingredient. Rather, the fixed body ought to correspond to a less costly material: the "second earth" made from the menstruum itself once the excess fluid has been distilled away. This earth is repeatedly soaked in the double "zaybeth" menstruum, and the distillation repeated "vntill all be dryed vp and fixed together."[76]

Maria's original saying has therefore been unpacked to give a process that maps exactly onto Ripley's own recipe for the vegetable stone, and accords with the fifteenth-century debate concerning the high cost of the *prima materia*. Ripley even adopts the same description of his imperfect body, "which by masters is called sericon," in the *Medulla*.[77] The rejection of gold as a material principle in favor of a base metal, coupled with Guido's stricture against excessive opening of purses, is also typical of Ripley. The relationship between the two works was certainly not lost on Norton, who noted, beside the reference to sericon, that this "agreth with G.R. Medulla."[78] For this Elizabethan reader, Ripley's authorship of the *Bosome Book* must have been self-evident.

74. Harley 2411, fol. 40r: "Corpus est Adrop de quo fit illud quod à magistris vocatur Sericon. Aqua est menstruum, de ipso tractum, quod ex duplici spiritum consistit; si Aceti et sui ipsius, quae sunt duo Zaybeth id est 2° Mercurij."

75. Sloane 2175, fol. 149v.

76. Ibid., fol. 149r.

77. Discussed in chap. 2, above.

78. Sloane 3667, fol. 125v.

THE NAME OF THE ALCHEMIST

The contents of the *Bosome Book* reveal that it was compiled with a particular aim in mind: to assemble theoretical and practical authority for a sericonian alchemy derived from pseudo-Lullian texts. It so happens that the results are highly compatible with the alchemy of George Ripley, particularly the *Medulla* and *Concordantia*, but also with items encountered in other fifteenth-century compendia. There seems no reason to doubt that the *Book* was an authentic product of the late fifteenth century, or that Ripley himself was the author and commentator of much of its content, if not all.

The issue remains of whether those contents were personally gathered by Ripley, or by another compiler—presumably one with access to the canon's own books. Both the Latin version (Harley 2411) and Norton's translation (Sloane 3667) record many occurrences of the initials "G.R." throughout the text, while Ripley's name is also attached to a number of individual treatises. Such instances are not unheard-of: for instance, Ripley's initials are appended to a set of verses, "Gaudeat artista," in one late fifteenth-century manuscript.[79] However, in this context the initials suggest the assignation of authorship by another scribe, rather than Ripley's own signature.[80] If the *Book* were Ripley's own, the lavish use of name and initials seems peculiarly redundant. Here, Gawin Smith's anonymous copy of the *Book* in Ashmole 766 (mentioned above) provides an important counterpoint to Harley 2411 and the Norton translation. In this copy the title does not attribute the collection to a named authority, raising the question of whether Ripley's connection with the *Book* was actually as overt as Norton's references suggest.

Ripley's presence is certainly less conspicuous in Ashmole 766. His name is absent from both the title and the explicit of the *Concordantia*, while many incidences of his initials recorded in Norton's translation are also omitted. However, the initials "G.R." still recur at several of the same points throughout the manuscript—for instance, in the *Notable Rules* and *De ignibus nostris*.[81] Some works are explicitly attributed to "George Ripley Cannon."[82] Furthermore, the manuscript preserves several epigrams containing Ripley's name at the same points as in Harley 2411 and Sloane 3667. It seems that we

---

79. CCC 336, fol. 4v. See Rampling, "Establishing the Canon," 196. The verses are CRC 14.

80. The initials are not present, for instance, in the version of "Gaudeat artista" in Trinity O.8.9, which seems to have shared a common exemplar with CCC 336.

81. Sloane 3682, fols. 41r, 43v.

82. The *Compendium* and *Somnium*: Sloane 3682, fols. 50v, 52v.

must accept these features as integral to the *Bosome Book*, the initials per-haps having been added by an early, even still fifteenth-century, owner.

The epigrams offer further persuasive evidence for Ripley's compilation of the *Book*. One provides a straightforward rearrangement of the alche-mist's name: "George gives you all his writings in a brief space to weigh up. Turn round piR, add Lay."[83] Besides yielding the name "Riplay," this for-mula puns on the alchemist's profession by including the Greek word "pir" (πυρ), or "fire." Not all are so straightforward. The longest is a Latin quatrain, containing two riddles:

> Maturus iacuit Geor ista Gius breviter dat
> Illi qui verso cognomine non variatur
> Cuius in hoc certe constantia notificatur
> Hinc datur ex arte, sibi quod sic dignificatur.[84]

Although Norton omitted this epigram from his own transcription, and Howes chose to record it in untranslated Latin, the first line was at some point translated to accompany another text extracted from the *Bosome Book*, the *Practise by Experience*: "George died when he was of ripe yeares, and he giueth these preceptes breefly."[85] Here, the reference to Ripley's death implies that he could not have composed the epigram himself. Such a loose rendering misses what is evidently another pun on Ripley's name, as we see when we adjust the translation:

> Ripe *Geor* lay down, this *Gius* briefly gives,
> To him who does not change when his surname is reversed,
> Whose constancy in this is surely shown,
> Hence that which is thus dignified, is given him by art.

What "Gius" briefly gives is a clue to the epigram: since the name "Geor-gius" has been divided, George's surname must also have been bisected. Sure enough, the answer is hidden within "Maturus iacuit"—"Ripe-lay," or Ripley.

In both Harley 2411 and Ashmole 766, the name "Maram" is jotted down beside the epigram. This palindrome clearly satisfies the second portion of the riddle: the constancy of Maram's name in either direction makes him a

83. Harley 2411, fol. 68r: "Compendere brevi spacio sua cuncta Georgi | vs dat scripta tibi verte piR adde Lay."

84. Ibid., fol. 40v.

85. Ashmole 1485, fol. 88r.

worthy successor to Ripley's teachings. This Maram was perhaps the compiler of the original *Bosome Book*, or may even have inherited the precious codex from Ripley himself. Either way, the epigram implies a personal relationship with the canon, and may account for the lavish scattering of Ripley's initials throughout the *Book*.

This inquiry must fall short of identifying the palindromic Maram, although Norton offers one suggestion. In the preamble to his *Key*, he includes among the English adepts "Marram Bishop of Yorke to whome Riplie wrote his Medulla."[86] George Neville, Archbishop of York, counted "Wharram" among his titles, but the names are clearly not identical, whereas Maram already existed as an English surname of the period; it was also a town in the diocese of Lincoln. Perhaps in this regard Norton allowed speculation to carry him too far. As one of the great conundrums of alchemical literature, it seems fitting that the *Bosome Book* should leave us with this final mystery.

NORTON AND PRACTICAL EXEGESIS

Despite the peculiar circumstances of the *Book*'s resurrection, and even in the absence of an exemplar, we can conclude that Samuel Norton possessed a genuinely fifteenth-century manuscript containing authentic writings of Ripley. This was probably compiled by the canon himself in the context of a strong, rhetorical push for his own brand of sericonian alchemy—the kind of push that might have been associated with a patronage suit. Contemporary audiences, able to view the actual book, seem to have accepted its provenance without question, and were swift to make copies of their own.[87]

The *Book*'s contents varied in popularity, and two Latin poems, the *Visio* and the *Somnium*, were particularly successful. The former is known to modern readers almost exclusively in its English translation, as *The Vision of Sir George Ripley*.[88] This famous poem, in which the death of a toad provides an allegory of the alchemical work, is actually Norton's translation of the original verse, "Pervigil in studio, nocturno tempore quodam."[89] The *Somnium*, an even more exotic allegory, describes a battle between Sol (gold) and the sericonian monster "Adropus," in which Sol at first perishes in the monster's

---

86. Getty 18/10, 15. Spelled "Maram" in Ashmole 1421, fol. 171v.

87. See CRC 32.

88. *TCB*, 374; discussed in chap. 9, below. Rabbards published the same translation in 1591.

89. Harley 2411, fol. 69r. Norton translates the incipit as "Whan busye at my booke I was vpon a serten nyghte"; Sloane 3667, fol. 149r.

flames, only to revive in a form that is, happily, resistant to fire.[90] Although never printed, this "Dream" was translated into English three times: by Norton and Howes, and also by the alchemically inclined theologian Edward Cradock in 1582, nine years after Norton first translated the *Book*.[91]

Some puzzles remain, for despite this transcription activity the extracts found in later manuscripts often vary significantly from the versions of the *Book* so far discussed. For instance, the *Book* contains a sericonian process, beginning "Recipe Adrop," that later acquired independent circulation in an English translation. In the new version, titled *George Ripley's Practise by Experience of the Stone*, the original recipe has been modified to accommodate several comments apparently based on empirical observation. This adaptation was transcribed by Dee and Robson, among others.[92]

While the source for the *Practise* can easily be identified as "Recipe Adrop," it is often accompanied in manuscript by a work somewhat harder to relate to the *Book*'s contents. This is a long, practical text, "The whole wourcke of the composicion of the stone philosophicall or greate Elixir, & of the fyrste solucion of the grosse Bodye," which for convenience I shall refer to as the *Whole Work*.[93] This item, in which the creation of the elixir is laid out in discrete and relatively comprehensible steps, was already circulating by the late 1570s, and by the early seventeenth century was completely integrated into the Ripley Corpus.[94] With the exception of the *Vision*, the *Whole Work* is probably the best-known item from the *Book*, even to the extent of assuming its title in print. It was published by William Cooper as *The Bosome-Book of Sir George Ripley* in 1683.[95]

The clear, practical orientation of the *Whole Work* is apparent from its incipit: "First take 30 pound weight of Sericon." Dissolved in vinegar, the sericon is evaporated to create a green gum, appropriately named the Green Lion: "which done our Sericon will be coagulated into a green Gum called our green Lyon, which Gum dry well, yet beware thou burn not his Flowers

---

90. Harley 2411, fol. 69v: "Sompnia ne cures, Dictum vulgare tenetur."

91. Philadelphia, University of Pennsylvania, Codex 111, fol. 43r: "Thus endeth [th]e dreame of Sir George Ripley Chanon of Bridlington. this was translated owte of laten verse [th]e 4th day of June anno 1582 by Mr doctor Cradocke."

92. "Practise," CRC 26.

93. Sloane 1095, fol. 75r.

94. CRC 35.

95. "The Bosome-Book of Sir George Ripley, Canon of Bridlington. Containing His Philosophical Accurtations in the making the Philosophers Mercury and Elixirs," in *Collectanea Chemica*, 101–21.

not destroy his greeness."[96] When the delicate gum is heated in a retort, it first yields the familiar faint water, then the white fume. Finally, stronger heating results in reddish "drops like blood" that should be carefully conserved. Once the vessel has cooled, the practitioner will also find other useful products remaining in the glass. The neck of the vessel becomes encrusted with "a white hard Ryme" like the "Congelation of a Frosty vapour," which the writer likens to mercury sublimate.[97] Black dregs, or feces, also remain behind in the bottom of the vessel, and these are called the "Black Dragon."

Although the Dragon may look like a waste product, it provides grounds for an extraordinary chemical effect. The *practica* continues:

> Then take all the rest of the aforesaid black Feces or black Dragon, and spread them somewhat thin upon a clean Marble, or other fit Stone, and put into the one side thereof a burning Coal, and the Fire will glide through the Feces within half an Hour, and Calcyne them into a Citrine Colour, very glorious to behold.[98]

Although such an account is not found elsewhere in Ripley's attested writings, the instructions are sufficiently clear that it can be recreated in a modern laboratory, by the simple expedient of using litharge (yellow lead oxide) in place of red lead. When this "Black Dragon" is tipped onto a heatproof surface and ignited with a hot coal, the black lead rapidly transforms to orange-yellow, creating the promised effect of a "gliding fire"—a simple process in modern chemical terms, signifying no more than the reoxidation of finely divided lead back into litharge (fig. 11).

The ease with which this striking transformation can be reproduced would surely not have been lost on an early modern reader of the *Whole Work*. Evidently, the Ripleian recipe provides testable and dramatic results. It is therefore particularly ironic that the *Whole Work* is one of the few texts related to the *Bosome Book* for which Ripley's authorship may be categorically excluded. Indeed, the "testable" portion of the work does not come from the *Book* at all.

In both Harley 2411 and Norton's translation, the long commentary *Maria dicit* is followed by a practical text, *Separatio elementorum* (Separation of Elements). This begins, "When our red humour is distilled from sericon, it

96. Ibid., 102.
97. Ibid., 102–3.
98. Ibid., 104.

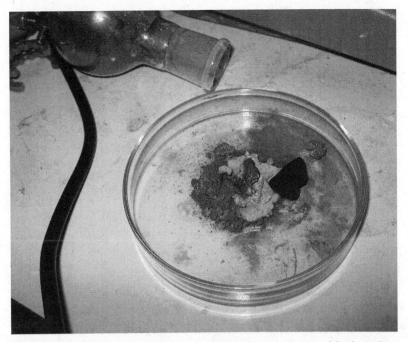

FIGURE 11. The "gliding fire" spreads outward from a hot coal. Photograph by the author.

should be rectified by *balneo* [i.e., water bath]," before briefly describing the extraction of a "most ardent spirit" from sericon.[99] The flammability of the spirit is gauged by a standard test: if a linen cloth previously dipped in the liquid is ignited, the spirit will burn without damaging the cloth. This clear spirit corresponds to "air," and the text goes on to describe the isolation of the remaining three "elements."

This material, including the flammability test, does not appear in the *Whole Work* until well into the process. Although the central part of the *Work* reproduces the *Separatio* fairly closely, supplementing Ripley's procedure with a few practical observations, the preceding text bears no relation to the *Book*, other than the reference to sericon in the opening line. What, then, is the source of the interpolated text? And how did it get there?

The puzzle is solved if we return to the preamble to Norton's *Key*. Here, Norton praises the accomplishments of English alchemists, singling out two in particular:

99. Harley 2411, fol. 43v: "Cum distillatus fuerit humor noster rufus à Sericone, rectificetur per balneum et trahatur tantum lentissimo calore spiritus eius ardentissimus."

[T]here is none that have deserved more Commendationes & honor, then have our owne Countrie men; of whome, I will name two, Ive, & George Riplie; whose worcks I Judge were by some divine providence of God left to the renuing of those excellent arts; that they should not bee hidden, & lie dead.[100]

The name of Ive is an unfamiliar one in studies of English alchemy. It is possible that he practiced medicine, since Norton explains how "Ive vnder Couller of Phisick, taught how to handle the base, and to exstract the Menstrue."[101] It is still Ripley, however, who merits the greatest praise:

[T]here fore I must needs say that Riplie was alonelie [th]e man: for beginning where Ive beginneth; hee ceaseth not; but plainlie sheweth, howe to beginne, how to Continue, & how to finishe & make perfect.[102]

Norton's statement that Ripley continues from where Ive leaves off is not merely rhetorical. In Sloane 3667, one of Ive's recipes is placed before the *Bosome Book*: "A notable worke of Sericon writen p*er* master Ive."[103] This work provides the missing segment of the *Whole Work*, including the distinctive reference to the "gliding fire" (table 7.2). The *Work* puts Norton's words literally into practice: the opening stages of the procedure are taken from Ive, and the remainder from the *Book*, while some additional details, including the time taken for the "fire" to spread, are apparently the result of practical observation.

The amalgamation of separate texts is a common feature of alchemical treatises throughout the fifteenth and sixteenth centuries, as illustrated by Lock's *Treatise*, among others. Yet the *Whole Work* brings a new dimension to this commonplace and commonplacing activity, demanding not only theoretical compatibility, but practical efficacy. Two distinct recipes are combined to produce a step-by-step procedure, with the implication that the greatest of these jumps—from Ive to Ripley—may be accomplished without stumbling. The consensus of the philosophers is simply assumed.

The source of these variant versions is almost certainly Norton himself. While Sloane 3667 represents Norton's original translation, the resourceful

100. Getty 18/10, 19.
101. Ibid. Although several "Ives" were members of the College of Physicians during the sixteenth century, it is not clear which of these, if any, corresponds to Norton's "Master Ive."
102. Ibid., 20.
103. Sloane 3667, fol. 115r.

TABLE 7.2. The *Work of Sericon* and the *Whole Work* compared

A notable worke of Sericon per Master Ive, Sloane 3667, fols. 115r-116r	The Bosome-Book of Sir George Ripley (London, 1683)
[T]take [tha]t Resedeue so blake & drye, & laye yt on astone abrode, or on Clyne Earth, & putte a Cole bernyng on [th]e on[e] syd of yt, & anon yt wyll set all on fire, glydyng thoroue yt, & Calcynyng yt, in [th]e 4ᵗʰ part of vn ower, into a yellowe Collour, bryght as [th]e bemes of [th]e sonne, which is great Wonder to be holde.	Then take all the rest of the aforesaid black Feces or black Dragon, and spread them somewhat thin upon a clean Marble, or other fit Stone, and put into the one side thereof a burning Coal, and the Fire will glide through the Feces within half an Hour, and Calcyne them into a Citrine Colour, very glorious to behold.

alchemist spent the three years between its completion and the composition of the *Key* both testing and adapting the *Book*'s material. A hint of this activity survives in the *Key*'s dedicatory letter, where he mentions his former plan to produce an edition of Ripley's *Bosome Book*, supplemented by his own practices and critical commentary:

> [W]*hi*ch bookes I soe provided, that it might Come to your Ma*ie*sties hands, not so much ffor the book it selfe, as for my owne severall practise; w*hi*ch should have beene there vnto annexed, with a right Censure & Judgement of proceedinge in the rest, ffor though some there are w*hi*ch I know have the same workes, yet have they failed in proofes, not by the fault of the Author, but by their owne follies.[104]

In the best tradition of alchemical authorship, much of what is important in this passage must be read between the lines. In soliciting patronage, Norton was concerned not to be dismissed merely as the passive mediator of Ripley's long-lost practice—a risk that must have arisen had he merely presented the queen with a translation of the *Book*. Rather, Norton emphasizes the value he has added to the text by subjecting Ripley's processes to personal, empirical examination, and then recording the results in the planned "annexe" to his translation. Potential rivals are dismissed with the observation that even those with access to the same works have failed to translate them into a workable practice.

At some point, Norton altered his plans. By writing the *Key* he opted to revise the original scheme, in order to offer Elizabeth "much more then that

104. Getty 18/10, 8.

booke Conteined, or my selfe at that time either knew or thought on, which since I have in practise found out."[105] "*That booke*"—the "annexe"—was no longer adequate for Norton's purposes. Yet relics of the earlier project survived. These include the complementary processes of Ive and Ripley, which Norton fused together to provide the *Whole Work*, while the Latin text "Recipe Adrop" became the *Practise by Experience*. These revised works may have started to circulate even before the *Key* was complete, as the early date of some copies suggests.

Norton's splicing together of recipes by Ive and Ripley is practical exegesis in action: grounded in and respectful to authority, yet irresistibly results-oriented. This pragmatic approach, coupled with Norton's repeated emphasis on his own experience, underlines the fact that he was seeking not just a licence, but investment. To this end he offers Elizabeth his writing, but also "the writers hand to performe it, yf your Highnes shall command."[106] Like Blomfild writing from the Marshalsea prison half a century earlier, Norton hoped that his ability to reconstruct fifteenth-century practices, proven by time (if not yet, entirely, by Norton himself), would convince the queen of his suitability to undertake more comprehensive assays.

Filling in the gaps in descriptive accounts was of tantamount importance to an alchemist who sought to reenact both the processes and the outcomes of Ripley's works for an exacting royal patron. Norton was himself aware that his proposal might be misconstrued. He explains that he has not yet pursued the work to its conclusion, having been advised by a certain friend "learned in the lawes" that such activity might fall within the compass of Henry IV's statute against multiplication. To reassure the queen and "remove suspect of sinister dealing," Norton undertakes not only to set out his processes clearly, but also to provide an estimate of the total cost of the work.[107] True to his word, he closes the *Key* with a list that sets out "such Charges as will rise in the accomplishing or performing of the whole art & science."[108]

These charges allow us to identify exactly how Norton envisaged his practice. His list begins with the ingredients for making sericon: "Red lead or minium in waight 280 [pounds]," which at four pence the pound is rather cheaper than the "280 Gallons of distilled vinegere" in which it is to be dissolved, at ten pence the gallon. A further 160 gallons of vinegar will be

105. Ibid.
106. Ibid., 155.
107. Ibid., 12.
108. Ibid., 156.

required for the second dissolution. These ingredients are to be prepared using three dozen "stone bodies to distill vinegere" at sixteen pence each, together with three dozen still heads (twelve pence each) and four dozen large receivers (sixteen pence apiece). £10 is assigned to cover the cost of brick and ironwork in making six furnaces, which Norton illustrates in diagrams appended to the list (fig. 12).

Having prepared the vegetable stone, Norton next requires ten pounds of crude mercury, at five shillings the pound, "to serve the minerall stone." The "materialls of the \mixed/ stone," comprising "Corrosive & Compound waters to dissolve the gold & silver," are not itemized, but simply rounded up to £4. Finally, Norton requires £14 for his most expensive item: four ounces each of gold and silver "for the Elixir of Life & firments of the stone," plus another £3 to cover the cost of "purging & beating thereof into foliate."

Such ingredients map exactly onto the processes described by Norton in the *Key* and in his other receipts, extracted and modified from Ripley's *Medulla* and *Bosome Book*. These materials embody Norton's plans to attempt a range of alchemical products: a vegetable stone distilled from lead and vinegar; a mineral stone sublimed from mercury; a mixed stone that fuses vitriol, saltpeter, and sericon; and an "elixir of life," for which—as Ripley warned the bishop a century earlier—only natural gold and silver will serve. In this list of charges, Norton itemizes both his claim to have penetrated the mysteries of the philosophers and his ability to personally recreate them for the queen.

REPLICATING SERICON?

If evidence were needed for the vitality of England's "native tradition" of sericonian alchemy during the latter part of the sixteenth century, the works of Ripley's Elizabethan disciples provide it in abundance. Practitioners like Norton, Blomfild, Charnock, Lock, and countless *anonymi* continued to treat pseudo-Lullian and Ripleian alchemy as materially and conceptually relevant to their own alchemical practice at a time when Paracelsian ideas were already circulating in England and causing serious disequilibrium in the Paris medical faculty. This relevance meant that authorities were not necessarily treated as monolithic texts, but as working documents apt for further experiment and adaptation. In the process of reclaiming Ripley's lost work, Norton did with the *Bosome Book* as Ripley had done with his own materials—supplementing and amalgamating earlier texts in the light of wide reading and personal, practical experience. The *Medulla*'s commentary on Lullian sources finds its analogue in Norton's *Key*, which also offers a fresh exposition and practical digest of an existing body of work.

The partly pseudepigraphic *Whole Work* is one outcome of this practical engagement with medieval texts. Rather than deliberate fraud, Norton's free use of earlier materials suggests a sincere attempt to reconstruct the laconic procedures excavated from manuscript sources, and regain the practical wisdom of past masters. His spliced recipes, backed up by experimental evidence that retrospectively justifies past authorities, suggests a new category of "forgery": one grounded on the conviction that medieval practices were both relevant and achievable, and that practical effects ultimately outweighed textual authenticity as a source of evidence.[109]

Yet was Norton any more successful in reconstructing Ripley's authentic practice than he was in reassembling the canon's lost *Bosome Book*? As modern attempts at laboratory replication (including my own) suggest, care must be taken when seeking to recreate past experiments, particularly those involving encoded and impure ingredients.[110] As we have already seen, pure minium does not readily dissolve in vinegar, suggesting that Norton may have had to modify his own ingredients in order to obtain the results he describes—results that can, however, easily be achieved by substituting another lead compound, litharge.

As it happens, early modern readers faced the same dilemma, and came to broadly the same conclusion. While Samuel Norton remained faithful to Ripley's likely intention by advocating the use of red lead, a different approach was taken by the French diplomat and translator Blaise de Vigenère (1523–1596). De Vigenère, inventor of a famous cipher, apparently deciphered Ripley's "sericon" without difficulty, but recognized that it was difficult to dissolve. He therefore turned to the more readily soluble litharge:

---

109. On early modern notions of forgery, see Anthony Grafton, *Forgers and Critics: Creativity and Duplicity in Western Scholarship* (Princeton: Princeton University Press, 1990). See also Didier Kahn and Hiro Hirai, eds., "Pseudo-Paracelsus: Forgery and Early Modern Alchemy, Medicine, and Natural Philosophy," *Early Science and Medicine* 5–6 (2020): 415–575.

110. On some of the methodological issues raised by attempting modern reconstructions of past alchemical processes, see particularly Lawrence M. Principe, "'Chemical Translation' and the Role of Impurities in Alchemy: Examples from Basil Valentine's *Triump-Wagen*," *Ambix* 34 (1987): 21–30; Principe, *Secrets of Alchemy*, chap. 6; and the collected essays in Hjalmar Fors, Lawrence M. Principe, and H. Otto Sibum, eds., "From the Library to the Laboratory and Back Again: Experiment as a Tool for Historians of Science," *Ambix* 63 (2016).

Some, as *Riply*, and others, have taken the *minium* of lead, but it is . . . of an uneasie resolution, as also ceruse & calcined lead. For my part I have found litharge, which is nothing else but lead . . . poure thereon distilled boiling vinegar, stirring it strongly with a staffe, and sodainly the vinegar will charge itself, with the dissolution of litharge.[111]

De Vigenère had good reason to be confident, for his substitution makes no appreciable difference in the early stages of the process described in the *Whole Work*—if anything, it simplifies the procedure. The resulting "gum" can be dry distilled to produce a white fume, exactly as described in the *Whole Work*. And, as De Vigenère goes on to explain, the dregs that remain in the distillation vessel (the residue that Norton called the Black Dragon) will still ignite:

That which remains in the *Cornue*, put burning charcoals upon it, and that will take fire as the match of a fusee: whence you may draw a fair secret: for as long as it feels not the air, it will not flame, and it may dissolve again with vinegar, to doe as before.[112]

The lighting of a match to a cannon fuse offers an evocative simile for the gliding fire described by Ive and Samuel Norton. More intriguing still is De Vigenère's statement that the Dragon will not ignite if "it feels not the air." This is one moment when a simple laboratory reenactment does help to shed light on the meaning of a text. The leaden Black Dragon is highly pyrophoric, to the extent that, if it is still hot when taken from the vessel, it will frequently "catch fire" on contact with air, even in the absence of a hot coal—an effect that De Vigenère also seems to have observed.

We might wonder why, if Norton indeed devised the *Whole Work*, he did not mention this striking property of the Dragon. The omission may of course relate to his choice of a different lead compound, such as minium. But it is also possible that he did not observe this effect because the instructions in the text, if followed correctly, make it far less likely that his dregs could oxidize on their own, by warning practitioners not to remove the receiver from the retort until "all things are cold."[113] Norton's advice in the *Key* on

111. Blaise de Vigenère, *A Discourse of Fire and Salt, Discovering many secret Mysteries, as well Philosophicall, as Theologicall* (London: Richard Cotes, 1649), 70; an English translation of Vigenère's posthumously printed *Traicté du Feu et du Sel* (Paris: Abel l'Angelier, 1618).

112. De Vigenère, *A Discourse of Fire and Salt*, 71.

113. "Bosome-Book of Sir George Ripley," 103.

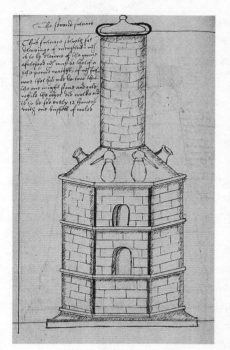
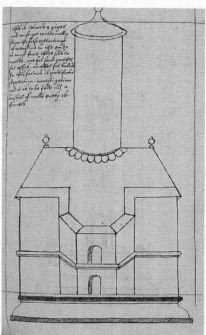

FIGURE 12. Furnace designs, copied by Thomas Robson from Samuel Norton, *Key of Alchemy*. Oxford, Bodleian Library, MS Ashmole 1421, fols. 218v-19r. By permission of The Bodleian Libraries, The University of Oxford.

the management of his furnaces would also ensure thorough cooling of the apparatus, reducing the likelihood that his Dragon would ignite on its own. In his notes to the furnace illustrations, he suggests using two furnaces for the process of drawing off the sericonian menstruum: "that the one might stand and cole while the other did worke" (fig. 12).[114]

Norton's and De Vigenère's differing responses to the same process offer a salutary reminder of both the diversity of approaches encountered in early modern Europe and the difficulty of reading even practically focused alchemical texts. While the *Bosome Book* offered Norton a variety of opportunities, in practice he chose not to present its contents in an unaltered form—an inevitable consequence, perhaps, of the challenges faced when reconstructing alchemical practices.

What does emerge from the manuscript record is the seriousness with which Elizabethan practitioners approached their written sources, as repos-

114. Ashmole 1421, fol. 218v.

itories of testable, practical information. Norton was aware of the historical importance of Ripley's lost book, but subordinated its antiquarian value to the practical information it contained, even to the extent of revising recipes where necessary. His respect for his source was tempered by a concern for results. After all, as a commonplace book rather than a coherent treatise, the *Book*'s contents did not always bear the stamp of Ripley's personal authority, as Norton learned after one ill-fated attempt at "separation":

> The proofe of which I full dearelie bought; ffor there by I lost all my quantite of white tincture in seeking of the Lunarie after that manner, ffor that which I found, thincking it to have been Riplies owne manner of Separation, was but a Note of Separation by Ripley taken out of the worcks of Hortulan[us].[115]

Setbacks aside, Norton was not easily disappointed in his investigations. Several years of research into the *Book*'s secrets seemed to increase rather than dim his enthusiasm for the canon's works. In his *Libri tres*, completed in 1599, Ripley is still one of the most-cited authorities. The canon's process for the vegetable stone even provides the basis for Norton's Book III, on "Saturn, or lead."[116] Adding a further turn to the wheel of the *Book*'s reception, in 1630 the York physician Edmund Deane (1572–ca. 1640) published the *Libri tres* in a series of eight short tracts, including Norton's sericonian process under the appropriate title of *Saturnus saturatus* ("Saturated Lead").[117]

Through such reformulations, Ripley's alchemy retained both its prestige and its relevance well into the seventeenth century. Yet his original writings now marched alongside an increasing number of close relatives, in the form of new, practical interpretations of the same source texts, some of which would be retrospectively assigned to his authorship. In this cycle of reinvention, Norton was simultaneously a casualty and a beneficiary of the canon's continuing fame. While his contribution to the *Whole Work* was ultimately eclipsed by the glow of Ripley's reputation, the efforts of his own mediator, Deane, granted him a measured afterlife in print—and with it, a foothold in England's alchemical history.

---

115. Ibid., 39.

116. Samuel Norton, "Liber ramorum tertius inceptis Aprilis die martis 16. anno domini 1599. De tabula Saturni siue plumbi ramus primus," Ashmole 1478, pt. 6, fol. 74v.

117. Deane's series begins with *Mercurius Redivivus, seu Modus conficiendi Lapidem Philosophicum* . . . (Frankfurt, 1630). On Deane, see Mandelbrote, "Norton, Samuel."

CHAPTER EIGHT
Home and Abroad

E. K. did open the great secret to me, God be thanked![1]

In April 1591, Edward Kelley, one of Europe's most famous living alchemists, was arrested in Bohemia. Up to this point the Englishman had pursued a spectacular alchemical career in the imperial capital of Prague and in Třeboň, an estate of the powerful Rožmberk family. Accounts of his remarkable transmutations had reached England, rekindling Elizabeth I's interest in alchemy and prompting Cecil to espouse Kelley's return to his native land. Kelley demurred, admitting to Cecil that his privileged situation abroad was unlikely to be matched at the English court. The Holy Roman Emperor, Rudolf II, a committed patron of alchemy, had validated Kelley's claims of noble Irish descent, presenting him with both an imperial knighthood and a seat on his privy council.[2] Kelley had also received several estates and villages from his patron Vilém of Rožmberk (1535–1592), the High Burgrave of Bohemia.[3] "I am not soe madd," he observed, "to runne away from my present honor and landes to shove for a newe."[4]

1. John Dee, diary entry for 10 May 1588, in *The Private Diary of Dr. John Dee, and the Catalog of His Library of Manuscripts*, ed. James Orchard Halliwell (London: Printed for the Camden Society, 1842), 27.

2. On Kelley's activities in Prague, see Evans, *Rudolf II and His World*; Michael Wilding, "A Biography of Edward Kelly, the English Alchemist and Associate of Dr John Dee," in *Mystical Metal of Gold*, ed. Linden, 35–89; Rafał T. Prinke, "Beyond Patronage: Michael Sendivogius and the Meaning of Success in Alchemy," in *Chymia*, ed. López Pérez, Kahn, and Bueno, 175–231; Vladimír Karpenko and Ivo Purš, "Edward Kelly: A Star of the Rudolfine Era," in *Alchemy and Rudolf II*, ed. Purš and Karpenko (Prague: Artefactum, 2016), 489–534.

3. Kelley received the estates of Liběřice and Nová Libeň from Rožmberk; Prinke, "Beyond Patronage," 183.

4. Edward Kelley to William Cecil, Lord Burghley, 10 August 1590, SP 81/6, fol. 65r.

With his arrest, Kelley apparently lost all that he had gained—not just his lands and his position at Rudolf's court, but his credibility in England as well. Questions over his alchemical prowess prompted his flight from Prague in an attempt to gain sanctuary with Rožmberk. Overtaken en route, he was incarcerated at the emperor's pleasure in the fortress of Křivoklát, or Pür-glitz. His arrest caused a sensation: rumors flew, and in a flurry of correspondence between Prague and England, Burghley's intelligencers threw out varied and contradictory reasons for Kelley's sudden, yet oddly unexplained, fall from grace. Some suggested that the Englishman was heavily in debt; that he had slandered the emperor, or even attempted to poison him. Others thought that Rudolf had learned of attempts by Elizabeth and Burghley to lure Kelley away from Prague, and imprisoned the goldmaker to keep him from absconding.[5]

At first glance, Kelley's failed enterprise is not obviously a story about England. During the later sixteenth century, alchemists of many nations gravitated to Bohemia, a state that, although Catholic, nonetheless tolerated other confessions to a marked degree, and boasted some of alchemy's most generous and attentive supporters in the persons of Rudolf II and his highest-ranking magnate, Rožmberk.[6] In that context, Kelley's rise and fall is just another example of the breakdown in patronage relationships that Nummedal has studied in the context of Europe-wide suspicion of alchemical fraud, or *Betrug*. Convicted *Betrüger* often arrived in states from abroad, laden with endorsements from previous patrons and promising profitable returns from their chemical techniques. In many cases, these men (and occasionally women) were probably sincere in their anticipation of successful results, only to be confounded by the physical limitations of their practice, the strict contracts that bound them to unattainable goals, and the serious penalties for failure.[7]

Just as Kelley's patronage enterprise is not obviously English in character, neither is it—again, at first glance—obviously a significant episode in the history of books and reading. Unlike his influential contemporaries Michał

---

5. The reasons for Kelley's arrest have been extensively discussed in the historical literature. For a summary of contemporary explanations, see Wilding, "Biography of Edward Kelly," 65–72. For patrons, the risk of abscondion abroad was taken seriously: Kelley's case has its parallel in Burghley's decision to confine Cornelius de Lannoy in the Tower of London (pp. 229–30, above); see also Nummedal, *Alchemy and Authority*.

6. The conditions that made Prague a center of both religious tolerance and alchemical production are discussed in Evans, *Rudolf II*; and the essays collected in Purš and Karpenko, *Alchemy and Rudolf II*.

7. See Nummedal, *Alchemy and Authority*, chaps. 4 and 6.

Sędziwój, or Sendivogius (1566–1636), and Michael Maier (1568–1622), also clients of Rudolf II, Kelley's later reputation rested on his fame as a master practitioner rather than as a writer of philosophical treatises.[8] Few of his writings were ever printed, and those that did appear under his name are highly derivative in nature, apparently offering little in the way of conceptual or practical innovation. When contemporaries later recalled Kelley's exploits, they lingered less on his literary remains than on his bravura practical demonstrations: most famously, an experiment in which, in the presence of the emperor himself, he produced the mercury of gold (*mercurius solis*) in the space of fifteen minutes.[9]

Yet Kelley's renowned practice, and even his choice of experiments, were deeply grounded in the history of English alchemy. He assimilated this history through his reading, taking advantage of access to one of Europe's greatest scientific libraries: that of his famed associate, John Dee. At a time when most alchemica, including almost everything attributed to English authorities, still circulated scribally, Dee's alchemical manuscripts (many of which accompanied the English party to Bohemia) offered a valuable resource for a budding practitioner. But Kelley acquired more than practical knowledge from his reading; he also learned how to reinvent himself as a philosopher within a distinctive national tradition. Just as his material success depended on proving descent from a noble Irish house, so his credibility as a philosopher required him to craft an authoritative genealogy for his practice.[10] He therefore set about inventing a new origin story for Ripley's sericonian alchemy, grounded on the authority of a still earlier adept, the tenth-century Archbishop of Canterbury, Saint Dunstan.[11]

Even far from home, Kelley provides one of the most remarkable examples of the English reception of late medieval alchemy. His Bohemian exploits show how medieval sources could be repurposed by a practicing alchemist seeking to appropriate the authority of a national tradition in support of his own patronage aspirations. Kelley did not merely read Ripleian texts; he reconstructed Ripleian practices, wrote commentaries on Ripleian works, and presented Ripleian works as gifts. As we shall see, he also used

---

8. Prinke suggests that Sendivogius's success rests partly on his strategy of claiming that he had not made the stone himself, but had acquired it from another adept; Prinke, "Beyond Patronage."

9. As attested by Matthias Erbinäus von Brandau, *Warhaffte Beschreibung von der Universal-Medicin* (Leipzig: F. Lanckisch, 1689), 13, 79, 92; cited in Evans, *Rudolf II*, 226.

10. Kelley's patent of nobility depended on him proving his descent from the Irish house of Imaymi: see Prinke, "Beyond Patronage," 182–83.

11. On the historical Dunstan, see Nigel Ramsay, Margaret Sparks, and Tim Tatton-Brown, eds., *St Dunstan, His Life, Times, and Cult* (Woodbridge: Boydell, 1992).

forged, pseudepigraphic tracts to further promote the authority of his English sources and, by extension, his own expertise. Above all, he exploited the relative unfamiliarity of new finds like Ripley's *Bosome Book* as novelties liable to intrigue his peers and catch the eye of patrons. In his hands, alchemical reading became the foundation for building a career.

As an alchemical reader and writer, Kelley's activities have been largely overshadowed by those of his renowned associate and traveling companion, Dee. His manipulation of alchemical history has also largely gone unnoticed because, like much of his reading matter, the testimony survives not in print but in manuscript. In this chapter, I marshal the evidence for Kelley's alchemical practice, using his own, previously neglected writings, many of which have not been identified before. These materials, copied during his lifetime or shortly after his death, were closely studied by alchemical aficionados based in Prague and elsewhere, including Kelley's assistants, correspondents, patrons, and—during his imprisonment—custodians. Most significantly, these include a codex from Rudolf II's library, now held in Leipzig University Library, that preserves a series of philosophical treatises and practical tracts prepared by Kelley for the emperor during the period of his incarceration at Pürglitz.

These "prison writings" offer a new view of Edward Kelley: not as the one-dimensional charlatan familiar from later histories, nor as a secondary player in the much better-known story of John Dee, but as an alchemical reader and exegete whose success rested, at least in part, on his skill at translating text-based descriptions into replicable effects. Kelley's Bohemian reconstructions breathed new life into old texts by demonstrating the ongoing practical significance of medieval authorities, including the contents of Ripley's recently uncovered *Bosome Book*. Like Samuel Norton, he also understood the necessity of putting his own stamp on these early materials. Indeed, Kelley went even farther than Norton in devising and exploiting "new" texts associated with prominent English adepts like Ripley and Saint Dunstan. In a curious reversal of fortune, Kelley's celebrity eventually served to raise the profile of his English forebears both in England and overseas.

ENGLISH ALCHEMY ABROAD

As English alchemists at home rejoiced in the alchemical achievements of their medieval forebears, alchemy on the Continent did not stand still. From editions of Paracelsus to the compilations of Italian "professors of secrets," alchemical material flowed into England in print and manuscript throughout the sixteenth and seventeenth centuries. Nor did these writings travel alone, as continental practitioners also traversed the English Channel to seek positions in London or farther afield. Cornelius de Lannoy, Burchard Kranich, and Giovanni Battista Agnello were among those who succeeded in winning serious support in England, but many other strangers revealed an interest in English alchemy. For instance, Michael Maier visited England in 1611–16 and later translated Thomas Norton's famous *Ordinal* into Latin prose; he also collaborated with the English iatrochemist Francis Anthony on a Latin edition of Anthony's celebrated treatise on potable gold.[12]

This traffic was not unidirectional. Medieval English authorities like Bacon and Dastin had long enjoyed Europe-wide reputations, but in the second half of the sixteenth century the writings of later adepts like Norton, Ripley, John Sawtrey, and Du Wes were introduced to continental audiences through the circulation of correspondence, books, and practitioners.[13] While Agnello was quoting extensively from the *Compound* in his own treatise dedicated to Elizabeth I, his compatriots could also have read Ripley's *Medulla* back in Venice, in a retranslation into Latin of Whitehead's Englished *Marrow*.[14] It is this "Type II" version of the work, rather than Ripley's original, that would eventually be printed in Frankfurt in 1614.[15]

Notwithstanding Maier's translation of Norton, among English adepts it was Ripley whose works achieved the greatest success in mainland Europe. Most of the core corpus was already circulating in Latin in France and Italy by the early 1570s, although only the *Philorcium* appeared in its original Latin. The *Compound* was translated by 1571 as the *Liber duodecim portarum* (Book of the Twelve Gates), and the *Epistle to Edward IV* not long

12. Michael Maier, *Tripus Aureus, hoc est, Tres Tractatus Chymici Selectissimi* . . . (Frankfurt: Lucas Jennis, 1618); Lenke et al., "Michael Maier," 5–9.

13. For instance, Kassel Landesbibliothek, MS 40 chem. 47, includes English translations of "The Booke of John Sawtre" (fols. 42r-65v) and the *Dialogue* of Giles Du Wes (fols. 73r-100v), as well as an English *practica*, *Elixir vini*, sometimes associated with the Ripley Corpus (fol 103r-v). The Kassel collection is discussed further below, pp. 314–16.

14. Florence, Biblioteca Nazionale Centrale, MS Magliabechiano XVI. 113, 10v: "Explicit Medulla Philosophye . . . Exemplata anno. 1566 Venetiis" (CRC 16.10). Agnello's treatise was printed as Giovanni Battista Agnello, *Espositione sopra un Libro intitolato Apocalypsis spiritus secreti* (London: John Kingston for Pietro Angeliono, 1566); later translated as John Baptista Lambye (i.e., Agnello), *A reuelation of the secret spirit: Declaring the most concealed secret of alchymie*, trans. Richard Napier (London: John Haviland for Henrie Skelton, 1623).

15. The Type II *Medulla* was published in *Opuscula quaedam Chemica. Georgii Riplei Angli Medvlla Philosophiae Chemicae. Incerti avtoris canones decem, Mysterium artis mira brevitate & perspicuitate comprehendentes . . . Omnia partim ex veteribus Manuscriptis eruta, partim restituta* (Frankfurt am Main: Johann Bringer, 1614). On the translation history of the text, see CRC 16.

after.[16] An established and trusted authority, Ripley was esteemed among continental alchemists both for his clarity of exposition and for the practical utility of his processes. The Saxon alchemist, physician, and schoolmaster Andreas Libavius (ca. 1550–1616), a sharp critic of contemporary practitioners, considered the late medieval "Hermetic" alchemy of Lull and Ripley superior to more recent Paracelsian works in practical efficacy and consistency of terminology.[17] In his monumental *Syntagmatis arcanorum chymicorum: tomus [primus] secundus. . .*, Libavius provided a commentary on the *Liber duodecim portarum* that ran to thirty-seven folio pages, in which he allowed the English canon to be either "the first of the best who have written of chrysopoeia . . . or to be not far from the first."[18] Ripley was also acclaimed for his medical prowess. The French royal physician and defender of chemical medicine, Joseph Du Chesne, ranked the English canon among the foremost "doctors and philosophers" who worked to uncover the universal medicine; the others being Raymond Lull, Roger Bacon, John of Rupescissa, and another pseudo-Lullian commentator, Christopher of Paris.[19]

The strength of this reputation generated an audience for new works by the celebrated adept. European demand helps account for the appearance of several "new" treatises attributed to Ripley during the late sixteenth century, which first appear on the Continent. Among them were two items later published in Ripley's *Opera omnia chemica* of 1649: the *Viaticum, seu varia*

16. The *Liber duodecim portarum* in Paris, Bibliothèque Nationale, MS Lat. 12993 is dated 1571 (CRC 9.45). Bologna, Biblioteca Universitaria, MS 142 (109), vol. 2 (see CRC 9.42), part of the Caprara collection, probably compiled in France shortly after 1570, includes the Latin *Epistle, Liber duodecim portarum, Philorcium, Medulla* (Type II), *Pupilla*, and *Terra terrarum* (the Latin translation of an anonymous poem, "Take Erth of Erth"). *De Mercurio* was published with the *Liber duodecim portarum* in Nicolas Barnaud, ed., *Quadriga aurifera . . .* (Leiden: Christophorus Raphelengius, 1599). The *Liber* was also printed in Penot, *Dialogus* (1595); Combach, *OCC* (1649). Penot's and Barnaud's editions were subsequently included in Lazarus Zetzner's *Theatrum chemicum*, ensuring wide diffusion throughout the seventeenth century; *TC*, 2:114–25 and 3:797–821, respectively.

17. On Libavius, see Bruce T. Moran, *Andreas Libavius and the Transformation of Alchemy: Separating Chemical Cultures with Polemical Fire* (Sagamore Beach, MA: Science History Publications, 2007).

18. Andreas Libavius, "Analysis Dvodecim Portarvm Georgii Riplaei Angli, Canonici Regularis Britlintonensis," in *Syntagmatis arcanorum chymicorum: tomus [primus] secundus . . .* (Frankfurt, 1613–15), 400–36, on 400: "Noster Riplaeus . . . videatur inter optimos, qui de chrysopoea scripserunt . . . primus, aut non procul à primis esse."

19. Joseph Du Chesne, *Ad Veritatem Hermeticae Medecinae ex Hippocratis veterumque decretis ac Therapeusi, . . . adversus cujusdam Anonymi phantasmata Responsio* (Paris: Abraham Saugrain, 1604), fol. [a.v]r: "Huiusmodi interpretes fuerunt Lullus, Rogerius Baccho, Riplaeus, Rupecissa [*sic*] Cristophorus Parisiensis, acplerique alij magninominis ac celeberrimi Medici & Philosophi."

practica (Viaticum, or Various Practices) and the *Clavis aureae portae* (Key to the Golden Gate).[20] Yet these enticing new tracts are merely permutations of earlier works attributed to Ripley—the *Bosome Book* and the *Accurtations of Raymond*, respectively. Their appearance illustrates not just the extent of continental interest in the English canon, but also the value still to be gained from harnessing his practical agenda in the cause of winning patronage and new audiences.

While the *Viaticum* is fairly straightforwardly composed of extracts from the *Book*, the *Clavis* is the outcome of repeated cycles of forgery.[21] It is actually a doctored version of an earlier text, *De lapide philosophorum*, attributed to Saint Dunstan, Archbishop of Canterbury. This Latin treatise, which I shall hereafter refer to as the *Work of Dunstan*, is itself a translation of the Middle English *Accurtations*, from which all references to Ripley's chief authorities, Raymond and Guido, have been removed—presumably in order to strengthen the case for an earlier origin, and Dunstan's authorship.[22] The *Work* thus represents a deliberate attempt to craft an "ancient" authority from a later source. As for the *Clavis aureae portae*, it offers a spectacular case of third-generation pseudepigraphy, for now it is Dunstan's turn to have his name stripped from the title, to be replaced by a spurious opening paragraph in which "Ripley" improbably reasserts his authorship.[23]

On the face of it, these treatises look like typical examples of alchemical pseudepigraphy of the kind we have already traced through the pseudo-Lullian corpus. However, the fact that the earliest manuscript copies of both works first appear in east-central Europe during the late 1580s is highly suggestive. Their arrival coincides with Kelley's stay in Třeboň, at exactly the moment when he was developing his own reputation both as a master practitioner and as the heir to an English tradition of alchemical philosophy. As we shall see, his entire practice seems to have rested on processes found in the *Bosome Book* and the *Work of Dunstan*, writings that were unknown in Bohemia prior to the arrival of the English party, and that Dee and Kelley played an important role in disseminating.

---

20. *OOC*, 337–65 and 225–94, respectively.

21. On the *Viaticum*, see CRC 31; Rampling, "Dee and the Alchemists," 504. On the evolution of the *Clavis*, see CRC 7; Rampling, "Alchemy of George Ripley," chap. 4.

22. In this I follow the example of Dee, Kelley, and other early modern readers who typically refer to the treatise as the *Work of Dunstan*. For manuscripts, see *CRC*, on 141.

23. *OOC*, 226.

THE RISE OF EDWARD KELLEY

It is paradoxical that one of the most famed scryers of Europe eventually achieved material success on the basis of his practical rather than his spiritual accomplishments. Kelley is still best known in scholarly literature for his role in mediating Dee's famous "conversations with angels," which has led to his being studied primarily as an appendage of Dee rather than as a practitioner in his own right.[24] Yet Kelley deserves study on his own merits, not the least as an accomplished client who consistently impressed prospective patrons with his abilities, whether in the field of alchemy or angel magic.[25]

Little is known of Kelley's early life. Born in Worcester, he may have trained first as an apothecary, although the details remain obscure. By the age of twenty-seven he was a practicing medium. Kelley's ability is first recorded during the famous series of scrying sessions—or, as Dee termed them, "actions"—that took place at Dee's house in Mortlake between 1581/2 and 1586.[26] During the actions, Kelley gazed into a glass and described the spirits and other figures that he claimed to perceive there, while Dee put questions to the spirits, and then recorded the responses that Kelley relayed.[27]

The angelic prophecies had religious and political ramifications that also intrigued some of Dee's patrons, who occasionally attended the sessions. One of these was Olbracht Łaski, palatine of Sieradz, whose pretensions to the throne of Poland received angelic encouragement during his visit to England in 1582. Rising debts and a seeming lack of patronage opportunities

24. On Kelley's role in the angel conversations, see especially Harkness, *Dee's Conversations with Angels*; also Clulee, *Dee's Natural Philosophy*; Parry, *Arch-Conjuror*; Stephen Clucas, "John Dee's Angelic Conversations and the *Ars notoria*: Renaissance Magic and Mediaeval Theurgy," in *John Dee: Interdisciplinary Studies in English Renaissance Thought*, ed. Stephen Clucas (Dordrecht: Springer, 2006), 231–73. Clucas's essay also includes a helpful overview of historiographical attitudes toward the scrying sessions, and the difficulty of parsing Dee's spiritual "actions" as historical events.

25. Although Kelley is typically presented as a charlatan in historical literature, for a more nuanced view, see Harkness, *Dee's Conversations with Angels*; Kassell, "Reading for the Philosophers' Stone"; Prinke, "Beyond Patronage."

26. Dee previously worked with another scryer, Barnabas Saul, in 1581. He later consulted two more: his own son, Arthur Dee, and Bartholomew Hickman; see Harkness, *Dee's Conversations with Angels*, 16–25.

27. Dee's records are preserved in Sloane 3188, 3189, and 3191; British Library, MS Add. 36674; and Ashmole 1790. Parts of the angel conversations were later published by Meric Casaubon, *A True and Faithful Relation of What Passed for Many Years Between Dr John Dee and Some Spirits* (London: Garthwait, 1659), based on a transcription of British Library, MS Cotton Appendix XLVI, pts. I–II.

at home convinced Dee to join Łaski's entourage and return with him to the Continent.[28] Dee, Kelley, and their families left England on 21 September 1583, traveling first to Cracow and then to Prague.

Prague was an obvious destination for men of Dee's and Kelley's interests and expertise. Rudolf II's interest in alchemy and natural magic made Bohemia one of the major European centers for the patronage of occult topics during the late sixteenth century, with some 200 alchemical practitioners employed across his territories—support on a scale that made Prague an objective for alchemists across Europe.[29] Even for those without expectations of imperial patronage, Prague's environment of relative religious toleration helped to establish the city as a crossroads for practitioners of many nations.[30]

Arriving in Prague on 1 August 1584, Dee, Kelley, and their families lodged with Tadeáš Hájek, a physician with close connections both to the court and to the many alchemical practitioners active in Bohemia.[31] Yet Dee's failure to personally impress the emperor, coupled with concerns over his conjuring activities, meant that he was unable to capitalize on Rudolf's interest in the very areas in which he specialized. The hostility of the papal nuncio, Germanus Malaspina, led to the exile of the English party in May 1586, and only Rožmberk's support eventually persuaded the emperor to relent. The party was granted permission to settle on the burgrave's estate at Třeboň, some eighty miles south of Prague, where they arrived on 14 September 1586.

The Englishmen were not the first visitors with alchemical interests to benefit from Rožmberk's patronage, or to set up their practice under his protection. The burgrave employed as many as fifty practitioners across his various estates, engaged in both transmutation and alchemical medicine.[32] Kelley rapidly grasped the scale of opportunity available to a skilled and

28. As Stephen Pumfrey has pointed out, by the standards of his time Dee was relatively successful in obtaining court patronage, even if the results fell short of his expectations; Stephen Pumfrey, "John Dee: The Patronage of a Natural Philosopher in Tudor England," *Studies in History and Philosophy of Science* 43 (2012): 449–59.

29. On Rudolf's patronage, see Evans, *Rudolf II*, chap. 6, and the collected essays in Purš and Karpenko, *Alchemy and Rudolf II*.

30. See Rampling, "Transmission and Transmutation."

31. Ivo Purš, "Tadeáš Hájek of Hájek and His Alchemical Circle," in *Alchemy and Rudolf II*, ed. Purš and Karpenko, 423–57. As Purš notes, Hájek's status has given rise to some confusion in previous literature, where he is sometimes erroneously characterized as personal physician to one or other of the Hapsburg emperors, or even as Rudolf II's "examiner of alchemists." In fact he provided medical care to some of the servants at court.

32. Ivo Purš and Vladimír Karpenko "Alchemy at the Aristocratic Courts of the Lands of the Bohemian Crown," in *Alchemy and Rudolf II*, ed. Purš and Karpenko, 47–92, on 59.

enterprising alchemist in this milieu, as well as the difficulty of distinguish-ing his own practice in a teeming alchemical marketplace. His expertise had also reached the point where he could distance himself from close collabo-ration with Dee, particularly since Rožmberk was apparently less interested in political prophecies than in chrysopoeia. Associations with necromancy had, furthermore, resulted in more harm than good for the English party. Kelley was certainly astute enough to realize that the angelic actions served as a brake rather than a spur to his own prospects, and he reoriented his activities accordingly.

The decline of the angelic conversations and the increasing status of Kel-ley's alchemical practice marked a shift in the relationship between the two Englishmen, as Kelley moved out of Dee's shadow to assume the role of mas-ter alchemist. Reports from Dee's diary show that Kelley set up his labora-tory in the gatehouse of Rožmberk's castle, often assisted by other practi-tioners, including Dee himself, who advised not just on theoretical matters, but on practical issues such as the most suitable shape of bricks for his fur-nace.[33] It was Kelley rather than Dee who took the lead in establishing and exploiting connections, regularly traveling to Rožmberk's other estates in Silesia and Bohemia, where he widened his circle of fellow practitioners and potential patrons, some of whom visited Dee and Kelley in turn.

One of the best-documented of these contacts is Nicolaus Mai, or Maius, another of Rudolf's councillors with alchemical interests, who was later appointed to the office of *Appellationsrat*, prefect of the imperial silver mines at Joachimsthal.[34] Mai was also a poet, and Kelley encouraged him to bend this skill toward the task of retranslating England's own leading alchemical versifier, Ripley. At Kelley's suggestion, Mai transformed the *Compound* into elegiac verse: a form more appropriate for a sophisticated courtly setting than the crude Latin translation already circulating in Europe by the 1570s.[35]

33. In a diary entry dated 28–29 October 1587, Dee noted that Kelley's associate Jan Kapr "did begin to make furnaces over the gate &c.: and he used of my round bricks"; Dee, *Diaries of John Dee*, 231.

34. On Mai, see Evans, *Rudolf II*, 209–10, 216; Olivier, "Bernard G[illes?] Penot," 609–10; Telle, *Parerga Paracelsica*, 176–77; Wilhelm Kühlmann and Joachim Telle, eds., *Alchemomedizinische Briefe, 1585 bis 1597* (Stuttgart: Franz Steiner, 1998), 13. Mai was already corresponding with Dee by late 1586, and actually visited Třeboň for a few days in January 1589, by which time Kelley's practice and reputation were well established; John Dee, *Diary*, in Ashmole 488, fols. 88v, 125r. Dee's so-called diary consists of marginal notes to the *Ephemerides coelestium motuum*, ed. Johannes Anto-nius Manginus (Venice, 1582).

35. Mai's translation, "Georgii Riplaei, canonici angli, XII. Portarum liber, elegiaco carmine edi-tus a Nicolao Maio," survives in two manuscripts. The first, Biblioteca Apostolica Vaticana, MS Reg. Lat. 1381, is described in Beda Dudík, *Iter romanum: Im Auftrage des Hohen Maehrischen Landes-*

Although this poetical undertaking served Mai's own aspirations (the translation was, naturally, dedicated to the emperor), it also points to growing interest in English alchemy at the Bohemian court in the wake of Kelley's arrival. Mai alludes to this attention in his verse dedication to Rudolf, noting that the ignorant among his contemporaries attack the writings of ancient philosophers as barbarous, or even dismiss the truth of *chymia* entirely, believing "the books of Ripley vain dreams."[36] While the emperor defends alchemical philosophy against calumny by supporting its practitioners, Mai seeks to counter these criticisms in a different way, by offering up Ripley's wisdom in a more elegant form. Furthermore, he does so on the advice of another Englishman—"Kelley, than whom no one is more excellent, ordered this work to be turned into Latin verse."[37] The message is clear enough: whatever the shortcomings of the English tongue, the value of English practice should not be underestimated.

It is a message that Kelley himself underscored by contributing a prefatory verse to Mai's translation. Punning on the shared meaning of *maius* and *magis* (greater), Kelley complimented his friend while reaffirming Ripley's Englishness as both practitioner and poet:

Kelley to the Reader
Whatever the crowd of philosophers has gathered in order,
Ripley in his father tongue has given to song:
Maius fashions the same into Latin with his pen,
Hence Ripley is esteemed—but Maius even more (*magis*).[38]

Such friendly overtures suggest an exchange of compliments between colleagues at Rudolf's court, but Kelley also produced presentation volumes

---

ausschusses in den Jahren 1852 und 1853, pts. 1–2 (Vienna: F. Manz, 1855), 228. The second, Kassel Landesbibliothek, 4° MS chem. 68, belonged to the alchemical collection of Moritz, Landgrave of Hesse-Kassel; see below, pp. 315–16.

36. Kassel Landesbibliothek, 4° MS chem. 68, fol. 4v: "Hic, veterum damnans doctissima scripta Sophorum, Cum nihil adsequitur, barbara scripta vocat; Ille, velut falsam Chymiae despicit Artem, Riplaei*que* libros somnia vana putat."

37. Ibid., fols. 4v–5r: "Causa mihi duplex: prior est; intelligis Artes, Philosophis*que* faves, Artifices*que* foves. Altera; Kellaeus, quo non praestantior alter, Hoc Latiis numeris vertere iussit opus."

38. Ibid., fol. 5v:

Kelleus LECTORI
Cuicquid philosophum congesserat ordine Turba,
patria Rypplaei carmine Lingva dedit:
Haec eadem Maius, calamo facit esse Latino,
Hinc notus Rypplai est, Notior ille magis.

on his own behalf. Książnica Cieszyńska SZ DD.vii.33 is a copy of one such volume: a substantial compendium of texts originally compiled by Kelley in 1589, and later copied in Prague between May and July 1592 by Jan Kapr of Kaprštejn—one of Kelley's regular associates, who assisted him in his laboratory in Třeboň and may also have served as his amanuensis.[39] Kelley intended the collection as a gift for Karl of Biberštejn (1528–1593), a Silesian official and imperial councillor who also had metallurgical interests, twice serving as master of the Bohemian mint.[40] A note from Kelley reveals that he presented it to Biberštejn in the same month in which his own noble status was formally ratified by the emperor:

> Edward Kelley wrote this book out of kindness and love for his most assured friend, the noble lord Karl von Biberstein, 2 August in the year 1589: whom he wishes to have known as his adopted philosophical son, and to be esteemed above all other mortals.[41]

The wording of the dedication imbues the gift with added significance. In adopting the mint-master as his "philosophical son," Kelley assumed the role of both master and tutor, a relationship embodied by the transfer of alchemical knowledge—specifically, the wisdom of an earlier English adept, Ripley. The Cieszyń manuscript contains seven Ripleian works, including Latin translations of the *Compound* and the *Epistle to Edward IV*, and the original "Type I" *Medulla*, *Pupilla*, and *Philorcium*. In addition to these well-known works, the collection includes the *Work of Dunstan*, as well as a newcomer to the Ripley Corpus: the *Clavis aureae portae*.

Such a trove of English knowledge served as a fitting gift from an English adept to his student in philosophy. Yet although Kelley was active in disseminating alchemica among his friends, it is not immediately clear how these gifts relate to his burgeoning reputation as a practitioner. For that, we must

39. Cieszyń, Książnica Cieszyńska, MS SZ DD.vii.33, fol. 146v: "Descripta per Johannem Carpionem Pragae die 14 Maij Annorum 1592." An administrator responsible for Rudolf's vineyards, Kapr appears frequently in Dee's diaries under the name of John Carpe or Johannes Carpio. His office was that of "perkmistr hor viničných" or "Bergmeister der Weingarten": Kühlmann and Telle, *Alchemomedizinische Briefe*, 165–66, 168; Rafał Prinke, personal communication.

40. Biberstein was *Landeshauptman* to the Duchy of Głogów. I am most grateful to Rafał Prinke for consulting several Polish records on my behalf.

41. Książnica Cieszyńska SZ DD.vii.33, fol. 119v: "Eduardus Keleus Hunc librum in graciam et amorem Magnifici domini Caroli de Bibeistaynn Amici sui integerrimi fecit. Annor[um?] 1589 Augusti 2°. Ipsumque filium suum philosophicum adoptat Praeferendum Etiam semper omnibus mortalibus merito suo existimat."

turn to the early evidence for Kelley's practical engagement with his source material: in particular, his adoption of the same sericonian alchemy as that presented in his favorite sources, the *Bosome Book* and Ripley-derived *Work of Dunstan*.

SYMBOLS OF TRANSFORMATION

Dee was already accumulating manuscripts of Ripleian texts long before Kelley entered his life. That he viewed them not just as antiquities but as useful practical guides appears from a series of alchemical experiments carried out at Mortlake between 22 June and 6 October 1581. At one point, Dee describes extracting more than ten ounces of "quick mercury" from a sublimate, "by my diligence in pressing the soft stuff betwene my fingers partly: and by washing it in destilled vineger." He also took care not to overcharge the mercury with vinegar, "as Riplay in philortium warnes of"—a reference to a passage in the *Philorcium* where Ripley advises practitioners to dissolve their mercury gradually, in order not to "incontinently suffocate it with water."[42]

Dee's careful cross-referencing between authoritative texts and his own observations points to an important dimension of early modern alchemical practice: the expectation that texts could guide practice because they described earlier iterations of the same procedures. By the same token, Dee's efforts to make the stone could be viewed as attempted reconstructions of Ripley's previous experiments. The assumption of continuity between earlier recipes and later attempts was aided by authorities' descriptions of specific effects—alchemical markers that gave practitioners confidence that they were following their instructions correctly. The "gliding fire" described in the Ripleian *Whole Work* is a particularly striking example of such a physical marker, or "token."[43]

Kelley had access to Dee's books, but he may also have learned from Dee's engagement with his authorities, for the same close correlation between text and practice is evident in Kelley's alchemical writings. Yet Kelley soon acquired a reputation in practice that far outstripped that of his associate. His Bohemian success is partly attributable to his skill at persuading audiences that he had reconstructed in his own practice the "tokens" laid down

42. Bodleian Library, MS Rawlinson D.241, fol. 3r; *OOC*, 200: "Trahe aquam de eodem lapide per alembicum, & cum illa aqua solve lapidem per alembicum, & cum illa aqua solve lapidem infinitum, sed solve per modicum & aqua non suffoces incontinenti, quia si rectè regatur cum aquae pinta una facies, si volueris, aquae quantitatem infinitam."

43. See pp. 274–75, 281, above.

by past authorities. His rapid manufacture of the *mercurius solis* in the presence of the emperor offers the most spectacular example, but, as we shall see, this was not the only effect that Kelley learned to recreate.

Dee's diary entry for 8 February 1588 records how that night Kelley invited him to his gatehouse laboratory to witness his replication of an alchemical experiment:

> Mr. E.K. at 9 of the clock after none sent for me to his laboratory over the gate: to se[e] how hee distilled sericon, according as in tyme past & of late he h[e]ard of me out of Riplay.[44]

Although Ripley described many experiments using sericon, we can pinpoint the exact process that Kelley was attempting to reconstruct, thanks to another record of the event preserved in Harley 2411. The only surviving Latin version of the *Bosome Book*, this seems to have been transcribed from Dee's own copy, as appears from the initials "J.D." added to several marginal notes. Signing his notes was a characteristic feature of Dee's annotation style, but it was rare for him to share the credit for an observation. Here, the addition of Kelley's initials signals his role in the proceedings:

> I saw the same on 8 February 1588 (new style) in Třeboň in Bohemia. From 2 lb. of sericon dissolved in distilled vinegar, and by means of spirit of wine cleansed of much sediment, came 4 oz. of red wine or oil. J:D. E:K.[45]

The note is appended to a process for a sericonian solvent titled *Magna philosophorum corrosiva* (The Great Corrosive of the Philosophers). The significance of this observation is underscored by the fact that Dee recorded it in both a signed annotation and a diary entry, suggesting that he viewed Kelley's result as an important step forward. On closer examination, however, the effect is as much the outcome of alchemical reading as of practical skill, since, like most of the *Book*'s content, "The Great Corrosive" cannot be reproduced in practice without an understanding of what sericon means.

Ripley's process begins much like any other late medieval recipe for a mineral acid or for corrosive sublimate. First, the practitioner should grind

---

44. Dee, *Diary*, fol. 111r.

45. Harley 2411, fol. 55r: "Ego idem vidi Anno 1588 feb: die 8. novo stilo Traeboniae in Bohemia Ex 2 lib Sericonis dissoluti in [vinegar] distillato, et per spiritum vini purificatum a multis faecibus 4 oz erant vini rubicundi sive olei. J:D. E.K." Another note, "J:D. Quae tantummodo vna est," appears on fol. 18v.

a pound of dry, yellow vitriol with half a pound of saltpeter, then place it in a flask and heat strongly to draw out an *aqua fortis*. However, ordinary saltpeter is evidently not intended here, since the term is immediately glossed as "salt of sericon."[46] The substitution of this sericonian salt for ordinary saltpeter transforms the nature of the resulting solvent, which therefore cannot be regular *aqua fortis*. Rather, it is a philosophical menstruum that, when sublimed, yields a strange, crystalline "mercury":

> When our menstrual "mercury" ascends from the sericon by the violence of the fire, a certain part of it is found cleaving to the side of the flask after the complete distillation and cooling of the glass, like salt and of a crystalline appearance. . . . And the form of this earth is like mercury sublimed, and therefore shines brightly. . . . This secret I learned through practice: G[eorge] R[ipley], as God is my witness.[47]

With his characteristic emphasis on practical experience, Ripley goes on to describe the special properties of his crystalline residue. First, it is a fixed matter that is "apt to receive any kind of form," as he determined by subjecting it to further experiments. For instance, he describes testing its fixity on a piece of heated glass. As long as the glass remained hot in the fire, he found that the earth could not be separated from it, and remained white. Another sample, tested on iron, turned yellow before his eyes, and remained fixed.[48] One of the intriguing aspects of this process is that Ripley repeatedly tells his readers what they should expect to see at different points in the process. In practice, this feature may have assisted with the "reproducibility" of the

46. Ibid., fols. 54v-55r: "*Recipe* Vitriol Romani dessicati in glaucum colorem super ignem lentum Libram vnam, salis petri nostri .i[.e]. salis sericonis bene desiccati lib: ser conterantur fortier in mortariolo enaeo donec optime incorporentur et tunc destilletur primo fantastica aqua lento igne et evacuetur, deinde fortissimo igne aqua fortis || quae ad opus solutionis et putrefactionis servetur &c." The unusual description of the vitriol as yellow in color may also suggest that another substance is intended.

47. Ibid., fol. 55r: "Quando [mercurius] noster menstrualis ascendit a Sericone per violentiam ignis, quaedam pars ipsius adhaerens lateribus vasis quasi salina et crystallina reperitur post distillationem completam et vasis infrigidationem, et illa terra crystallina, est materia fixa & apta ad recipiendam quamcumque formam. Colligatur ideo & servetur. Et terrae illius figura est quasi figura argenti vivi sublimati, et ita resplendet. . . . Haec secreta in practicando didici. Haec G:R: teste Deo." A similar recipe appears in the *Whole Work*, suggesting that it was also adapted by Samuel Norton.

48. Ibid., fol. 55r: "Ego viso hoc admirabar et probavi illius fixationem super peciam vitri, quam donec canduit ignivi et terra mansit adhaerens quasi inevellibilis \vitro:/ et permansit in albedine sua. Aliam autem particulam terra probaui super ferrum ignitum, et illa particula citrinavit coram oculis meis et mansit."

experiment. The description of specific effects in alchemical texts aided replication by allowing practitioners to correlate their own observations with
the written instructions of their authorities. From Robert Greene's mercurial
tree to the sericonian "gliding fire," such signs and tokens served as observable signposts, reassuring reader-practitioners that they were on the right
path.[49]

Dee seems to have experienced the same sense of confidence during Kelley's distillation of sericon, when he observed a mercury-like residue form
around the neck of the glass. For Dee, this was a clear sign that they had interpreted Ripley's salt of sericon correctly—an accomplishment that he marked
by sketching a flask at the end of the recipe, with a dotted line to indicate
the white residue. He added the date, 1588, and an explanatory note: "So in
a circle aboue [th]e matter was the cleare matter lyke [mercury]" (fig. 13).

Yet it is not clear how we should read Dee's claim that he "saw the same"
during this experiment. Although his annotation refers to a red oil, this color
is not actually mentioned in the *Book*'s recipe, which instead describes a
golden oil. Since the color red does not appear in the original text, Dee and
Kelley may have been working from other sources, and possibly even other
practices within the *Bosome Book*. From Dee's perspective, however, they
were recreating essentially the same effect, and hence the same practice, as
that seen by Ripley himself.

This ability to produce striking chemical effects, which seemed actually to
bring the descriptions of his authorities to life, was one of the most successful techniques in Kelley's armory. Since the reconstruction of these tokens
was based on successful reinterpretation of encoded texts, it followed from
Kelley's demonstrations that he also possessed the appropriately "philosophical" combination of wisdom, revelation, and technical expertise. Such
evidence helped to convince audiences, from Dee to Rudolf, of the promise of his approach. It simultaneously validated the authority of his medieval
sources—in this case, Ripley.

Throughout 1588, Dee's terse diary entries present his associate as an
alchemist on the cusp of a major practical breakthrough. Just a few months
after the sericon experiment, on 10 May, he exclaimed in his diary that Kelley
had opened "the great secret" to him. The revelation marked a further shift
in their relationship, and a sign of Dee's gradual demotion from patron to

49. A good example of such a "token" is the mercurial tree reconstructed by Lawrence Principe; Lawrence M. Principe, "Apparatus and Reproducibility in Alchemy," in *Instruments and
Experimentation in the History of Chemistry*, ed. Frederic L. Holmes and Trevor H. Levere (Cambridge, MA: MIT Press, 2000), 55–74.

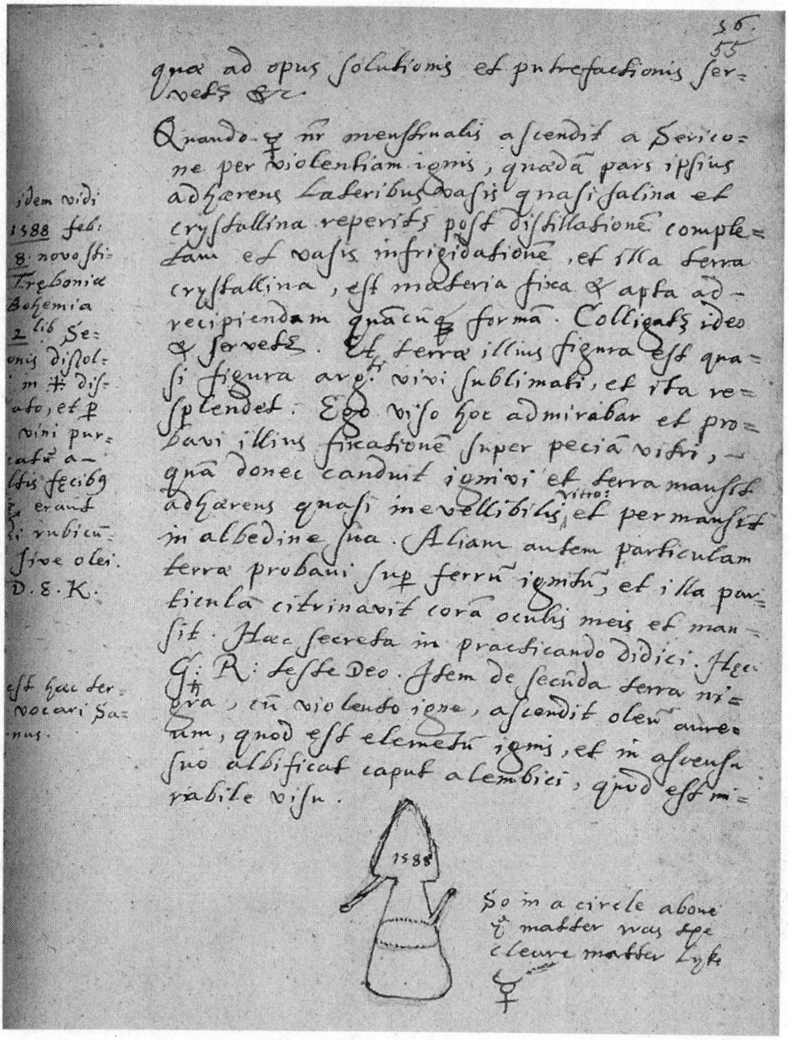

FIGURE 13. Drawing of the mercurial residue observed in a flask after distilling sericon, copied from an original sketch made by John Dee in February 1588. The marginal note above is signed "J.D. E.K." © The British Library Board, MS Harley 2411, fol. 55r.

assistant. If Dee found his associate's alchemical disclosures significant, he was not alone; the fruits of Kelley's practice also impressed Rudolf, who summoned him to Prague. From Kelley's perspective, the change in his circumstances also indicated the end of Dee's usefulness, and the dissolution of their partnership. In March 1589, disheartened by his lack of fortune and under increasing pressure from Rožmberk to leave Třeboň, Dee departed from Bohemia with his family on the long journey home.[50] Despite repeated overtures from Elizabeth I, encouraging him to return, Kelley remained behind.

For two more years, Kelley enjoyed the advantages of fame and success, planting the seeds for legends of his alchemical prowess that would continue to flourish in Europe long after his demise. In practice, his ready production of gold was financed by loans rather than transmutation. Rumors of debt, coupled with Kelley's unwillingness to conduct decisive trials of his expertise, seem to have contributed to his arrest and imprisonment, instigating a dramatic new phase of his alchemical career. Yet a man of his mettle could not tranquilly resign himself to incarceration. Although disgraced and imprisoned, Kelley continued to manipulate texts, people, and even alchemical history in order to win his freedom.

THE PRISON WRITINGS

Edward Kelley was not the first alchemist to attempt to write his way out of trouble, as we have seen from the examples of Richard Jones and William Blomfild, both accused of conjuring. Over the course of the sixteenth century, a large quantity of alchemical material, ranging from philosophical treatises to substantial collections of receipts, was produced in prisons. In Bohemia, the alchemist Bavor Rodovský of Hustiran (1526–1592) prepared treatises and translations for Rožmberk during the 1580s while imprisoned for debt in Prague.[51] In England, debt-ridden alchemists also saw strategic composition as a way of escaping their troubles. As we have seen, Francis Thynne, jailed for debt, offered his alchemical services to Burghley, to whom he had previously dedicated alchemical works.[52] More prosaically, the mer-

---

50. The circumstances precipitating Dee's return are discussed in detail in Parry, *Arch-Conjuror*, chap. 17.

51. These include the *Kniha Dokonalého umieni chymiczkého* (The Book of the Perfect Art of Chemistry) in Leiden Universiteitsbibliotheek, MS Vossianus Chym. F.3. This compilation of Rodovský's translations of Latin works into Czech includes Ripley's *Medulla alchimiae*, retitled "Wybrane gadro z Hermesowe filozofie Sepsane Skrze Cztihodneho pana Girzika Ryplea, kanovnika w Englandu" ("Extracted Marrow of the Philosophy of Hermes, Compiled by Honourable Mr George Ripley, Canon in England"); Voss. Chym. F.3, fol. 159r. The text is identified only as "Commentaire au philosophe Hermès" in P. C. Van Boeren, *Codices Vossiani chymici* (Leiden: Universitaire pers Leiden, 1975), 10. The manuscript was later owned by Vilém's brother, Petr; it is inscribed with Petr's motto, *Contra spem in spe* (ibid., 9).

52. On Thynne, see pp. 251–52, above.

chant Clement Draper may have copied and translated alchemical texts (including a rare English translation of Raymond's *De secretis naturae*) as a means of generating income from his cell.[53]

Yet these attempts all differed from Kelley's situation in a key respect: none of these men was originally imprisoned for alchemical fraud. Kelley's incarceration, however, was probably the direct outcome of his failure to convince Rudolf of his alchemical competence.[54] Under the circumstances, the onus lay even more heavily on the English alchemist to prove his skill—a challenge that he answered by preparing a series of treatises that laid out his own approach to the theory and practice of alchemy, now preserved in a single codex.

Formerly in the library of Rudolf II, Leipzig Universitätsbibliothek MS 0398 is a thick, quarto manuscript comprising more than 400 folios of alchemical content in Latin and German.[55] The most substantial items are a series of treatises authored by Kelley (although not transcribed by him), several of which address the emperor directly. Most but not all of this material is written in the same clear, formal secretary hand, although several items have been added later, including one dated 1600, several years after Kelley's death. The manuscript must therefore represent a later attempt to compile copies of the Englishman's "prison writings," supplemented by the compiler's own additions and attempted expositions.

That Kelley was able to write at all shows just how far his situation improved after his initial arrest, when he was confined within a cramped cell. During this time Kelley may have relied on his patron, Rožmberk, to intercede with the emperor, and there is evidence that Rožmberk did succeed in securing better conditions for his client.[56] Kelley lost this source of support when his powerful protector died on 31 August 1592—a loss that coincided with a seeming olive branch from the imperial household.

53. Sloane 3707, in the hand of Clement Draper. On Draper's alchemy and note-taking practices, see Harkness, *Jewel House*, chap. 5.

54. The discovery that Kelley was financing his lavish expenditure through borrowing and pawning jewels, rather than through transmutation, coupled with his reluctance to present himself at court for a final trial of his expertise, are likely factors in Rudolf's disillusionment; see Prinke, "Beyond Patronage," on the evidence for this version of events.

55. This manuscript has not previously been associated with Kelley's activities during the term of his imprisonment. I am very grateful to Rafał Prinke for first alerting me to the catalogue entry for this item, which suggested possible connections with Ripley and Kelley.

56. Vienna, Österreichische Nationalbibliothek, Sammlung von Handschriften und alten Drücken, Cod. 8964 [Fugger-Zeitungen 1591], fol. 641r; cited in Karpenko and Purš, "Edward Kelley," 513.

On 8 February 1592, one of Rudolf's chamber servants, Hanuš Heyden, wrote to the Castellan of Křivoklát, Jan Jindřich Prolhofer of Purgersdorf (d. 1604). Heyden was seeking Prolhofer's assistance in extracting alchemical secrets from Kelley, including his process for potable gold, as well as the meaning of symbols encountered in books confiscated from Kelley's laboratory.[57] By recruiting the castellan as intermediary, he also opened a channel of communication with the English alchemist.

This approach marks the moment when Kelley took his destiny back into his own hands. The earliest dated item in the codex was finished less than a month later: an untitled treatise in sixteen folios, with the heading "Edward Kelley composed this book in prison at Pirglitz in the year 1592, 1 March, in Bohemia."[58] The content is primarily philosophical rather than practical, as suggested by the incipit "Plurimum quidem in dies in scholis ("Often indeed, every day in the schools"). In it, Kelley teases two of his favorite themes: the transformation of one element into another, and the identity of his mysterious prime matter.[59] A few months later, "in the fourteenth month of his misfortune, in prison at Pirglitz"—that is, in June 1592—Kelley submitted a much longer treatise, *De chymicis oratio* (An Oration on Chymists), an extravagant piece of rhetoric embellished with phrases from classical poets, the goal of which is to distinguish true philosophers from frauds.[60] This was followed by a series of elaborate, Ramist tables, prefaced by a separate letter to Rudolf.[61]

Throughout these writings Kelley's tone swings between arrogance and despair, although he never gives an inch on the subject of his own innocence. Thus he concludes the *Oratio*'s prefatory letter with a passionate appeal to Rudolf, "on bended knees and with all submission, that Your Majesty might at last have pity on me, the most innocent of men, and well deserving."[62] In the introduction to the tables he condemns those enemies who have falsely accused him, and urges the emperor to "desist from destroying a most inno-

---

57. Karel Pejml, *Dějiny české alchymie* (Prague: Litomyši, 1933), 57; cited in Ivo Purš, "Rudolf II's Patronage of Alchemy and the Natural Sciences," in *Alchemy and Rudolf II*, ed Purš and Karpenko, 139–204, on 195n192. Unfortunately the whereabouts of the original letter are currently unknown.

58. Leipzig 0398, fol. 31r: "Eduardus Kelleus hunc librum Conscripsit Pirglitz in Carcere Anno 92. Mart j° In Boeme."

59. Ibid., fols. 31r-48r. Incipit: "Plurimum quidem in dies in scholis, de prima illa Materia seu Matre rerum, ab Aristotelis sectatoribus est disputatum."

60. Ibid., fols. 51r-106r.

61. Ibid., fols. 112r-22v.

62. Ibid., fol. 51r: "Addo etiam has preces genibus flexis & omni submissione, vt mei, hominis innocentissimi & de se benemeriti, Maiestas tua tandem misereri velit."

cent man . . . with such bitter anger."[63] The preface ends on a heartrending note from "Kelley, the most afflicted of mortals."[64] Yet just a few lines earlier we find him bargaining, none too subtly, for better treatment. He regrets that he cannot provide the tables for gold and silver (the most crucial part of the process), since he lacks adequate writing materials:

> If more paper had been available, I would have completed the remaining golden tables of Sol and Luna. And if at any time the grace of Your Majesty will smile upon me with favor, I will make you the master of this wonderful science. Finally, I will repeat this again: have mercy on me, Caesar, according to your great mercy; on me, I say, the most innocent of men, and your servant.[65]

In these early treatises, we find Kelley balancing philosophical exposition with a strong emphasis on the factors that distinguish him from *Betrüger*: his knowledge, his probity, and his godliness. This is particularly the case with the *Oratio*, the message of which is aptly summarized by its conclusion, where Kelley pointedly quotes a late antique epigram: "You who are serious, be free from fraud, and you who are ignorant, believe in the learned."[66]

From this text we learn that Kelley was probably under pressure to produce faster practical results. In one significant passage, he reminds the emperor that the successes of past authorities were not achieved overnight. He points to the many years of study necessary to acquire the secret, even for great adepts: thus "Roger Bacon was fifty-seven years old before he descended to the field of this study; three lustres [i.e., fifteen years] had passed before he attained the *scopus*."[67] Arnald of Villanova and Raymond

---

63. Ibid., fol. 112v: "& tandem cessa hominem innocentissimum . . . tam acerba indignatione perdere."

64. Ibid., fol. 113r: "Afflictissimus mortalium, Kelleus."

65. Ibid., fol. 112v-13r: "Si plus chartae adesset, reliquas Solis & Lunae aureas perfecissem Tabulas; Et si aliquando Mag^{tis}. tuae mihi affulserit gratia admirandae scientiae Dominum faciam. Deni*que* hoc iterum repetam: Miserere mei Caesar, secundum magnam misericordiam tuam, mei dico innocentissimi hominis, & serui tui."

66. Ibid., fol. 106r: "Fraude carete graues, ignari credite doctis"; translation from Jacob Handl, *The Moralia of 1596*, ed. Allen B. Skei (Middleton, WI: Madison, A-R Editions, 1970), pt. 1, 16. This is a quotation from the *Carmina duodecim sapientum* (fourth or fifth century AD), a collection of late antique epigrams that Handl set to music in his *Moralia*, published in Prague in 1589: Jacobus Gallus Carniolus, *Quatuor vocum Liber I. Harmoniarum Moralium . . .* (Prague: Georgius Nigrinus, 1589). Given the timing, Kelley may have encountered the text through this source.

67. Leipzig 0398, fol. 79r: "Bacon quinquaginta septem annorum erat, ante*quam* in campum hujus descenderat studij; Tria elapsa sunt lustra, scopum prius*quam* attigit." The sentence suggests

Lull also spent many years striving for the stone before they succeeded. Kelley then goes on to list other practitioners who eventually attained the art after long study: John Garland, Hortulanus, Ripley, Isaac Hollandus, Dunstan, Brixham, and Bernard of Treviso.

This catalogue is interesting not only as an appeal for continuing patience, but as a record of Kelley's reading. English practitioners dominate in this selection, including Kelley's own preferred authorities, Ripley and Saint Dunstan. The inclusion of the little-known alchemist "Brixham" also indicates a link with Dee's library. Dee owned a late fourteenth-century English manuscript that contains several recipes attributed to the mysterious Brixham, suggesting that Kelley had at some point studied that volume for practical information.[68] As a personal model, however, Kelley devotes more attention to Ripley's younger contemporary Thomas Norton—suggesting that, for all his practical reliance on religious authorities like Dunstan and Ripley, he sought to identify himself with Norton as his equal in both wisdom and social status: a gentleman and fellow knight (*eques auratus*), learned in all the sciences.[69]

Perhaps the most remarkable feature of Kelley's campaign is its apparent success. Unlike many other disgraced practitioners of the sixteenth century, Kelley was released from prison, obtaining his freedom in the autumn of 1593. Although he never regained his former level of intimacy with the emperor, he was able to return to his estates and resume the life of a gentleman. Did Rudolf simply relent, or did the Englishman find other means of demonstrating his usefulness?

As Kelley well knew, it was not enough merely to claim innocence and assert his place among the philosophers; he had to show beyond question that his practical expertise also merited that status. Kelley's attention to his English forebears in fact prepares the groundwork for a more practical exposition of his alchemy, which draws heavily on his long-standing authorities, Ripley and Dunstan. As we move through the volume, and through the term of his imprisonment, it becomes clear that Kelley was not left sitting idly in his cell, but was permitted to resume his practice, now under the supervision

---

a literary echo of Horace, *Odes* 3.1: "Descendat in campum" (signifying the Campus Martius). One lustre usually indicated a period of five years, although it was sometimes used to denote four: Ovid, *Fastorum libri sex: The Fasti of Ovid*, vol. 3, *Commentary on Books 3 and 4*, ed. and trans. James George Frazer (Cambridge: Cambridge University Press, 1929), 47–48.

68. Ashmole 1451, e.g., on fols. 37v, 40v, 41r, 57v.

69. Leipzig 0398, fol. 79r-v: "Inter hos Norton (.Eques auratus ille || doctissimus in omni scientiarum genere Vir.) assus est, vigenos bis totos dies sibi ad mysteria & hujus artis secretiora a Magistro peritissime discendum vix quidem satisfecisse."

of his jailor, Prolhofer. Furthermore, Kelley's constant citation of English alchemists was not merely rhetorical. Sericonian alchemy, pursued with a view to both transmutation and medicine, provided the basis for his prison practice—and, we must suspect, his eventual liberty.

DAMNED EARTH AND NATURAL FIRE

Imprisoned in Křivoklát, Kelley lost no time in cultivating the castellan. Clues to his interactions with Prolhofer appear in his last two treatises in Leipzig MS 0398: the *Responsum ad Interrogata* (Response to Questioning) and *Syntagma philosophicum* (Philosophical Syntagma), only the second of which is actually written for Rudolf. The former is unusual in being addressed to Prolhofer, whom Kelley enthusiastically hails as "Excellent man, Lord Captain."[70] Despite the ominous title, the *Responsum* is not a legal deposition but a philosophical treatise, which Kelley perhaps hoped would evoke the authoritative set of questions posed to Morienus by his royal pupil, Prince Khālid.

In the *Responsum*, Kelley draws together several of the threads encountered in his earlier treatises. He alludes to the Ramist tables he prepared for the emperor, claiming that knowledge of these will help expound the process further. Sure enough, in his preface to the tables Kelley introduces two substances familiar from the pseudo-Lullian tradition: the natural fire and the damned earth (*terra damnata*). The damned earth, which also appears in the Lullian *Testamentum*, is the mighty "instrument" that opens the gates of matter, allowing the innermost parts of bodies to be revealed and intimately examined. Like a sailor who cannot acquire knowledge of other peoples without his ship, "so the philosopher, unless he has this damned earth, will perceive little or nothing of the most important secrets in other metals."[71] This chemical instrument is evidently a solvent capable of dissolving precious metals, but it is not made from alums and salts, like *aqua fortis*. "No alum, sal niter, nor vitriol belongs here," brags Kelley. "The Damned Earth alone is the doorkeeper, the only master, which can open up these mysteries

---

70. Ibid., fols. 316r-22v, on fol. 316r: "Magnifice Vir, Domine Capitanee."

71. Ibid., fol. 112r: "Sicuti Nauta, vectrice naue ad alias rapitur oras, sine qua neque aliarum gentium, neque tantae aut voluptatis aut vtilitatis cognitionem acquireret; Ita Philosophus, nisi hujus Terrae damnatae habuerit, amplissimorum in caeteris metallis arcanorum parum aut nihil percipiet. Haec enim est vnica illa ductrix, sate sanguine Diuum Caesar, qua bis nigra videre tartara licet; Haec Cymba ferruginea sola subuectat Acherontis ad vndas; Haec lata illa via, quae campos pandit Elysios."

of nature."[72] This mysterious solvent springs from "blood"—that is to say, it is drawn from an imperfect metal, sericon.

In the best tradition of alchemical *dispersio*, Kelley reveals the identity of this metal in another text, the *Responsum*, where he first introduces the term "sericon." He assures the castellan that he has worked "over the course of eight years' great labor, watchfulness, and expense" to purify the elements of the stone and the elixir, elements that must be "continually augmented with raw sericon, so that there will be no need for any other solvent."[73] Although this process sounds Ripleian in character, he attributes it to an even earlier English authority, Saint Dunstan.[74] The archbishop's importance is underscored by the fact that Leipzig MS 0398 includes two copies of the *Work of Dunstan*: the Latin original and a German translation. Both versions are annotated, while several of the notes to the German copy are signed "J.D.," suggesting that they were copied in turn from one of Dee's manuscripts.[75]

Although grounded in fifteenth-century sources, the prison writings provide the philosophical trappings for what seems to have been a busy practical program. They also reveal a previously unknown facet of Kelley's imprisonment in Křivoklát: his return to practice under the supervision of Prolhofer, who reported on his progress back to Prague. Echoes of these reports live on in Leipzig MS 0398 in the form of unattributed fragments that refer to Kelley in the third person, suggesting that Prolhofer was corresponding with another high level intermediary, addressed only as "Your Highness."

One such account relates to the Englishman's famous process for the mercury of gold: a process titled "Kelley worked the impregnation of common mercury with *mercurius solis* in such a way."[76] The writer reports that Kelley first sublimed and congealed his solar mercury with a "calcined oil" before grinding it to powder with common quicksilver. In the space of a single night, the combined mercuries sublimed together into the top of the vessel.

72. Ibid.: "Hoc grande illud Instrumentum, quo fores, claustra, quo ipsa denique omnium corporum penetralia panduntur, deteguntur & lustrantur intima. Nihil alumen, sal Nitrum, nihil victriolum hic habet loci, sola Terra Damnata Janitor, solus Dominus, qui haec naturae mysteria pandere potest."

73. Ibid., fol. 316r: "sunt mea propria, octo annorum laboris magnis, vigilijs et sumptibus honeste Comparata; Elementa nimirum Lapidis & Elixiris depurata, In quibus et per quae soluitur & multiplicatur Lapis, quae etiam perpetue cum crudo augentur Sericon, ita vt numquam sit opus aliquo alio dissoluente."

74. Ibid.: "Haec igitur se multiplicant, vt scribit Dunstain."

75. John Dee's own transcription of the *Accurtations*, with his annotations, is preserved in London, Wellcome Library, MS 239.

76. Leipzig 0398, fol. 299r: "Impregnationem [mercur]ij communis cum [mercur]io [sola]ri tali Kelleus fecit modo."

The sublimate was then gathered by Kelley, who reserved part for further use in his work, giving the remainder to the writer, "to be sent back to His Most Sacred Majesty in Prague."[77]

Another recipe apparently acquired from Kelley is a secret possessed by his former patron: the "Oil of Lord Rosenberg." Extracted from litharge of gold, this is said to be "that same gum with which [Rožmberk] was used to multiply and incerate his medicine."[78] The medicinal oil expels poison and pestilence, but it can also be used for "the great work of the Chymists" (that is, transmutation) when mixed with gold prepared using the fire against nature, or *aqua fortis*—provided, of course, that the volatile product does not escape the vessel.[79] This process sounds remarkably similar to Ripley's account of the multipurpose vegetable stone in the *Medulla*, including the reference to the oil's volatility. It is no coincidence that Leipzig MS 0398 also contains a copy of the *Medulla*, accompanied by a commentary on the vegetable stone that focuses explicitly on the dissolution of sericon in the "sharpest humidity of grapes"—the latter presumably written by either Kelley himself or an official attempting to make sense of his process.[80]

The fact that Kelley's treatises and *practicae* do not merely cite English texts but are actually accompanied by copies of those works testifies to the centrality of these authorities in his campaign. The alchemy of Dunstan and Ripley provides the philosophical and practical core of the last of his dated treatises: the *Syntagma philosophicum*, completed on 20 September 1593, less than a month before his release. This is the only one of Kelley's Pürglitz tracts ever printed, albeit posthumously, as one of "Two Excellent Tracts on the Philosophers' Stone" published in 1676.[81] There it is renamed

77. Ibid.: "Itaque totus iste [mercuri]us vnius noctis tempore ascendit sublimatus; Quem [libra]tum ita vase suo exemit, nec non 6. lotonum quantitate de eo, pro vsu necessario operis sui reseruata, reliquam partem omnem nobis dedit, Sacrissimae suae Maiestati Pragam remittendam."

78. Ibid., fol. 299v: "Oleum illud Domini Rosenbergij est illud Gummi, cum quo solitus erat suam multiplicare & incerare Medicinam."

79. Ibid.: "Res lixata auro, vel potius aurum viuum, cujus vsus in peste, Venenem & grauioribus contagijs abigendis apparebit satis. Ex illo gummi, conjuncto cum auro praeparato & liquato prius in igne contra naturam (.id est, Aqua forti.) deinde ab eodem igne separato, & in puluisculum redacto, magnum opus Chimicum praestari potest, saltem si includatur."

80. Ibid., fols. 408v-10v: "Praxis Lapidis Vegetabilis ex Medulla Alchimiae Georgij Ripley." The text opens with the line on the humidity of grapes, and a reference to the relevant passage in the *Medulla*, earlier in the manuscript: "*Recipe*. Acerrimum Vuarum humiditatem &c (Vide supra fol. 390. lin. 14)."

81. Published as "Edouardi Kellaei Via Humida, sive discursus de menstruo vegetabili Saturni. E Manuscripto," in Edward Kelley, *Tractatus duo egregii, de lapide philosophorum* (Hamburg: Schultze, 1676), 43–96. The contents of the book are translated by Waite in *The Alchemical Writings of Edward Kelly*, trans. Arthur Edward Waite (London: James Elliott, 1893); in this chapter I provide my own translations based on the text of Leipzig 0398.

"The Humid Way, or, a Discourse on the Vegetable Menstruum o[
taken from a Manuscript": a title that recognizes the text's strong s[
focus. Leipzig MS 0398 may have been the exemplar for the printe[
although the latter omits all reference to Kelley's imprisonment, ...
includes interpolations that do not appear in the original—an unattributed
passage from the *Medulla*, and several concluding recipes that, although by
Kelley, have been plucked from elsewhere in the manuscript.[82]

In the *Syntagma*, Kelley explains the nature of the imperfect metal in his
fullest elaboration of sericonian doctrine, as a work based on Saturn, or lead.
The power of lead is introduced using an analogy between the metals and
the orbits of their corresponding planets; thus the work begins with Saturn
because this is also the outermost planet, "within whose circle the spheres of
the others are naturally encompassed." Just as the orbit of Saturn must con-
tain those of the inner planets, so the metalline water drawn from Saturn's
metallic analogue, lead, must include the properties of the other metals. It
follows that lead is the only metal whose menstruum will dissolve the rest.[83]

Both Kelley's orbital analogy and the theory of metals that underpins
it are grounded in late medieval sources that present alchemy as a "lower
astronomy" (*astronomia inferior*), in which the seven metals map onto the
seven Ptolemaic planets.[84] By the sixteenth century, this idea was widespread. Long before he met Kelley, Dee had developed the idea of "inferior astronomy" in his *Monas hieroglyphica* (1564), in which his eponymous
figure, the *monas*, expressed analogies between heaven and earth.[85] Yet
Kelley's own vision of concentric planetary spheres is closer to that imagined almost a century earlier by Ripley, in the form of the wheel diagram
appended to the *Compound*:

82. The concluding text of the *Via Humida* (pp. 82–93) is composed of recipes extracted from
Leipzig 0398, fols. 4r-v and 5r-v, and a further process on the use of antimony that I have not identified. It also includes several passages adapted from Ripley's chapter on the animal stone and prefatory verses in the *Medulla*.

83. Leipzig 0398, fol. 348r: "Huius vero initium operis, Saturnum, quippe cuius circulo
sphaerae aliorum complectuntur omnes assumpsisse: cuius cum uirtute plumbum productum sit
et illud ipsum aeque metallorum in se essentiam continere omnem uoluit . . . plumbum illis etiam
singulis qua menstrualis erit."

84. On alchemy as *astronomia inferior*, see Ruska, *Turba Philosophorum*, 80; Joachim Telle,
"Astrologie und Alchemie im 16. Jahrhundert: Zu den astroalchemischen Lehrdichtungen von
Christoph von Hirschenberg und Basilius Valetinus," in *Die okkulten Wissenschaften in der Renais-
sance*, ed. August Buck (Wiesbaden: O. Harrassowitz, 1992), 227–53, on 238–40; Newman and
Grafton, "Introduction," in *Secrets of Nature*, on 18; Rampling, "Depicting the Medieval Alchemical
Cosmos."

85. On Dee's use of the term, see Nicholas H. Clulee, "Astronomia inferior: Legacies of Johannes
Trithemius and John Dee," in *Secrets of Nature*, ed. Newman and Grafton, 173–233.

Our heaven this Figure called is
Our table also of the lower Astronomy.[86]

Kelley evidently paid attention to the wheel, since a Latin translation of these verses is also appended to the end of the *Clavis aureae portae* that he previously sent to Biberštejn.[87] Yet Ripley's influence goes deeper than the borrowing of a neat analogy; Kelley also draws on the theory of metallic generation expressed in the Ripleian *Accurtations of Raymond*, and subsequently appropriated by the *Work of Dunstan* and the *Clavis*. As we saw in chapter 3, this paradigm rests on the assumption that metals exist on a continuum, whereby lead gradually "ripens" into gold through a slow, natural process of mineral vegetation. Lead thus has the potential to grow into all the other metals. Its very crudity implies that it also possesses the most potent "vegetable" power, making it the most appropriate solvent for the rest.

Kelley's innovation is to add a hierarchical component. The lead-based menstruum serves to dissolve Jupiter, or tin, the next "contiguous" metal, to produce a new solvent—a process that continues through the metals in increasing order of maturity. Next, for instance, a solvent is made that will dissolve iron, followed by copper. Finally a solvent is obtained that can dissolve gold and silver, thereby yielding the "great menstruum" for transmutation.

Yet if Saturn is already the outermost planet, how is it to be dissolved in turn? Kelley explains: since no metal is cruder than Saturn, its proper solvent cannot be drawn from another metallic source. It must therefore arise from the vegetable rather than the mineral kingdom:

Truly, there is nothing worthier with respect to the form of Saturn, and participating more with [the nature of] metallic bodies, than that which is in vegetable things. Therefore the instrument for dissolving Saturn shall be made from some vegetable thing . . . and this thing ought to agree completely with lead in its own properties.[88]

---

86. *TCB*, 117. See also Rampling, "Depicting the Medieval Alchemical Cosmos."
87. Książnica Cieszyńska SZ DD.vii.33, fols. 60v-63r.
88. Leipzig 0398, fol. 348v: "Est igitur solutio corporis alicuius actio quaedam per appetitus uel innatae sympathiae leges, inferioris classis sibi simile, in habitum uirtutis suae proprium dirigens. Verum dignior respectu saturni forma, maximeque cum corporibus metallicis participans, nulla alia est, quam quae uegetabilibus inest, fiet igitur instrumentum saturnum dissipans ex re aliqua uegetabili . . . et illud suis proprietatibus plumbo omnino conuenire oporteat."

Although minerals and vegetables differ in species, a vegetable substance may still share the properties of lead "through appetite or laws of natural sympathy"—in this case, through the crude, unripe nature shared by lead and vinegar.[89] Since sweetness is associated with ripeness, it follows that an "unripe" solvent will be sour; thus "the subject for the proper dissolution of lead must be, of its own nature, a kind of vegetative, vinegary water."[90] Kelley's reasoning leads us to a familiar formula: the dissolution of lead in distilled vinegar. This conclusion is hardly surprising given that Kelley's compendium also includes the sericonian *Medulla*, which describes exactly this process. But the manuscript also includes plenty of other hints that Kelley's practice tends in this direction, such as a process for "The Ancient Way of Distilling Saturn," which describes how "Ancient philosophers were accustomed to dissolve lead in vinegar" in order to make a gum.[91]

All these examples reveal the influence of Ripley's *Bosome Book* on Kelley's practice, but none does so more profoundly than a recipe found toward the beginning of Leipzig MS 0398. Here, Kelley reveals another of his favorite tokens—the secret of the "gliding fire" that we have already encountered in Samuel Norton's adaptation of Master Ive. Kelley describes how, after drawing a menstruum out of the sericonian gum through distillation, a black earth remains behind:

> Which done, immediately break the glass, a little above the lute which covers its base. In this way the black earth will be kindled of its own accord, and calcine itself marvellously; which secret even the Philosophers would never commit to writing: they said only that our stone is able to calcine, wash, dissolve, perfect, and multiply itself. Once this earth is made to kindle like a live coal, it should be stirred by the worker several times with some iron rod, so that all of its parts may be well and perfectly calcined.[92]

89. Ibid., fol. 348v: "per appetitus uel innatae sympathiae leges."

90. Ibid., fols. 348v-49r: "ex quibus collectis sequitur, subiectum natura sua ad dissolutionem hanc plumbj idoneam aquam || acetosam quandam uegetabilem esse oportere." This assumption can be compared to the more unusual association of sweetness with corruption in the fifteenth-century *Tractatus brevis*, discussed above, pp. 122–24.

91. Ibid., fol. 6r: "Antiquus Modus distillandi Saturnum. Antiquiores Philosophi Soliti erant plumbum per acetum sic soluere."

92. Leipzig 0398, fol. 4v: "Quo facto rumpatur protinus vitrum, paulo altius luto illo, quo fundus ipse tegitur. Hac ratione nigra illa terra sua Sponte incendetur, seque Calcinabit mirifice. Quod secretum ne quidem Philosophi litteris Committere Voluerunt unquam, saltem dixerunt: lapidem nostrum semetipsum Calcinare, abluere, dissoluere, perficere, & multiplicare posse. Dum uero

Although this is clearly the same effect as that set down in the *Whole Work*, Kelley's procedure differs in one important respect. While Kelley mentions using a coal to ignite the Black Dragon, he claims that the powder may also "be kindled of its own accord." This is, in fact, exactly the effect that Blaise de Vigenère would later allude to in his own revision of the process—and one that, as we have seen, can be easily reproduced under modern conditions, simply by opening the retort before it has had an opportunity to cool.[93]

This effect also sheds light on Kelley's probable role in adapting the *Work of Dunstan*. Copies of *Dunstan* include a very similar description, which is not present in the original *Accurtations*, but has apparently been inspired by the same token of the gliding fire—an assurance that the black dregs remaining after the distillation of the Green Lion will of "ther owne accord . . . be Calcined into a most yellowe earth."[94] By retrofitting an earlier text to accommodate one of his own tokens, the canny Englishman here took a hand in manipulating alchemical history, reinventing the tenth-century archbishop as a sericonian alchemist whose work prefigured that of Ripley by half a millennium.

How better to set the stage for a dramatic "reconstruction" of Saint Dunstan's ancient practice than by writing a known, replicable effect into a forged medieval treatise? Such methods seem characteristic of Kelley, who throughout his career exhibited remarkable skill at producing exactly the results his audiences expected, whether in alchemical experiments or angelic conversations. The crafting of the *Work of Dunstan* offers one clue to his success, while underscoring the continued relevance of fifteenth-century practices in early modern Europe. Even in Kelley's Bohemian fortress, English sericonian alchemy reigned supreme.

LEGACIES OF DEAD ALCHEMISTS

Kelley's good fortune was not long lived: only a few years after his release he was imprisoned again, this time in the fortress of Most. Records of his final days are blurry; possibly he took poison after a failed escape, as one contemporary account suggests.[95] Probably he sought once more to retrieve his

---

Terra haec instar Viui Carbonis incensa fuerit, ab artifice est aliquoties mouenda ferrea aliqua Spatula, ut omnes illius partes bene & perfecte calcinentur." The same text, with some minor alterations, was later appended to the *Via humida*; Kelley, *Tractatus duo egregii*, 83.

93. See p. 281, above.

94. English translation from Robson's copy of the *Work of Dunstan* in Ashmole 1421, fol. 151v.

95. Karpenko and Purš, "Edward Kelley," 521.

position by writing to the emperor. The first of the "Two Excellent Tracts" to appear in print is presented as a work written during his second imprisonment, suggesting just such an attempt.[96] It consists of a series of commonplaces on Kelley's old themes: the action of the elements, the maturity of metals, and quotations from medieval texts.

Kelley's influence, and his relationship with medieval English alchemy, did not end with his death. His former associates now picked over the bones of his written legacy. Among them was Nicolaus Mai, who seems to have remained in touch with Kelley after his disgrace and imprisonment, later composing an epitaph for his widow, Joan.[97] Mai also had access to at least some of Kelley's papers, including his prison writings—material that he took to heart when seeking patronage on his own behalf.

Despite his prestigious role as overseer of mining operations at Joachimsthal, Mai continued to cultivate connections with other high-placed advocates of the alchemical art. On 22 January 1603, he wrote to one of Europe's most prominent supporters of alchemy, Moritz, Landgrave of Hesse-Kassel, to acknowledge receipt of Moritz's letters, "dearer to me than any gold." Replying to Moritz, he confesses his relief that his earlier attempts at "philosophizing" have not offended the landgrave but, on the contrary, incited his interest. Moritz has gone so far as to send his chamberlain, John Eccelius, to meet with Mai and encourage him to return with them to Kassel, "so that I might go straight to Your Highness with them and say in your ear those things illustrious and most desired by Your Highness."[98] Although Mai has to excuse himself from attending in person because of press of business, he has conferred privately with the landgrave's messengers in the meantime

---

96. Kelley, *Tractatus duo egregii*, 3–40.

97. Susan Bassnett, "Absent Presences: Edward Kelley's Family in the Writings of John Dee," in *John Dee*, ed. Clucas, 285–94, on 290. Kelley's stepdaughter, the poetess Elizabeth Jane Weston, known as Westonia, dedicated a series of poems to him: Elizabeth Jane Weston, *Parthenica*, vol. 1 (Prague: Paulus Sessius, [1606]).

98. Nicolaus Mai to Moritz of Hesse-Kassel, 22 January 1603, Kassel Landesbibliothek, 2° MS chem. 19, fol. 273r: "Accepi, Illustrissime et Clementissime Princeps, Cels[itudinis] Tuae literas, quouis auro mihi cariores, quae me non tantum sollicitudine quadam levarunt: sed etiam de pristinâ erga me meosque clementia et voluntate confirmarunt. Dolui enim vehementer, et veritus sum, ne Cels. T. in luctu gravissimo, Philosophicis meis inscius appellassem, et simul offendissem. Nunc vero cum intelligam, gratum Cels. T. fuisse officium meum, et partem lacrymarum abstersam esse, dupliciter gaudeo. Caeterum quod per ablegatos fratrem meum lucam et Joannem Eccelium Cels. T. cubicularium, clementer me salutare, mecumque agere jusseris, ut vel cum iis ad Tuam Cels. recta eam, et quae habeam praeclara maximeque Cels. T. exoptata, in aurem dicam." On Moritz, see Moran, *Alchemical World of the German Court*.

regarding his philosophical studies.[99] To further placate Moritz's curiosity, Mai announces that he will set down his own opinion concerning the matter of the philosophers' stone.

The screed that follows is instantly familiar. In fact, it has been lifted almost verbatim from Kelley's *Syntagma philosophicum*, beginning with the role of Saturn in the planetary hierarchy, and the need for a vegetable solvent. Mai's appropriation extends even to substituting his own name for that of his authority. For instance, Kelley concludes the *Syntagma* on an enigmatic note, quoting from Virgil's *First Eclogue*:

> But do not say too much, Kelley; for already smoke ascends in the distance from the roofs of the houses, and the shadows of the hills begin to lengthen.[100]

Mai finishes his exposition with the same phrase—but replaces "Kelley" with "Maius."[101]

It seems fitting that Kelley's techniques of invention and appropriation should eventually be applied to his own writings, and directed to a similar end: currying the favor of a prince. Yet, in a further twist, Mai's borrowing was ultimately advantageous for Kelley's posthumous reputation, by allowing material associated with Kelley to circulate abroad, and even to reach the printing press.

When Prague was sacked by Swedish troops at the culmination of the Thirty Years' War, Rudolf's alchemical library did not escape. Many of Rudolf's and Rožmberk's books, including a presentation volume of Mai's *Liber duodecim portarum*, were borne away from Prague as booty.[102] It is

---

99. Ibid., fol. 273r-v: "Venissem cum ablegatis ipse, nisi me Caesaris iussa ad Comites Mansfeldenses legatum, in negotio quodam arduo, moram non ferente, avocarent. . . . Interim cum ablegatis de studiis Philosophicis familiariter contuli, ad quaesita Eccelii aperte respondi, et quanta mihi sit Auxilio divino spes melioris fortunae, multis argumentis demonstravi."

100. Leipzig 0398, fol. 352r:

> Sed parce Kellee nimium procedere ripae.
> Nam iam summa procul, uillarum culmina fumant,
> Maioresque cadunt, altis de montibus umbrae.

101. 2° MS chem. 19, fol. 276v: "Sed parte Maje nimium procedere ripae . . ."

102. Dudík, *Iter romanum*, 228. Dudík suggests that the manuscript may have been part of the Rožmberk collection in Prague (see also Evans, *Rudolf II*, 210 n1). It eventually reached Rome via the royal library of Queen Christina of Sweden, which included alchemical manuscripts from the collections of Rudolf II and Rožmberk. On the fate of Christina's alchemical books see Frans Felix Blok, *Contributions to the History of Isaac Vossius's Library* (Amsterdam: North-Holland, 1974); on the diffusion of books and manuscripts from Rudolfine Prague, see Nicolette Mout, "Books from

presumably thanks to Mai's overtures to Moritz that some records of his relationship with Kelley escaped the purge, eventually coming to rest in the archives of the princely court of Hesse-Kassel. Mai's manuscripts include, besides a copy of his Ripley translation, a separate volume packed with Ripleian texts, annotated by Mai himself.[103] The latter also records precious traces of Kelley's Bohemian activities: a recipe heard from Kelley's own lips; a testimonial to Kelley's noble Irish lineage, dated Galway, 10 March 1593; and several extracts apparently taken from Kelley's letters (one dated Prague, 20 June 1587).[104]

Decades later, this cache came to the attention of Moritz's former physician Ludwig Combach (1590–1657), himself a devotee of chemical medicine. Combach was surprised to find an entire collection of writings by Ripley, some of which, he concluded, must have been translated by the famous Kelley himself. He lost no time in publishing Kelley's epistolary fragments, touting their connection with "Councillor Mai" as evidence for their authenticity.[105] In 1649, he followed up this volume with an edition of twelve of Ripley's attributed works, using Mai's manuscript as a major source.[106] In total, this manuscript includes eight of the twelve texts included in the *Opera omnia*

Prague: The Leiden *Codices Vossiani Chymici* and Rudolf II," in *Prag um 1600: Beiträge zur Kunst und Kultur am Hofe Rudolfs II*, ed. E. Fučíková (Freren: Luca Verlag, 1988), 205–10; Astrid C. Balsem, "Books from the Library of Andreas Dudith (1533–89) in the Library of Isaac Vossius," in *Books on the Move: Tracking Copies through Collections and the Book Trade*, ed. Robin Myers, Michael Harris, and Giles Mandelbrote (London: Oak Knoll Press, 2007), 69–86.

103. *Liber duodecim portarum*, 4° MS chem. 68. The Ripleian collection is in Kassel Landesbibliothek 4° MS chem. 67, which also includes one of Mai's Latin verses, "Ænigma M. Nicolaii Maii" (fol. 183r).

104. Ibid.: "Ex ore EK. *Recipe* [mercur]ium [Jov]is et pone in crucibulum super tripodum" (fol. 141r); "ex epistola K: 20 Junii anno [15]87 Prahae data" (fol. 181v); "Datum Galuiae vrbis huius prouinciae principalis .X. die Martii, anno ab incarnatione Dominj M.D.XCIII" ("Testimonium Eduardi Kellaei Angli," fol. 143v). The latter text is reproduced in Karpenko and Purš, "Edward Kelley," 534; English translation on 505–6. See also Rampling, "Dee and the Alchemists," 503.

105. Ludwig Combach, *Tractatus aliquot chemici singulares summum philosophorum arcanum continentes, 1. Liber de principiis naturae, & artis chemicae, incerti authoris. 2. Johannis Belye Angli . . . tractatulus novus, & alius Bernhardi Comitis Trevirensis, ex Gallico versus. Cum fragmentis Eduardi Kellaei, H. Aquilae Thuringi, & Joh. Isaaci Hollandi . . .* (Geismar: Salomonis Schadewitz for Sebaldi Köhlers, 1647), 31–33. See also *OOC*, 11: "Insequentes tractatus parvuli . . . cum fragmentis Kellaei . . . ex codice ms. Domini Nicolai Maij, Augustiss. quondam Imperatoris Rudolfi II. &c. Consiliarij, excerpti sunt."

106. 4° MS chem. 67 provides the primary exemplar for his edition of the *Liber*, and is the second of two manuscripts used in preparing the *Medulla* (the other is 4° MS chem. 66).

chemica; six of which, including the *Clavis aureae portae*, had previously appeared in Kelley's gift to Biberštejn.[107]

Combach's edition appeared at a timely moment for adherents of pseudo-Lullian alchemy, as the authority of the corpus was under attack. Just the previous year, Hermann Conring (1606–1681), professor of natural philosophy, medicine, and law at the University of Helmstädt, had condemned *De secretis naturae* as a work "full of follies and vanities."[108] By publishing the works of Ripley, one of Europe's most eminent interpreters of Lullian alchemy, Combach rose to Raymond's defense. In his edition, the English canon's writings provide necessary links in a temporal chain that connects one of the most celebrated medieval theorists, Lull, to one of the most lauded early modern practitioners—Ripley's own countryman and commentator, Kelley.

Combach may have been responding to contemporary polemics, but his case rested on the activities of a network of practitioners based in Bohemia some sixty years earlier; a circle linked to Kelley, and distinguished by its interest in the English canon of Bridlington. It is through such tortuous routes that Ripley's "Collected Chemical Works" came to be printed for the first and only time; not in the canon's native land, but in a wider European context where adherents of *chymia* marshaled the alchemical histories of all nations in defense of their science.

RIPLEY'S RETURN

Kelley's Bohemian enterprise, from written patronage suits and presentation volumes to the content of his laboratory practice, is characterized to a remarkable extent by his use of late medieval English authorities. Yet although works like Ripley's *Bosome Book* informed Kelley's alchemical activities, the benefits of the association were not one-sided. The connec-

---

107. The *Liber 12 portarum, Medulla philosophiae chemicae, Clavis aurae portae, Pupilla Alchemiae, Terra terrae philosophicae, Viaticum seu varia practica, Cantilena*, and *Epistola ad Regum Eduardum*. Of the remaining four texts printed by Combach, three (*Liber de Mercurio & Lapide philosophorum, Philorcium Alchymistarum*, and *Accurtationes & practicae Raymundinae*) are found in Combach's second major exemplar, 4° MS chem. 66. I have not identified an exemplar for the remaining item, the *Concordantia* (although this work is mentioned by title in 4° MS chem. 67, fol. 133v). It is also likely that Combach had access to additional exemplars for at least some of the texts named above.

108. Hermann Conring, *De Hermetica Ægyptiorum vetere et Paracelsicorum nova medicina liber unus* (Helmstedt: Hemming Müller, 1648), 382: "Et vero iam supra demonstratum est, librum de Quinta essentia Raimundi plenum esse ineptiarum ac vanitatum." On this and other anti-Lullian critiques, see Rampling, "Transmission and Transmutation," 493–95.

tion with a successful and charismatic practitioner, Kelley, attracted new interest in the author upon whom Kelley himself relied, smoothing the passage of Ripley's English works through the courts and presses of the empire. In this international and transgenerational conference of philosophers, one authority supported another—yet Combach still chose to publish Kelley's own writings before those of his authority, Ripley.

Back in England, Dee, Kelley, and Ripley would remain inseparably bound in print. Raph Rabbards, magistrate and frustrated engineer, published Ripley's *Compound* in 1591; it was the first time that an English vernacular alchemical work had been printed in its original language.[109] In his dedication to Elizabeth I, Rabbards hailed the achievements of English alchemical philosophers, "especially M. Doctor *Dee* in his *Monas Hyeroglyphica*," praising their "depth of learning *Theoricall*."[110] Despite this lip service to theoretical acumen, he also hinted at the results that might be obtained if the work "were yet executed by any experienced practitioner," before expressing his own willingness to work in this capacity on the queen's behalf.[111]

This image of the alchemist as an "experienced practitioner" able to resurrect the art of past adepts is also an apt description of Edward Kelley's self-presentation in the years prior to his fall. Although Rabbards does not mention the disgraced Englishman directly in his preface, he does include a verse attributed to him: "Sr. E. K. concerning the Philosophers Stone, written to his especiall good friend, G.S. Gent."[112] The initials suggest a connection to Dee's friend, "Mr Gawyn Smithe gentleman," the same royal projector who

109. Rabbards describes how he, "hauing these fortie yeares amongst many other most commendable exercises and inuentions of so warlike Engines, founde out diuers deuises of rare seruice, both for Sea and land," only to lose the credit to "ignoraunt persons . . . [who] vainely arrogated the inuention vnto themselues." Rabbards, "Epistle dedicatorie," Ripley, *Compound*, sig. A3v.

110. Ibid, sig. A4v. The prefatory poems include one by "J.D. gent: in praise of the Author, and his Worke" (Ripley, *Compound*, sig. *2r), sometimes attributed to Dee, apparently on the basis of the initials; Peter J. French, *John Dee: The World of the Elizabethan Magus* (London: Routledge & Kegan Paul, 1972), 82n2. Given Rabbards's earlier singling out of the *Monas*, it is entirely plausible that Dee is the author: he sometimes wrote English verses, and his brief *Testament* was later included in the *TCB*; see below, p. 324.

111. Rabbards, "Epistle dedicatorie," *Compound*, sig. A4v.

112. Ripley, *Compound*, sig. *3r-v. The same poem appears, dated 1589, in one of a group of manuscripts connected to the Kelley circle in Prague: Copenhagen, Royal Library, GKS 242, 300–301. There it is titled "The praise of vniti for frendships sake made by a stranger to furder his frende his Conceyts. 1589," signed "Sir Edward Kelle." See Jan Bäcklund, "In the Footsteps of Edward Kelley: Some MSS References at the Royal Library in Copenhagen Concerning an Alchemical Circle around John Dee and Edward Kelley," in *John Dee*, ed. Clucas, 295–330. It is quite likely that Rabbards was unaware of Kelley's arrest at the time of printing: the scandal in Prague broke only twelve days before the *Compound* was entered in the Stationer's Register on 12 May.

commissioned Howes's translation of the *Bosome Book*.[113] If so, we can add Smith to the growing list of would-be adepts who received philosophical correspondence from Kelley. As fellow engineers with a taste for alchemy, Smith and Rabbards may also have been acquainted: a possible indicator of the route by which Kelley's poem reached the English press.

The host of synchronicities surrounding the publication of Ripley's works provides ample evidence for the vigorous, scribal transmission of early modern alchemical texts, even in the age of print. The first editions of the *Compound*—by Rabbards, Penot, Nicolas Barnaud, and Combach—did not emerge in isolation, but lay enmeshed within webs of communication, authority, and patronage: a Pan-European network in which English practitioners like Dee and Kelley were enthusiastic and influential participants. This network now survives only in fragmentary form: in friendly dedications, marginal notes, and the appearance of particular works in unexpected places—most strikingly, in Kelley's earnest petitioning for his freedom. Such clues guide us to the routes by which Ripley's masterpiece attained a level of success that Dee, indifferently successful petitioner to a host of European monarchs, might well have envied: written for an English king, printed for an English queen, and translated for a Holy Roman Emperor.

---

113. Discussed above, pp. 262–63. The poem's link to Gawin Smith receives some tentative support from the appearance of "Smith" among the deleted names identified by Bäcklund, "Footsteps of Edward Kelley," 299, in the margins of another manuscript in the Copenhagen "Kelley" group: GKS 1727 4° (ca. 1593–95). See Rampling, "Dee and the Alchemists," 506. The *Bosome Book* also left its mark on Rabbards's edition. The *Compound* is prefaced by "The Vision of Sir George Ripley, Chanon of Bridlington" (*Compound*, sig. *4r), Samuel Norton's English translation of the *Visio*, excerpted from the *Book* (CRC 32); see below, pp. 340–44.

CHAPTER NINE

Antiquity and Experiment

The Subject of this ensuing Worke, is a Philosophicall account of that Eminent Secret treasur'd up in the bosome of Nature; which hath been sought for of Many, but found by a Few, notwithstanding Experience'd Antiquity hath afforded faithfull (though not frequent) Discoveries thereof.[1]

When Elias Ashmole (1617–1692) published the *Theatrum Chemicum Britannicum* in 1652, he trusted that the work would whet the appetites of his English contemporaries for their own national tradition, while also preserving its most cherished fruits.[2] The *Theatrum* marked the first substantial collection of English alchemical poetry in print, if we discount the handful of verses added to Rabbards's 1591 edition of the *Compound*. Its contents reinforced the pantheon of English authorities that had taken shape over the course of the previous three centuries: a chorus of adepts whose practical ingenuity, as much as their semilegendary origins, arose from the permutations of manuscript transmission and the accumulated insights of generations of reader-practitioners.

Ashmole's project coincided with another remarkable chemical enterprise that took English alchemy as its anchor text: the composition by George Starkey (1628–1665) of a new alchemical corpus attributed to a mysterious American adept, Eirenaeus Philalethes. Starkey is now well known to historians of science, both for his role in tutoring the young Robert Boyle in chemistry and for the success of alchemical works written under his

---

1. Elias Ashmole, "Prologomena," *TCB*, sig. A2r.

2. The leading source on Ashmole remains C. H. Josten, ed., *Elias Ashmole: His Autobiographical and Historical Notes, His Correspondence, and Other Contemporary Sources Relating to His Life and Work*, 5 vols. (Oxford: Oxford University Press, 1967). See also Vittoria Feola, *Elias Ashmole and the Uses of Antiquity* (Paris: Librairie Blanchard, 2012); Bruce Janacek, *Alchemical Belief: Occultism in the Religious Culture of Early Modern England* (University Park: Pennsylvania State University Press, 2011), chap. 5.

pseudonym.[3] During the 1650s, he assumed the persona of Philalethes to pen what would become a hugely influential series of commentaries on Ripley's works, covering the first six gates of the *Compound*, the *Epistle to Edward IV*, and the *Vision*—all published by Raph Rabbards some sixty years earlier.[4] These commentaries preserved Ripley's authority even as the sericonian paradigm of multiple stones increasingly gave ground to proponents of different practical and theoretical positions.

Both the *Theatrum* of Ashmole and the commentaries of Philalethes pay homage to the tradition of English alchemical verse exemplified by Ripley and his late medieval peers.[5] But their presentation, and the evidence they relied upon, otherwise look very different. For Ashmole, the relics of English alchemy had value beyond practice alone, as evidence for an ancient tradition of British wisdom (including knowledge of magic) traceable back to the Druids.[6] To recover this tradition meant locating and collating the "Collected Antiquities" of the past—the tradition preserved in manuscript.[7] While Starkey also studied past texts, his interest was resolutely practical. His source base included the evidence of his own trials and experiments, set down in notebooks that recorded his commitment to making and selling alchemical products.[8] The two projects also differed in the nature of their rhetoric. Where Ashmole called for the rescue of England's neglected alchemical legacy through preserving and publishing texts, Starkey presented himself (through Philalethes) as a practicing philosopher who actually possessed the secret of transmutation: physically embodying the tradition not in the form

3. The main source on Starkey's life and work is Newman, *Gehennical Fire*; on his experimental and reading practices, see also Newman and Principe, *Alchemy Tried in the Fire*.

4. Of these, "Sir George Ripley's Epistle, to King Edward Unfolded" was printed (without Starkey's consent) in Samuel Hartlib's *Chymical, Medicinal, and Chyrurgical Addresses* (London: G. Dawson for Giles Calvert, 1655), and a commentary on the *Compound*'s "Recapitulation" in Eirenaeus Philalethes, *A Breviary of Alchemy; or a Commentary upon Sir George Ripley's Recapitulation: Being a Paraphrastical Epitome of his Twelve Gates* (London: for William Cooper, 1678). The Ripley commentaries, including "The Vision of Sr George Ripley, Canon of Bridlington, Unfolded," were collected in *Ripley Reviv'd: or An Exposition Upon Sir George Ripley's Hermetico-Poetical Works* (London: William Cooper, 1677–78).

5. In addition to his expositions of Ripley's verses, Starkey published an alchemical poem of his own, *The Marrow of Alchemy*—a title that evoked Ripley's famous treatise of the same name. Eirenaeus Philoponus Philalethes, *The Marrow of Alchemy, Being an Experimental Treatise, Discovering the Secret and Most Hidden Mystery of the Philosophers Elixer. Divided Into Two Parts* (London: A.M. for Edward Brewster, 1654).

6. *TCB*, sigs. A2v-A3r.

7. Ibid., sig. A4v.

8. Now edited as George Starkey, *Alchemical Notebooks and Correspondence*, ed. Lawrence M. Principe and William R. Newman (Chicago: University of Chicago Press, 2004).

of a text or book, but through the reproduction of the stone itself. In the writings of Philalethes it is the practice rather than the text that constitutes the true antiquity.

At first glance, Ashmole and Starkey seem to offer two visions of English alchemy: a seventeenth-century divergence between what we might think of as "antiquarian" and "experimental" modes. Yet we should be wary of assuming separate readerships for alchemical history and alchemical practice. Throughout the century, the writings of medieval alchemists continued to inform the living practice of their seventeenth-century successors, including natural philosophers of the stature of Robert Boyle and Isaac Newton.[9] Conversely, the compilations of antiquarians were often shaped by their own practical commitments. The challenge of distinguishing between modes becomes more difficult still as we plunge back into the manuscript record, including Ashmole's own sources for the *Theatrum*. Whether text, practice, or history, the matter of alchemy never stayed fixed for long.

THE BRITISH CHYMICAL THEATER

The activities of Ashmole and Starkey did not take place in a vacuum, for the mid-seventeenth century witnessed a sharp spike of interest in alchemy, related to the expansion of the English printing industry.[10] After remaining relatively stable during the previous decade, the number of English books printed on chemical topics increased tenfold during the 1650s, a level maintained almost until the end of the century.[11] While striking in themselves, these figures—based on the tallies of the London printer William Cooper between 1673 and 1688—are further inflated by the volume of continental

9. See particularly Newman and Principe, *Alchemy Tried in the Fire*. Newton's manuscript notes are packed with extracts from medieval sources, which he also studied in print. For instance, he owned Ripley's *OOC* (Cambridge, Trinity College Library, NQ 10.149), and copied extracts from the *Medulla*, *Pupilla*, and *Clavis aureae portae* into Cambridge, King's College Library, MS Keynes 17. On his copy of the *TCB*, see note 124 below.

10. On the political, economic, and technological factors behind the expansion of English print, see James Raven, *The Business of Books: Booksellers and the English Book Trade, 1450–1850* (New Haven: Yale University Press, 2007).

11. Lauren Kassell, "Secrets Revealed: Alchemical Books in Early-Modern England," *History of Science* 48 (2011): 1–27 and A1–38, on 1. Kassell's data is based on William Cooper, *A Catalogue of Chymical Books Which Have Been Written Originally or Translated into English*, printed with W. C. Esquire, *The Philosophical Epitaph* (London: William Cooper, 1673). Cooper printed new versions of the catalogue as *A Catalogue of Chymicall Books. In Three Parts* (London: William Cooper, 1675), and *The Continuation or Appendix to The Second Part of the Catalogue of Chymical Books* (London: William Cooper, 1688).

publications, printed in Latin and European vernaculars, that became available to English readers over the course of the seventeenth century. These printed books still constituted only a portion of the alchemical literature available to interested readers. They were supplemented by countless manuscripts, both new and old, that continued to circulate at different levels of English society—a world of scribal publication that included texts hand-copied from printed books, and printed books that became manuscripts through the annotations of their users. Even the *Theatrum* falls into that category, as Ashmole continued to annotate and amend his personal copy long after the formal date of publication.[12]

Ashmole's *Theatrum*, like other compendia of the seventeenth century, is grounded in practices of scribal copying and compilation, but it also blends several distinct genres of print. On the one hand, it is an edited compilation of alchemical texts: a genre recently embodied by Zetzner's multivolume *Theatrum Chemicum*, a continental endeavor to which Ashmole's collection provides a distinctively anglophone response.[13] On the other, the texts are supported by extensive bio-bibliographical notes of the kind gathered in the catalogues and indexes assembled by English antiquaries such as John Leland and John Bale, both of whom Ashmole mined for information on his chosen authors.[14]

Aspects of both genres shape Ashmole's self-presentation in the *Theatrum*, as both English antiquary and alchemical philosopher (if not, by his own admission, one who had yet embarked "*Effectually* upon the *Manuall Practise*").[15] The *Theatrum* is prefaced with the author's engraved portrait bust, while Ashmole's name is embellished with the patriotic sobriquet "Mercuriophilus Anglicus"—the English lover of Mercury.[16] The name was apt enough for the editor of a volume intended to preserve the relics of England's alchemical past: those precious manuscripts that might, as Ash-

12. Ashmole added notes to an interleaved copy of the *TCB* prepared for that purpose, now bound in two separate volumes as Ashmole 971 and 972.

13. On Zetzner and the *TC*, see Gilly, "On the Genesis of L. Zetzner's *Theatrum Chemicum*"; Kahn, *Alchimie et paracelsisme*, 112–21.

14. Ashmole singles out Leland and Bale in *TCB*, sig. A2v, and frequently references their writings in his concluding annotations. On Ashmole's use of antiquarian sources, see Feola, *Elias Ashmole*.

15. *TCB*, sig. B2v.

16. This marks a retreat from Ashmole's earlier authorial modesty. In May 1650 he published alchemical treatises by Arthur Dee and Jean D'Espagnet under an anagram of his name, James Hasolle; Arthur Dee, *Fasciculus chemicus, or, Chymical collections: expressing the ingress, progress, and egress of the secret Hermetick science, out of the choisest and most famous authors . . . whereunto is added, the Arcanum, or, Grand secret of hermetick philosophy*, ed. and trans. Elias Ashmole (London: J. Flesher for Richard Mynne, 1650).

mole reminded his readers, have been lost to the ravages of time, "but that my *Diligence* and *Laborious Inquisition* rescued them from the *Jawes* thereof."[17] In his oft-quoted "Prologomena," he further deplored the "great Devastation of our *English Libraries*" wrought by willful iconoclasm and blind neglect.[18]

Ashmole's tone, so far divorced from that of Elizabethan patronage suits, reflects the temporal distance that allowed even devout Protestants in his time to lament the despoilment of books and buildings occasioned by the dissolution.[19] The fact that alchemical books were circulating outside religious houses long before the 1530s, and that that many of Ashmole's own sources for the *Theatrum* were produced by merchants rather than monks, does not dilute either the narrative of loss or the specificity of the English context he invokes. The tradition of alchemical history preserved in his manuscripts belonged to a pre-Reformation world in which monks and kings were intrinsic components of English society, and philosophers might still cultivate personal relationships with English princes. It is a fantasy captured in the vernacular poems of Ripley and Norton, and engagingly acted out in the verses of Charnock and Blomfild, all of which Ashmole published in the *Theatrum*.

In interregnum England, a realm without monks or kings, the art's long associations with monkish adepts and English royalty carried particular freight. As a former Royalist officer, banned from living within twenty miles of London, Ashmole already belonged to a different world from that celebrated in the pages of alchemical histories. It is telling that the first alchemical manuscript we know for certain he owned was Henry Harrington's treatise dedicated to King Charles I: "by the grace of God Emperour of greate Brittanye, and Kinge of France and Ireland Defender of the ancient Catholique faith."[20] Ashmole received it as a gift from the surgeon Nicholas Bowden, whom he earnestly thanked in a verse dated 8 July 1649, promising to follow the directions of the text until he achieved the elixir:

And stubborne Nature to Obedience wrought:
Then to make Gold 't shalbe a thing of nought.[21]

17. Ashmole, "Prologomena," *TCB*, sig. B3v.
18. Ibid., sig. A2v.
19. On early modern "nostalgia" for the monastic past, see Margaret Aston, "English Ruins and English History: The Dissolution and the Sense of the Past," *Journal of the Warburg and Courtauld Institutes* 36 (1973): 231–55; Harriet K. Lyon, "The Afterlives of the Dissolution of the Monasteries, 1536–c. 1700" (PhD diss., University of Cambridge, 2018).
20. Ashmole 1459, pt. 1, fol. 4v.
21. Ibid., fol. 26v.

From these early beginnings, Ashmole's collecting gathered pace. Since his own library of alchemica was still developing when he commenced work on the *Theatrum*, he relied to a large extent on copies owned by acquaintances, transcribing not just their original contents but also later emendations and additions. These included Ripley's apologia from the *Bosome Book*, in which the canon urged readers to disregard his experiments carried out before 1470. Ashmole found the passage in one of John Dee's books and jotted it down as a kind of frontispiece to his own Ripleian compendium, Ashmole 1459, noting its distinguished provenance: "This I mett with in a \vellum/ Manuscript of Dr. Dees; & written before Riplies 12 gates."[22] Beneath, he added Ripley's Latin explicit to the *Compound*, taken from the same manuscript of Dee's, which he later printed with his own English translation.[23]

Ashmole raided the waste spaces of such association copies for traces of distinguished former readers. In the *Theatrum*, the marginal flourishes of Dee and Charnock carry equal billing with late medieval texts of the type they once annotated. One of Ashmole's main exemplars for alchemical verse, the fifteenth-century Harley 2407, was previously owned by Dee, who added an English poem, the *Testament*, to a half-empty page. Ashmole included it in the *Theatrum*, later annotating his own copy to affirm that it was indeed "Coppied from Dr Dee's owne hand."[24] Another coup for Ashmole was the discovery of a large cache of Thomas Charnock's own annotations in Trinity O.2.16. Few alchemists made such enthusiastic use of waste space as Charnock, who crammed the margins of fifteenth-century manuscripts with notes, including his near-obsessive reflections on the number of circulations required to make the stone. Ashmole harvested these, transcribing Charnock's annotations as well as removing several pages (with or without the owner's permission is unclear) from the original manuscript.[25] In his own copy in Ashmole 1441, he mimicked Charnock's signature and initials, even

---

22. "Ex Libro Collectaneorum G[eorg]ij R[ipley]," Ashmole 1459, fol. 27v. Ashmole referred to this retraction in *TCB*, 456.

23. *TCB*, 193.

24. "Testamentum Johannis Dee philosophi sum[m]i ad Joannem Gwynn transmissum/ 1568," in Harley 2407, fol. 69r-v; printed in *TCB*, 334, with Ashmole's note in Ashmole 972, 334.

25. The pages in Charnock's own hand from Trinity O.2.16 are now in Ashmole 1441, 85–88 and 98. With the exception of the notes on p. 98, these were published by F. Sherwood Taylor, "Thomas Charnock," *Ambix* 2 (1946): 148–76, on 160–62. Ashmole's transcriptions from Trinity O.2.16 are added afterward: Ashmole 1441, 99–104; published by Taylor, "Thomas Charnock," on 162–63. Although Taylor was not aware of the Trinity College manuscripts, he rightly speculated that Charnock's original books would eventually come to light (ibid., 176).

attempting to preserve the mise-en-page of the original manuscript, proudly recording in the *Theatrum* that these relics were "Coppied from Charnock's own handwriting," or "Fragments scattered in the wast places of an Old Manuscript, written with T. Charnock's own Hand" (figs. 14–15).[26] Torn away from their original setting, however, Charnock's enigmas no longer speak to their fifteenth-century anchor text, but only to one another. For knowledge of his practical sources and how he read them, we must still return to the Kentish alchemist's own manuscripts.[27]

GENEALOGIES OF PRACTICE

On 7 December 1650, Ashmole came to Ripley's *Bosome Book*. His source was not the original *Book* but a Latin copy transcribed by an Elizabethan reader, Thomas Mountfort.[28] Mountfort's version was not, however, an accurate facsimile of Ripley's famous compendium, since he chose to conceal the practical content of his sources by replacing the names of ingredients with his own idiosyncratic set of symbols. The effect was to dramatically transform the text, simultaneously obscuring the original terms used by his authority and imposing new interpretations upon their meaning. For instance, in a marginal note to *De ignibus nostris*, Mountfort offered his interpretation of Ripley's two opposing fires: "The fire against nature is [Mercurius]," and "Natural fire is [water] of our [Sol]."[29] But what kind of "mercury" did Mountfort intend, given that Ripley's original text actually equated the fire against nature with vitriol, rather than quicksilver? Without comparing Mountfort's notes to a less obviously adulterated copy, such as Harley 2411, the oddness of his reading would go unnoticed. In copying his transcription, Ashmole was thus recording more than a medieval antiquity; he also silently absorbed Mountfort's late sixteenth-century interpretation of the text.

The themes of experiment and antiquarianism fuse in the books of

---

26. *TCB*, 425.

27. Discussed above, pp. 236–37.

28. I have been unable to identify Mountfort with certainty. Possibly he is the physician who served seven times as censor to the College of Physicians; Norman Moore, "Moundeford, Thomas (1550–1630)," rev. Patrick Wallis, *ODNB*. Several collections of his transcriptions of medieval monastic sources, now Ashmole 1406 (pt. 4) and 1423, later entered Ashmole's library. Also in Mountfort's hand is Bodleian Library, MS Rawlinson B.306, parts of which seem to have been copied from Richard Walton's manuscript, Ashmole 1479, suggesting a London connection.

29. Ashmole 1459, pt. 2, 5: "Ignis contra na*turam* est .[Mercurius]. . . . Ignis naturae est .[aqua]. [Sol]. nostri."

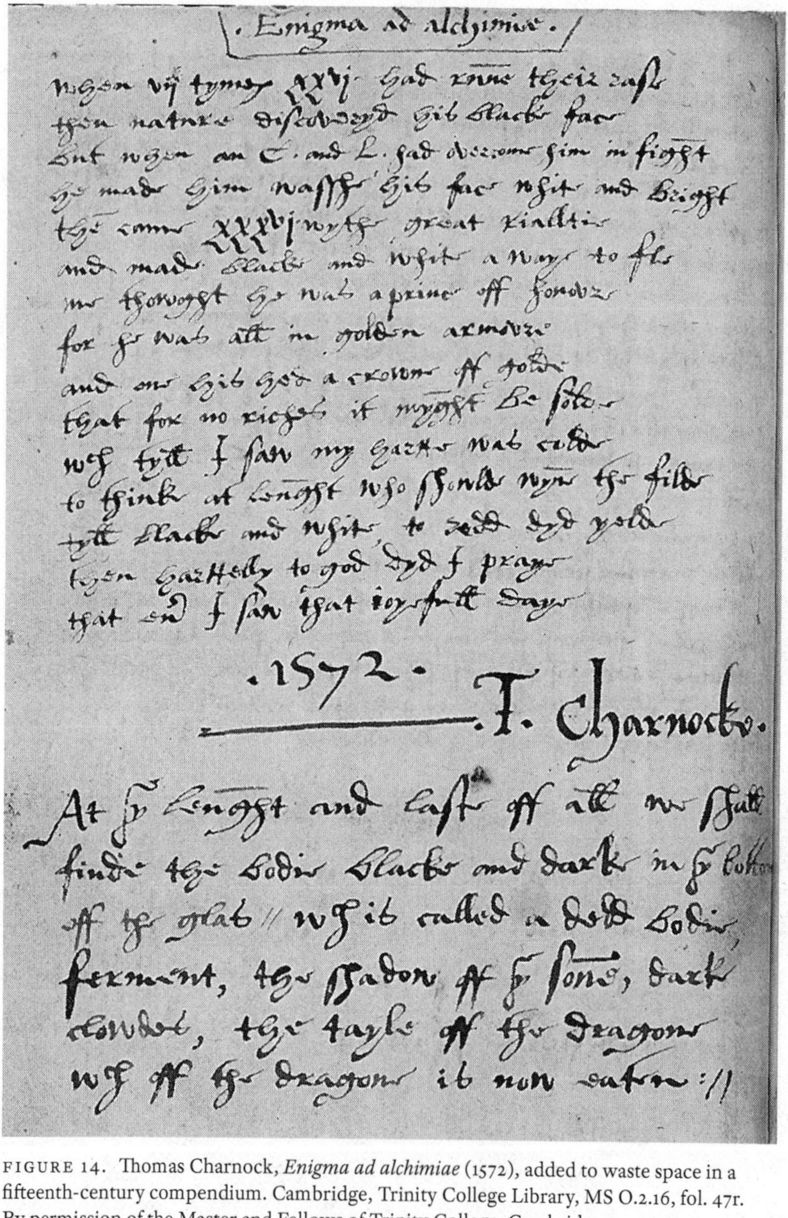

· Enigma ad alchimiæ ·

when vij tymes xxvj had runne their cast
then nature discovered his black face
but when an C. and L. had overcome him in fight
he made him wasshe his face white and bright
the came xxxvj wyth great riallti
and made black and white away to fle
me thought he was a prince off honour
for he was all in golden armour
and on his hed a crowne off golde
that for no riches it myght be solde
wh tyll I saw my harte was colde
to think at length who sholde wyn the filde
tyll black and white to redd dyd yelde
then hartely to god dyd I pray
that ons I saw that ioyfull daye

·1572· T. Charnocke.

At the length and last off all we shall finde the bodie blacke and dark in the bottom off the glas // wh it is called a dedd bodie, ferment, the shadow off the sonne, dark clowde, the tayle off the dragon wh off the dragon is now eaten ://

FIGURE 14. Thomas Charnock, *Enigma ad alchimiae* (1572), added to waste space in a fifteenth-century compendium. Cambridge, Trinity College Library, MS O.2.16, fol. 47r. By permission of the Master and Fellows of Trinity College, Cambridge.

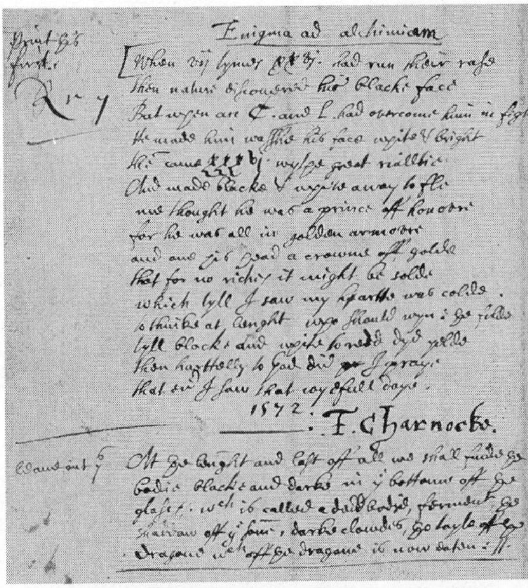
Print this first.

Enigma ad alchimiam

When vij tymes xxvj had run their rase
then nature discovered his blacke face
but when an C. and L. had overcome him in fight
he made him wasshe his face white & bright
then came xxxvj wythe great riallti
and made blacke & white away to fle
me thought he was a prince off honour
for he was all in golden armour
and one his head a crowne off golde
that for no riches it might be solde
which tyll I saw my hartte was colde
to thinke at length who should wyn the filde
tyll blacke and white to redd dyd yelde
then hartely to god did I pray
that ever I saw that joyfull daye.

1572. T. Charnocke.

Leave out ye

At the lenght and last off all we shall finde the bodie blacke and darke in ye bottome off the glasse: wch is called a dead bodie, the shaddow off ye sun, darke cloudes, the tayle of the dragone wch off the dragone is now taken :

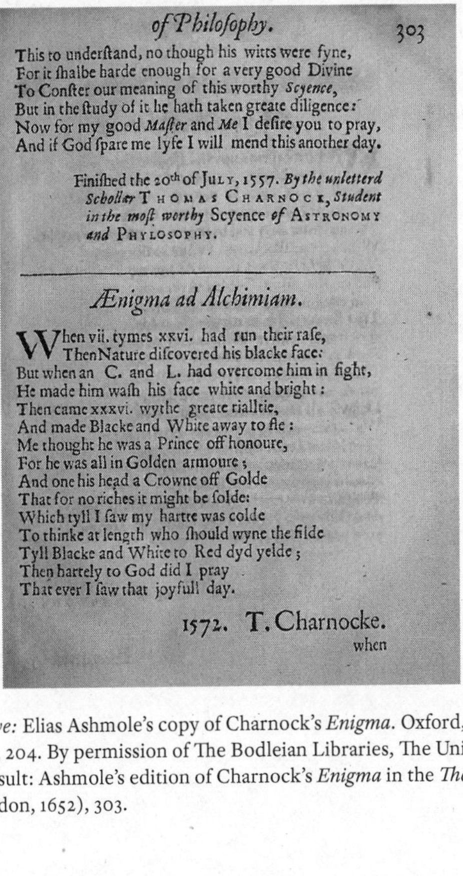
of Philosophy. 303

This to understand, no though his witts were fyne,
For it shalbe harde enough for a very good Divine
To Conster our meaning of this worthy *Scyence*,
But in the study of it he hath taken greate diligence:
Now for my good *Master* and *Me* I desire you to pray,
And if God spare me lyfe I will mend this another day.

Finished the 20th of July, 1557. *By the unletterd Schollar* Thomas Charnock, *Student in the most worthy* Scyence *of* Astronomy *and* Phylosophy.

Ænigma ad Alchimiam.

When vii. tymes xxvi. had run their rase,
Then Nature discovered his blacke face:
But when an C. and L. had overcome him in fight,
He made him wash his face white and bright:
Then came xxxvi. wythe greate rialltie,
And made Blacke and White away to flee:
Me thought he was a Prince off honoure,
For he was all in Golden armoure;
And one his head a Crowne off Golde
That for no riches it might be solde:
Which tyll I saw my hartte was colde
To thinke at length who should wyne the filde
Tyll Blacke and White to Red dyd yelde;
Then hartely to God did I pray
That ever I saw that joyfull day.

1572. T. Charnocke.

when

FIGURE 15. *Above:* Elias Ashmole's copy of Charnock's *Enigma*. Oxford, Bodleian Library, MS Ashmole 1441, 204. By permission of The Bodleian Libraries, The University of Oxford. *Below:* The end result: Ashmole's edition of Charnock's *Enigma* in the *Theatrum Chemicum Britannicum* (London, 1652), 303.

compilers who, like Mountfort, were practitioners as well as readers—
alchemists who read as a guide to both practice and their own history. Ashmole's diligence in collecting and copying the relics of this active readership
therefore preserved more than just a set of writings. His efforts salvaged precious evidence for the descent of books from owner to owner, and texts from
copyist to copyist: a genealogy of ownership in which children do not always
exactly resemble their parents. The result is a kind of generational "nesting"
effect, as the books of one authority passed from one set of hands to another,
the marks of each encounter inscribed upon their pages. Thus Giles Du Wes,
as we have seen, lovingly compiled and repaired medieval manuscripts in
the opening years of the sixteenth century, only for Robert Greene to copy
texts and acquire books from him, and Robert Freelove from Greene. Some
of their manuscripts came to Dee in turn, and many more of Dee's books
came to Ashmole, who also acquired, through different routes, original manuscripts in the hands of Du Wes and Greene, as well as copies of Freelove's
translations.[30] This was not merely a process of transmission; as books and
texts changed hands, their contents also changed through the philological
and practical interventions of successive readers.

This is most obvious in the case of a reader-practitioner like Mountfort,
who salvaged lost collections of monastic receipts while simultaneously con-
cealing (and, in time, obliterating) their original form behind a set of symbols, unlocked by a key. The results offer tantalizing references to some of
Ripley's contemporaries, among them "that famous Channon of Walton:
Rowland Greye" and "Sir Brian Goodricke Cannon of Glocester a man of
famous and singular memorie"—men with a contemporary reputation for
alchemical expertise, now all but forgotten.[31] The most extensive is a "trewe
coppy" of a parchment book, written in 1437 by Sir George Marrow, a monk
of Nostell Abbey in Yorkshire, "now most faithfully written ouer ageyne word
for worde, this 20th of marche Anno. 1596" (fig. 16).[32] But despite his claims
of accuracy, Mountfort's copy is far from word-for-word. In these recipes, he
has stripped the text down to its most practical content, replacing the names
of ingredients, instruments, and processes with symbols of his own devising.
The extent of his intervention appears most startlingly when we compare

30. See pp. 330, 332, 340 below.
31. Ashmole 1406, fols. 240r, 238v. Walton Priory was a Benedictine house, and so did not have
canons; more likely Mountfort intends Waltham Abbey, the largest Augustine house in England.
32. Ashmole 1423, fol. 1v; 71. Presumably he intends the Augustine priory of Nostell, Yorkshire.

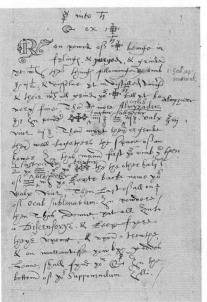

FIGURE 16. Thomas Mountfort's elaborate frontispiece to his copy of the *Book of George Marroe* [Marrow]. The names of several ingredients have been replaced by symbols, which Mountfort expanded using a key. Oxford, Bodleian Library, MS Ashmole 1423, title page and p. 1. By permission of The Bodleian Libraries, The University of Oxford.

Mountfort's version of Ripley's *Philorcium* to other extant copies.[33] Removing prefatory and discursive material, Mountfort skins Ripley's recipes to the bone, substituting his own symbols for the original Latin terms.

While the encoding of terms suggests that Mountfort placed a high value on their practical content, he may also have enjoyed the alchemical game of encipherment. Certainly he appreciated some of the more artful aspects of the *Bosome Book*. Copying out the distich "Maturus iacuit," with its play on Ripley's name, he added a punning verse of his own:

T: M: in the comend[ation] of G[eorge] R[ipley]
A noble George now thou hast
Greate secrets herein revealede
Which many frownes hath cost
By those that therein hath feylede.

33. Rawlinson B.306, fols. 66r–71v.

Rype mayst thou wel be termde
For ript vp hast thou the matter
Lay=men of thee may learne
The truth wherein others clatter.[34]

Mountfort's own copy has now disappeared along with the original *Bosome Book*: both his verse and his reading of Ripley live on in Ashmole's library of transcriptions.

Ashmole's antiquarian project was always a work in progress, endlessly extended as new material came into his hands. Five years after the *Theatrum* went to press, he encountered one of the impressive folio volumes compiled by Robert Greene, whom he had previously known only from Blomfild's slighting comment in the *Blossoms*. In his personal copy, Ashmole annotated his edition of Blomfild in light of the new information: "I haue seene a Manuscript in folio, being a Collecion of seuerall peeces of Chimicall Philosophy, all written with this Greenes hand, & his name subscribed almost at the end of euery Chapter." In 1657, the time of writing, this volume was "in the handes of deacon Goddard." Later, however, Ashmole updated the note with the information that the manuscript had since been "given to me by Colonell Sanchy"—presumably the Anabaptist Sir Jerome Sankey, a member of Parliament and former colonel in the parliamentary army.[35] Ashmole remained unaware of Greene's relationship with Du Wes, although at some point he acquired several of Du Wes's transcriptions of English alchemical poems; our only evidence that the librarian paid attention to vernacular as well as Latin sources.[36]

Ashmole's alchemical library expanded dramatically in 1669 with his purchase of books formerly owned by the clergyman and astrologer Richard Napier, alias Sandy (1559–1634), rector of Great Linford, from his greatnephew, Thomas Napier.[37] The value of the collection lay not just in Napier's own notes, purchases, and transcriptions, but in the large quantity of material he acquired from other reader-practitioners of the late sixteenth and early seventeenth centuries. These included books inherited from Napier's

---

34. Ashmole 1459, pt. 2, fol. 28v. On "Maturus iacuit," see p. 271, above.

35. Ashmole 972, fol. 320r. On Sankey, whose name was frequently spelled as "Sanchy" or "Zanchy," see A. J. Shirren, "'Colonel Zanchy' and Charles Fleetwood," *Notes and Queries* 168 (1953): 431–35.

36. Ashmole 1441, 89–95.

37. Jonathan Andrews, "Richard Napier (1559–1634)," *ODNB*.

own master, the astrologer-physician Simon Forman, and a number of manuscripts compiled by Thomas Robson, one of England's most avid collectors
of alchemica prior to Ashmole himself.

We know very little about Robson (alias Fletcher), one of the most distinctive and energetic alchemical scribes of early seventeenth-century
England.[38] He was clearly well-connected in alchemical circles, enjoying
access to a remarkable number and range of exemplary manuscripts in addition to Napier's, including books owned by such notable Elizabethan collectors and compilers as Forman, Christopher Taylour, and Robert Garland,
"practizioner in the arte spagericke."[39] From the libraries of his associates
Robson exported transcriptions in large quantities and often with considerable duplication, as in the case of two compendia with near-identical contents that suggest that he was preparing volumes either as gifts or for sale—
perhaps for the unknown "friend" whom he addresses at several points. As
an intelligencer of alchemical texts, Robson evidently strove to obtain copies
of desirable works as they became available. For instance, on 20 September 1606 he secured a new translation of that famous medieval testament of
Morienus's teachings to Prince Khālid, *De compositione alchemiae*, "beinge
truly written from the copye therof lattly beinge translated out of latynn into
englishe in londone, and nowe agayne written by me for my very good and
ashured freende."[40] He copied Howes's translation of the *Bosome Book* in the
same year, and Norton's *Key* in 1613.

Through Ashmole's acquisition of Napier's books, we can also trace some
of Robson's exemplary manuscripts. These include one that Robson

38. Robson reveals his alias in Ashmole 1424, fol. 50r, and Ashmole 1406, pt. 3, fol. 214r, where
he has added a note to an unknown recipient (unfortunately trimmed by the binder): "Your pore
servant to t[. . .] power Thomas fle[tcher]." Manuscripts written in Robson's hand include Sloane
1744 (which includes items dated 1602, 1604, and 1606); Ashmole 1394 (pt. 5), 1407, 1408 (pts.
3–4), 1418 (pt. 1 comp. November 1605–6; pt. 2 comp. 20 September 1606; pt. 3 comp. after October 1604), 1421 (includes items dated 1614 and 1615), 1424 (comp. 23 April 1614–April 1623); MS
Ferguson 133 (comp. 23 June 1606). Shorter pieces in Robson's hand are found in Ashmole 1441
(pt. 2). Robson's manuscripts and annotations in the Ashmole collection are detailed by Black,
Descriptive, Analytical, and Critical Catalogue; see index, 138.

39. Besides annotating Napier's manuscripts, Robson frequently credits his friend in his
copies—for instance, in Ashmole 1418, pt. 1, fol. 24r. He made several transcriptions from Ashmole
1486, formerly owned by Robert Garland, who added his name and monogram on 20 November
1596 (fol. 27r). On Robson's copying of Forman's manuscripts, mediated by Richard Napier, see
Lauren Kassell, *Medicine and Magic in Elizabethan London: Simon Forman: Astrologer, Alchemist,
and Physician* (Oxford: Clarendon Press, 2005), 229. Kassell has also noted the likely connection
between Robson and Taylour (personal communication).

40. Ashmole 1418, pt. 2, 110. The text is discussed in chap. 1, above.

described as "the old booke writt in the old hand before printe": a book that we can now identify as Ashmole 1486 (pt. 5), the Opinator's gift to Thomas Ellys, prior of Leighs.[41] Robson added his own annotations to this fifteenth-century source, besides marking up several items for transcription—copies of which resurface in his own compendium, now Ashmole 1421.[42] He also copied extensively from manuscripts belonging to Christopher Taylour, another enthusiastic compiler who may have prepared an alchemical collection for a female patron, Margaret Clifford, Countess of Cumberland.[43] Taylour's books supplied Robson with a plethora of useful English texts, including copies of the *Work of Sir Robert Greene*; Blomfild's *Practicke* (although neither Robson nor Taylour attributed the work to that irascible preacher); and a compilation of pseudo-Lullian quotations titled "Of the Menstrewes," possibly translated by Robert Freelove.[44] Robson also garnered translations of important works by foreign adepts, including Alexander von Suchten and Cornelius de Lannoy.[45]

Robson's compendia of English alchemists, like the rest of Napier's manuscripts, did not fall into Ashmole's hands on time to influence the contents of the *Theatrum*. Once secured, however, these storehouses of bio-bibliographical bounty prompted Ashmole to update his own book by hand. In Robson's copy of Norton's *Key*, for instance, Ashmole learned of Samuel's putative descent from Thomas Norton, as well as his assertion that Thomas had learned his alchemy not from Ripley, as was commonly believed, but

---

41. Ashmole 1421, fol. 44r.

42. Ibid., fols. 44r-62v, including diagrams of vessels and furnaces copied from Ashmole 1486, pt. 5. Robson's annotations appear in Ashmole 1486, pt. 5, on fols. 18r, 19v, 22r, 23r, and elsewhere.

43. The link is suggested by Penny Bayer, who notes that Lady Cumberland owned a manuscript (Kendal Archive Centre, MS Hothman 5, the so-called Margaret Manuscript) largely written in Taylour's hand; Penny Bayer, "Lady Margaret Clifford's Alchemical Receipt Book and the John Dee Circle," *Ambix* 52 (2006): 71–84.

44. "The Admonition of Sir Robert Greene," Sloane 1744, fols. 22v-29, copied from Ashmole 1492, pt. 9, 197–205; "Raymundus to kinge Edward off Woodstocke" (Blomfild's *Practicke*, from which Robson has removed the introductory petition and biographical content), Ashmole 1418, pt. 3, fols. 78v-81r, copied from Ashmole 1492, pt. 9, 151–54; "Of the Menstrewes," Ashmole 1418, pt. 3, fols. 68v-70v, copied from Ashmole 1478, pt. 2, fols. 97r-98v, a version that includes the colophon "Anno 1548 25° Augusti." The date would be appropriate for Freelove, who translated one of the preceding items in the collection: the pseudo-Baconian *The Privitie or Secrete of Avicen*, "translayted oute of latine into ynglysshe By Robert freelove the .15. Day of December Anno 1550" (fol. 96r).

45. Ashmole 1418, pt. 3, fols. 17r-30r ("Verbis honoris primyssis, sic incipit tractatum suum de antimonio, ad Dominum Johannem Babtistam de Sepache, Alexander a Suchsten"); 43v-47v ("Cornelius Alvetanus Arus Rodius of makeinge the divine elixar").

from another, unnamed master.[46] Napier's own diary offered an account of a projection apparently conducted by Edward Kelley himself, although Ashmole was disappointed to find the entry undated—"Had the yeare been also set downe, it might, perhaps, haue given better satisfac*cion*."[47]

The books in Ashmole's collection are intricately interconnected through a web of references that Ashmole himself, slowed down by law suits and lacking valuable contextual information (such as Du Wes's relationship with Greene), did not himself always discern. But these books do more than supply connective tissue between compilers of texts; they also show how readers and copyists translated their contents into practice.

As with Mountfort, Robson's labors as a compiler were not divorced from practical concerns. He was preoccupied with lead-based processes, supplementing the well-known tracts of Ripley and Raymond with lesser-known works, such as the anonymous *Tractatus brevis* and a brief tract on Adrop, both copied on 23 June 1606 from "the booke that Came from Mr Sandeys [i.e., Napier] of linforde."[48] Ashmole 1421, his most conspicuously sericonian collection, includes an entire recipe collection devoted to works on lead, "taken forth of the litle blacke boke."[49] To these he adds a process drawn from his own experience: "To drawe the salt of .[Saturn]. as my selfe did in the yeare 1607."[50]

Yet reading with a view to practice was a risky business, as Robson well knew. Elsewhere he observes that readers can be easily deceived by the multiplicity of possible processes, which cater to a diversity of stones rather than one thing alone. "Many men in thes days," he notes, "are given to Reed of this stone and to worke therin," not recognizing that their source texts are the product of different practitioners working at different times. Thus, although "very ancient bookes of Record" offer many ways for making the mineral stone, it does not follow that these can be interpreted the same way, or conflated into a unified process:

In the Reeding of thes bookes you haue many disscorses of the stone which to mens thinking at the first sight is very playne for [th]e proceeding therof,

46. Ashmole 972, fol. 286r: "Samuell Norton maketh Tho[mas] Norton's Master & G[eorge] Ripley two men, See his Praeface to his Clavis Alchimiae."

47. Ibid., fol. 318v.

48. Ferguson 133, fols. 1r-6v. On the *Tractatus brevis*, see pp. 122–24, above.

49. Ashmole 1421. The collection, marked by the symbol for Saturn (lead) at the top of each page, begins with "To make Red lead after Paracelsus" on fol. 2v and ends with "The worke of Sericon" on fol. 9v.

50. Ibid., fol. 5v.

but yf you shall take the Ruells of one stone and proceed to the working of
another, you shall never attayne to any liklyhod of the worke you so goe
aboute.[51]

Robson perceptively acknowledges that although the aim of alchemical
reading is to understand one text in light of another, such conflation may
produce results in practice that were unintended by either of the original
authorities. Yet his concern relates primarily to the mineral stone, for which
the largest number of processes exist. In the case of the vegetable stone, he
claims there are three authors who can be safely read in concert, because
they all describe the same process from different angles—thus "the one may
be a helpe to the other."

Robson's confidence reposes in a set of texts that should be very familiar
to us by now: the *Work of Dunstan*, Norton's *Key of Alchemy*, and the *Whole
Work* of George Ripley. Robson outlines his reasons for viewing these three
as compatible. Two of his authorities seem to have independently hit on the
same procedure: thus "Mr. dunstone and George Riply speake of none other
but of the vigitable." Ripley is also the ultimate authority for the *Key*, which
Norton himself "doth father vppon George Riply," since its contents are
"taken forth of his bosome booke." And the *Whole Work* is actually extracted
from the *Bosome Book*, as Robson delightedly records: "as I had it by great
fortune, the vigitable worke after Riplyes owne hand writing in his bossome
booke as the authore hath sett it downe."[52]

While Robson is correct in spotting similarities between these three monuments to sericonian alchemy, there is still plenty of irony in his selection.
As we have seen, the *Work of Dunstan* is a deliberate forgery based on the
Accurtations of Raymond, while the *Whole Work* is a synthetic text—in fact,
an example of exactly the kind of creative alchemical splicing that Robson
had earlier warned his readers against. From a modern perspective, all three
texts are essentially products of the sixteenth century, albeit derived from a
shared body of late medieval sources attributed to Raymond Lull, Guido de
Montanor, and George Ripley.

For early modern readers like Robson, however, pseudepigraphy mud-
died the waters; as a result, the clear similarities between the texts did not
necessarily point to a shared context of production. Rather, they implied
that the transfer of alchemical knowledge was a long-term process, in which

51. Ashmole 1424, pt. 2, 9.
52. Ibid., 10.

an ancient treatise of Saint Dunstan anticipated (or perhaps informed) the fifteenth-century work of Ripley, and subsequently of Ripley's Elizabethan disciple Samuel Norton. The wisdom of philosophers is transmitted down through the centuries, while maintaining its unified character—a character that can of course be tested in practice, through the physical process of making the stone.

This narrative is the very stuff of English alchemical histories. In reality, however, the *Secretum secretorum*'s celebration of one, transhistorical stone, found in all times and all places, is contradicted by the sheer multiplicity of methods recorded in treatises and recipe collections. Robson already acknowledged this in the case of the mineral stone, while holding out hope that the vegetable path, at least, might remain constant over time. Perhaps this optimism merely reflects his own view that he had identified the correct ingredient. In a marginal note to his own copy of the *Whole Work*, he offers a familiar solution: "You*r* Sericon is the matter drawne fro*m* Red lead with [vinegar] and then vapored."[53] By the first decade of the seventeenth century, however, this reading was already just one of many.

SUBSTITUTING SERICON

Alchemical practices could hardly remain stable when both their textual repositories and the skills required to read them were subject to such constant flux. Fortunately reader-practitioners had another source of information to turn to: experiment. The difficulty of construing textual sources could be ameliorated by practical experience working with chemicals, which offered readers an external means of evaluating written works, or at least of making educated guesses about their meaning.

This task was undoubtedly aided by the fact that the range of possible chemical operations available to the early modern practitioner was finite. The capacities of various metals and their products to dissolve in mineral acids, change color, or flee the fire were widely known. The tendency of certain metallic compounds to dissolve in distilled vinegar, known since ancient times, therefore provided one obvious reading for Ripley's sericonian receipts. Yet lead was only one among a range of substances that had this property, leaving open the door for inventive practitioners to explore alternative readings.

Accordingly, those seeking to replicate Ripley's processes looked else-

53. Ashmole 1418, pt. 2, fol. 26r.

where in their attempts to identify the precise nature of sericon. In a compilation of treatises and recipes related to the vegetable stone, Clement Draper added a marginal gloss to his own transcription of the *Medulla*, offering alternative readings of sericon: either ceruse of lead, or a "vittrioll made of most sharpe moysture of grapes to be verdigreace for yt is made of vineger and tarter."[54] While Draper's first suggestion takes account of sericon's long-standing association with lead (in this case, ceruse), his second proposes verdigris, an acetate of copper that—like minium—was in widespread use as a pigment. The alternative readings may here offer us a glimpse of Draper's own exchanges with fellow readers and practitioners both within and beyond London's King's Bench prison, where he compiled extensive alchemical notebooks and sometimes tested their contents over the course of his long confinement.[55]

Another Elizabethan alchemist, Thomas Potter, recorded treatises and recipes in an alchemical compendium compiled around 1579–80, now MSS Sloane 3580A and B. Like Draper, Potter seems to have been in contact with other alchemical enthusiasts, to judge from his success in accessing and collating copies of alchemical treatises, including Ripley's famous *Compound*.[56] In his own annotations, he identifies Ripley's Green Lion with a product drawn out of copper: "The lyon greene, is mercury of venus, which muste be calcyned with [gold]. & [silver]."[57] This reading may have influenced his speculations on the nature of sericon in other Ripleian works. For instance, he added a marginal note to the opening line of the *Whole Work* ("First take 30 pound weight of sericon"), glossing sericon as "a minium powder of metal," while leaving the choice of metal open: "copper. &c."[58]

Such metalline readings predominate in Elizabethan commentaries on Ripley's writings, although they did not go unchallenged. Ripley's reference to the "most sharp humidity of grapes" led other readers to speculate that his sericon denoted a vegetable ingredient rather than a metallic body. For instance, another popular Ripleian *practica*, "The Vegetable of George Ripley to the Bishop of York," probably started life as an attempt by a later reader to

---

54. Sloane 1423, fol. 29r.

55. Harkness, *Jewel House*, chap. 5.

56. On Potter's editing strategies, see Keiser, "Preserving the Heritage," 189–214; Rampling, "Depicting the Medieval Alchemical Cosmos," 75–76.

57. Sloane 3580A, fol. 142r; see also a similar note on fol. 144v. On Ripley's use of copper in the *Compound*, see Rampling, "Establishing the Canon," 205–6; chap. 3, above.

58. Sloane 3580A, fol. 214v. Potter also noted Ripley's use of sericon in the *Medulla*, although without recording his own interpretation of the term, on fol. 143v.

decode the *Medulla*'s process for the vegetable stone.[59] This reading starts with the manufacture of tartar from argol (the lees of wine), which is repeatedly distilled in *aqua vitae* to make a fiery corrosive capable of dissolving gold—a secret that is the "key of all the Science," for it unlocks the door to "the chamber of Dame Philosophie."[60]

Ripley did indeed use tartar in the *Medulla*, as a component of the wine-based solvent in which mineral sericon, "the body well calcined into red," is dissolved. However, the pseudo-Ripleian *Vegetable* omits sericon entirely from its discussion of the medicinal aspects of the vegetable stone, which requires only gold and silver (although an imperfect body, copper, is employed for the "lesser" work of transmutation). Ingeniously, the commentator suggests that the resulting vegetable stone is still "mineral" in nature because vines grow from the earth, and tartar is accordingly "mineral in the place of his generation."[61] The commentary becomes a fully fledged example of alchemical pseudepigraphy in several copies, where it is reframed as a letter from Ripley to George Neville, or, as one scribe suggests, to Edward IV.[62]

The tartaric reading shaped the interpretation of at least one Elizabethan reader—Humfrey Lock. Lock's own reading of the vegetable stone follows the tartaric line of the pseudo-Lullian *Epistola accurtationis* and pseudo-Ripleian *Vegetable* rather than the *Medulla*'s sericon. In a passage excerpted from the *Medulla*, he therefore replaces Ripley's sericonian gum, which is simultaneously vegetable *and* mineral, with a wholly vegetable product, black tartar:

> In the name of God, take the sharpeste humidity of [th]e grape, which is not the grape of the vine, but the black grape of Kotolory [i.e., Catalonia], which in this worke of the vegitall stone must be made into the manner of a gum.[63]

To reinforce this reading, Lock adds a passage on the making of the tartar-sharpened vinegar that seems to have been adapted from the opening of

---

59. The *Vegetable Work* is CRC 30. The treatise actually concludes by quoting from the *Medulla*, in a passage that describes "things which in this arte are named colorabely to deceave fooles"; Ashmole 1407, pt. 4, 27 (hand of Thomas Robson).

60. Sloane 288, fol. 95r.

61. Ashmole 1407, pt. 4, 25.

62. Sloane 3645, fol. 40r: "Scriptus ut opinor Regem Edvardum 4t: Angl."

63. Lock, *Treatise*, ed. Grund, in *Misticall Wordes and Names Infinite*, 224 (ll. 25–28). On Lock, see chap. 7, above.

the *Vegetable*.[64] While the results may not correspond to Ripley's intended practice, they do suggest that Lock was following the advice of the philosophers on how to read alchemically: in this case, by reading diverse texts against one another in order to extract the "true" method for the vegetable stone.

However, by far the most frequent interpretation of the alchemists' red lead was "antimony"—a term invariably used to denote ores such as stibnite (natural sulphide of antimony) rather than metallic antimony itself. The success of this reading reflects the fact that sixteenth-century practitioners were generally more interested in antimony than their medieval predecessors, and consequently more alert to possible references in their source texts.[65] The recommendation of antimonial compounds as medicinal purgatives in Paracelsian treatises, and their increasing use by medical practitioners throughout the sixteenth century, culminating in the notorious "antimony wars," provide one likely context for the increasing substitution of antimony in sericonian recipes.[66] Another is the value of antimony in metallurgy, as a means of purging gold of impurities. In the late sixteenth and early seventeenth centuries, the influential writings of Alexander von Suchten and the fictitious Basil Valentine also helped to secure antimony's place as a key ingredient in alchemical practice.[67] For instance, Simon Forman silently

64. Ibid., 225 (ll. 37–42); see also the commentary on these lines on 301. Grund discusses the *Vegetable Work* under the title *Thesaurus pauperum* and not as as Ripleian text; Grund, *Misticall Wordes and Names Infinite*, 52–54.

65. The major exception is *De consideratione quintae essentiae*, in which John of Rupescissa describes a medicinal remedy made by distilling antimony with the quintessence of wine; *De consideratione*, 88: "Scientia in extrahendo quintam Essentiam ab antimonio, que appellatur Marchasita plumbea." When antimony is mentioned in fifteenth-century English texts, including the *Compound*'s "Admonition" (where Ripley rejects it as an ingredient), it is often in relation to this passage from Rupescissa. For example, Sloane 3747, fol. 94r, includes an extract from *De consideratione* on the "quintessence of antemony."

66. The "antimony wars" (1566–1666) began with an attempt by the Paris medical faculty to ban the use of antimonial compounds in medicine; see Kahn, *Alchimie et Paracelsisme*.

67. On Suchten's use of antimony, see Alexander von Suchten, *Liber unus de Secretis Antimonii, das ist von der grossen Heimligkeit des Antimonii* (Strassburg, 1570); *Antimonii Mysteria Gemina* . . . (Leipzig, 1604); also discussed in Newman and Principe, *Alchemy Tried in the Fire*, 50–56. Basil Valentine was a fictive alchemical persona, probably created by "his" first editor, Johann Thölde; see particularly Basil Valentine, *Triumph-Wagen Antimonii . . . An Tag geben, durch Johann Thölden* . . . (Leipzig, 1604); discussed in Claus Priesner, "Johann Thoelde und die Schriften des Basilius Valentinus," in *Die Alchemie in der europäischen Kultur- und Wissenschaftgeschichte*, ed. Christoph Meinel (Wiesbaden: Harrassowitz, 1986), 107–18.

adapted one of von Suchten's antimonial recipes as early as November 1598, alloying antimony (which he called "Cako") with gold and silver.[68]

The reinterpretation of red lead as antimony is assisted by similarities between the two sets of procedures. Medicinal preparations of antimony often employed the vegetable solvents of wine, vinegar, and tartar, creating a clear, practical bridge between antimonial and lead-based practice.[69] In *De consideratione*, John of Rupescissa had described the preparation of a "quintessence of antimony" using spirit of wine, which he explicitly distinguished from the manufacture of sugar of lead, maintaining that the antimonial product was sweeter and better than ceruse dissolved in vinegar.[70] This procedural correspondence may have been deepened further by the perceived relationship between various "leaden" metals that were not always sharply distinguished in early chemistry, including antimony, marcasite, and lead—a correspondence noted by both Rupescissa and his later reader, Paracelsus.[71] A "red antimony" also existed in the form of kermesite, making "red lead" a plausible cover name for the brownish-red antimonial ore.[72] Whatever the deciding factor, the substitution of antimony for red lead helped shape the reception of Norton's *Whole Work*.

As we have seen, the prescription of "30 pounds of sericon" in the *Whole Work* originally signified minium. This reading was indeed accepted by early readers, including Potter, mentioned above, and the compiler of the Elizabethan manuscript Sloane 1095, who gives the term a recognizably Ripleian gloss: "Lead once it is burned is made of a red color, which by masters is

---

68. "Of Cako," Ashmole 208, fols. 78–93v; Kassell, *Medicine and Magic*, 176–86.

69. The archetypical antimonial product, emetic tartar (antimony potassium tartrate), is made using tartaric acid. For a table of antimonial preparations, see R. Ian McCallum, *Antimony in Medical History: An Account of the Medical Uses of Antimony and Its Compounds since Early Times to the Present* (Edinburgh: Pentland Press, 1999), 99–102.

70. John of Rupescissa, *De consideratione*, 90.

71. For instance, *Paracelsus, his Archidoxis comprised in ten books : disclosing the genuine way of making quintessences, arcanums, magisteries, elixirs, &c. . . .* , trans. J. H. (London, 1660), bk. 6, 82–83: "Gold and the Marcasite, Antimony and Lead, the which in their framing and Constelation, may be compared to each other mutually, but are neverthelesse Separated in Virtue." I here use the English translation of Paracelsus's *Archidoxa . . . Zehen Bücher* (Basel, 1570). It is unclear exactly what substance John of Rupescissa intended by antimony, which he describes only as "a leaden marchasite." However, he clearly differentiates it from lead compounds; *De consideratione*, 90. On the influence of Rupescissa's alchemy on the *Archidoxis*, see Dane Thor Daniel, "Invisible Wombs: Rethinking Paracelsus's Concept of Body and Matter," *Ambix* 53 (2006): 129–42.

72. Antimony sulphide is also easily turned from black to red, either by dry sublimation or by dissolving it in alkaline lixivia and precipitating with any acid; I am grateful to Lawrence Principe for pointing out this significant color change.

called sericon."[73] However, others clearly viewed it as a *Deckname* for antimony. John Dee, for instance, glossed sericon as "red leade," followed by the symbol for antimony, in the margin of his own early seventeenth-century transcription of the *Work*.[74] By 1683, the antimonial reading was sufficiently well entrenched to be silently incorporated into the version published by the London bookseller William Cooper: "30 pound weight of Sericon or Antimony."[75]

The difficulty of reading alchemical texts means that such substantive changes in interpretation and practice have to be inferred from relatively minor alterations to manuscript copies. Yet even small changes to a recipe could have serious practical implications for the outcome of a chemical operation.[76] Over a period of several centuries, textual accounts became uncoupled from their original traditions and were paired with new techniques— transformations unwittingly mediated by practitioners whose intention was simply to clarify or supplement their authority. Thus, although the sericonian paradigm of menstrua and vegetable stones continued to circulate during the seventeenth century, it faced increasing competition from other theoretical and practical models, as sericonian works were adopted and adapted by proponents of different methods.

TWO VISIONS OF ALCHEMY

We have already encountered Ripley's *Vision* as a popular component of the *Bosome Book*, translated by Samuel Norton in 1573/4 and subsequently published by Rabbards and Ashmole. This short poem describes the poet's vision of a red toad that expires after consuming the "juice of grapes":

> When busie at my booke I was upon a certeine night,
> This Vision here exprest appear'd unto my dimmed sight,
> A *Toade* full rudde I saw did drinke the juce of grapes so fast,
> Till over charged with the broth, his bowells all to brast.[77]

---

73. Sloane 1095 (1550–1600), fol. 75r: "Plumbum cum comburitur coloris rubei efficitur quod a magistris Sericon appellatur."

74. Ashmole 1486, pt. 5, 1. On Dee and sericon, see Rampling, "John Dee and the Alchemists"; and chap. 8, above.

75. *Bosome-Book*, 101.

76. For some examples, see Principe, "'Chemical Translation' and the Role of Impurities in Alchemy."

77. *TCB*, 374.

The remainder of the poem describes the death agonies and putrefaction of the toad, which first leaks "poysoned sweate," then a white vapor, and a golden humor. The rotting corpse passes through the various color changes of the work, from "colours rare" to white and finally red.

In the first chapter of his *Key of Alchemy*, Norton interprets this poem in light of his extensive knowledge of Ripleian alchemy, gleaned from the *Medulla, Concordantia,* and the amassed processes of the *Bosome Book*. The result is a sound sericonian reading of the toad and its solvent: "Vinneger commeth of the vine, & hath vertue ingressive . . . By this toad hee meaneth red Ledd that is Adrop or Minium or Saturne, or Capricorne or Rupescissus Antimonie."[78]

Although Norton here lists alternative terms for the red lead, we have already seen that his shopping list for Elizabeth I supports the traditional reading of minium and distilled vinegar.[79] For Norton, the death of the toad therefore alludes to the reduction of lead to its prime matter, by means of the vinegar's ingressive virtue:

> By which meanes, the bodie is now become no bodie, but brought, or reduced into the first matter, into a viscous matter, whereof it was in the bowells of the Earth ingendred. . . . Heere Riplies toad drinks so fast, that his Bowells be all burst, heere have wee made *spissum Liquidum*.[80]

Norton reduces the allegorical poem to a straightforward, practical procedure of the type already noted in the *Whole Work* and *Practise by Experience*. Even familiar tropes, such as the "faint water," are meticulously explained:

> Because in this solution wee have a great deale too much vinegere, which wee seeke not but rather vse as a meane to draw our gummie water, from the Lead; wee therefore place this water over a slow fire on a trevet, that the superfluous watrishnes of the vinegere, may be so evapored away that wee may find the extracted matter of lead drawne out by the vertue of vinegere.[81]

---

78. Getty 18/10, 31–32. Norton here interprets John of Rupescissa's "antimony" as red lead. Norton equates red lead with "capricorn" and minium elsewhere: "Lead also is by Rodagirius named Capricornus, & being burnt or Calcined they Call that Minium" (ibid., 23).

79. See pp. 278–79, above.

80. Getty 18/10, 32. This passage is omitted from Robson's transcription in Ashmole 1421.

81. Getty 18/10, 32.

Throughout the process, Norton emphasizes that the vinegar serves as a means of preparing the lead, rather than as an ingredient of the stone itself.[82] In this way, he avoids conflict with those philosophers who insist upon the stone's metallic nature. Perhaps we may infer from these caveats an increasing skepticism toward the use of vinegar in the later sixteenth century. As we have seen, within a few decades of Norton's exegesis, alchemical practitioners had access to a wide range of alternative methods, particularly within the area of transmutational alchemy, and accordingly approached Ripley from different directions.

Some seventy-five years after Norton's *Key*, another "Exposition" on the *Vision* was written by George Starkey. Since the source texts for all of his "Ripley" commentaries were available in Rabbards's edition, Starkey may not have realized that the *Vision* was based on Norton's translation of a Latin original. The circumstances of the *Bosome Book*'s Elizabethan resurrection, and the peculiar transmutation of much of its content, were perhaps similarly unknown to him. At all events, reconstructing Ripley's alchemy in its fifteenth-century context was not Starkey's goal. His exegesis marks a strikingly different approach from that of both Norton and Ripley himself.

Starkey begins his commentary in conventional style, by warning readers that Ripley's poem is a parable or enigma, "which the Ancient Wise Philosophers have been wont to use often in setting out their secrets."[83] That he differs from those earlier writers is clear from the outset, however, since he identifies the ruddy toad not as lead, but as gold. "To this," he pronounces, "Authors assent with one accord"—although some philosophers have deliberately denied the truth of this, "on purpose to deceive the unwary."[84]

Starkey is, in fact, employing the familiar technique of appropriating past authority in support of his own, signature practice: one based not on lead and vinegar, but on metallic bodies. In a passage stuffed with elaborate *Decknamen*, he expounds his meaning:

> The Juice of Grapes then, which is our Mercury, drawn from the Chameleon or Air of our Physical Magnesia, and Chalybs Magical, being circulated upon our true Terra Lemnia; after it is grossly mixed with it by Incorpora-

---

82. For instance, Getty 18/10, 31: "Vegetables are not vsed in the stone to give anie metallike vertue, but onlie to serve for preparation of metalls, That thereby the vertues may bee the better extracted; & yet vsing the selfe same reason, I would prove that some vegetable giveth ingression to metalls."

83. "Vision . . . Unfolded," 1.

84. Ibid., 2.

tion, and set to our fire to digest, doth still enter in and upon our Body, and searcheth the profoundity of it; and makes the occult to become manifest by continual ascension and descension: till all together become a Broth.[85]

Starkey here reads Ripley's "juice of grapes" as a cover name for mercury—not common quicksilver, of course, but the philosophical (or "sophic") mercury already described in his earlier work, the *Introitus apertus ad occlusum Regis palatium* (Open Entrance into the Closed Palace of the King).[86] Starkey is, in fact, reiterating a process that he presents throughout his writings, particularly in the Ripley commentaries. As William Newman has shown, these repeatedly describe the reduction of antimony ore with iron to produce the star regulus of antimony.[87] The regulus is combined with conventional quicksilver and distilled to yield the sophic mercury. This process also requires silver, although Starkey obscures his process by omitting all references to this fourth ingredient, both in the *Vision* commentary and elsewhere.

We can still identify the other three ingredients in his cryptic account. The sulphureous spirit of iron is disguised as "Chalybs," the same term he uses in the *Introitus*, while antimony, "our Physical Magnesia," elsewhere appears as "Magnes," or "our Magnet," since it can draw the spirit from the iron.[88] Together the two make regulus of antimony, the "chameleon" that should be circulated with gold (*Terra Lemnia*) until it dissolves and putrefies.[89] This process, Starkey tells us, reduces the gold "into a Powder, like to the Atoms of the Sun, black of the blackest and of a viscous matter."[90] Such language stirs recollections of the pseudo-Lullian "black blacker than black," as well as the black "unctuous humidity" that Ripley describes in the *Medulla*—another example of the repurposing of traditional cover names.[91]

Often throughout the commentaries Starkey constructs his own riddles, disguising his process with obscure language and imagery that may be

85. Ibid., 7.

86. Eirenaeus Philalethes, *Introitus apertus ad occlusum Regis palatium* (Amsterdam: Johannes Janssonius van Waesbergen, 1667).

87. William Newman deciphers in detail one such passage from Starkey's account of Ripley's first gate, "Calcination," in *Gehennical Fire*, 115–69, particularly on 125–33.

88. "De Chalybe Sophorum" and "De Magnete Sophorum" are described in chaps. 3 and 4, respectively, of the *Introitus apertus* (on 6–7); while the "Aër Sophorum" drawn from star regulus ("Chaos") appears in chap. 6 (on 9).

89. *Terra Lemnia*, or *terra sigillata*, a reddish earth gathered annually on the island of Lemnos, was a well-known remedy for snakebite.

90. "Vision . . . Unfolded," 8.

91. Discussed on p. 95, above.

decoded by those well versed in alchemical literature and practice. In the case of the *Vision*, the riddle comes ready-made. Since both the sericonian and the antimonial approaches require the dissolution of a metallic body in a solvent, the fatal thirst of the *Vision*'s toad serves the purposes of either. While Norton's interpretation is probably the more faithful reading, his motives in supplying it cannot be convincingly distinguished from Starkey's. In providing a commentary on the famous poem, each practitioner demonstrates his ability to decipher perplexing alchemical authorities, and consequently his own fitness to propagate the art.

THE MINERAL STONE REVISED

In Starkey's antimonial chemistry we encounter a radical departure from the sericonian method beloved of fifteenth- and sixteenth-century texts—and with it, a rejection of the ancient tradition of animal, vegetable, and mineral stones. This shift should not surprise us. Early modern thinking about matter developed in multiple directions throughout the later sixteenth and seventeenth centuries, driven by a move toward particulate theories of matter.[92] Corpuscular theories minimized concerns about the relationship between mineral and vegetable kinds: the very distinction on which the traditional division of *materia medica* (and the basis for pseudo-Lullian alchemy) depended. It is telling that the writings of Starkey, who essentially stripped Ripley's alchemy of its medicinal component by ignoring the vegetable stone, proved to be important influences on Robert Boyle, one of England's most active proponents of a corpuscular matter theory.[93]

What is more remarkable, perhaps, is that this shift occurs without any diminution in Ripley's authority as either philosopher or practitioner. In the preface to *Ripley Reviv'd*, Eirenaeus Philalethes praises the canon above all other authorities: "*Ripley* to me seems to carry the Garland." Later, he compares his single, mineral process to those "pitiful Sophisters" who "dote on many Stones, Vegetable, Animal, and Mineral."[94] Yet the approach rejected by Starkey is integral to Ripley's *Medulla*, alluded to in the *Compound*, and

---

92. On these developments and their relationship to alchemical theorizing and experiment, see particularly Christoph Lüthy, John E. Murdoch, and William R. Newman, eds., *Late Medieval and Early Modern Corpuscular Matter Theories* (Leiden: Brill, 2001); Antonio Clericuzio, *Elements, Principles, and Corpuscles: A Study of Atomism and Chemistry in the Seventeenth Century* (Dordrecht: Kluwer, 2000); Newman, *Atoms and Alchemy*.

93. Newman and Principe, *Alchemy Tried in the Fire*.

94. *Ripley Reviv'd*, 23.

a staple of many sixteenth-century commentaries, including Norton's *Key*. Ripley's authority remains, but the alchemy is no longer his, as the structures and metaphors of his familiar works are appropriated to serve new agendas. The practical impetus of the work stems not from the pseudo-Lullian tradition, but from Starkey's later continental authorities, particularly von Suchten.[95]

The influence of such antimonial readings shaped even the reception of the *Bosome Book*. As we have seen, sericon was glossed as antimony in later copies of the *Whole Work* (including Cooper's edition)—but one seventeenth-century reader went farther still, by reframing the entire text as an "antimonialist" tract. In Sloane 689, Norton's already-composite text undergoes a further cycle of digestion, now transmuted into the pseudepigraphic *Liber Secretissimus Georgii Riplei* (Most Secret Book of George Ripley).

Where the *Whole Work* begins with the dissolution of thirty pounds of sericon, the *Liber* introduces a new starting matter:

> Take our artificiall Antimony, but not of the Naturall Antimony as it comes out of the Earth, for that is too dry for our worke, & hath little or no humiditie, or fatnes in it, but take I say, our artificiall Antimoniall compound, which is abundantly replenished with the dewe of heauen & the fatnesse & vnctuosity of the earth.[96]

The commentator explains that the artificial antimonial compound is made from a "Mineral Trinity," referred to throughout as the "three noble kinsmen."[97] The original *Whole Work* provides a loose framework for this imported material, but relics of medieval alchemy nonetheless survive. During the distillation phase an "insipid & fainte water" is first drawn off, followed by "the white fume, which is called our aire."[98] These staple tokens of the sericonian approach are here turned to new ends, lending a glimmer of Ripleian authenticity to processes that otherwise share nothing in common with the canon's own writings, or even Norton's reworking. This ingenious appropriation is all the more striking given that the *Whole Work* is one of the most practical texts in the Ripley Corpus, lacking the kind of figurative

---

95. See Newman, *Gehennical Fire*, 135–41.
96. Sloane 689, fol. 20r.
97. Ibid., fol. 21r. Although the additions to the text are obscure, they may refer to an alloy of antimony, silver, and quicksilver, in which the latter serves to dissolve the regulus.
98. Ibid., fols. 22v-23r.

language encountered in the *Vision*, or even the *Compound*, that lends itself more readily to diverse interpretations.

Also evident in these late examples of Ripleian exegesis is a shift toward a more overtly chrysopoetic alchemy. The multipurpose vegetable stone, effective as both medicine and agent of transmutation, plays no part in these accounts, and is sometimes explicitly excluded, as in *The Golden Age: or, the Reign of Saturn Reviewed*, authored by the pseudonymous "Hortolanus Junior."[99] This book is essentially a florilegium of statements relating to the antimonial alchemy of Eirenaeus Philalethes, the work's principal authority. In the opening pages, "Hortolanus" lists erroneous approaches:

> Diana has superfluity of *Menstruums*, she hath Simple, Vegetable *Menstruums*, made of *Philosophical* Wine only, others of the Spirit of *Philosophical* Wine, and the hottest Vegetables, Herbs, Flowers, Roots, &c. being Oyly. Also, Simply Mineral *Menstruums* made of the matter of *Philosophical* Wine only, others of that and acid Spirits, as *Aqua Fortis*, Spirit of *Nitre* &c. Also, Mineral *Menstruums* Compounded of Vegetable, and Mineral *Menstruums* mixed together.[100]

Possibly the author was reacting to a recent publication on the "philosophical wine" by the Lithuanian chemist and scholar Johann Seger Weidenfeld, which compiled a variety of sericonian sources as *Four Books . . . Concerning the Secrets of the Adepts, or, of the Use of Lully's Spirit of Wine*.[101] In doing so, Hortolanus Junior rejects an entire tradition of alchemy descended from John of Rupescissa and Raymond's *De secretis naturae*: both the straightforward use of spirit of wine and "hot" plants, and the long-lived institution of compound waters. Ripley, who presented just such a recipe in the *Medulla*, avoids inclusion among the derided "Slipp-slop-Sawse makers" by virtue of his new relationship with Eirenaeus Philalethes, becoming, instead, one of the volume's major authorities. The *Compound* and *Epistle* are extensively cited, particularly in combination with Starkey's commentaries. This is not

99. *The Golden Age: or, the Reign of Saturn Reviewed, Tending to set forth a True and Natural Way, to prepare and fix common Mercury into Silver and Gold . . . An Essay. Written by Hortolanus Junr* (London: J. Mayos for Rich. Harrison, 1698).

100. *Golden Age*, 3–4.

101. Johann Seger Weidenfeld, *De secretis adeptorum sive de usu spiritus vini Lulliani libri IV. Opus practicum per concordantias philosophorum inter se discrepantium . . .* (London, 1684; re-ed. Hamburg, 1685), published in English as *Four Books of Johannes Segerus Weidenfeld Concerning the Secrets of the Adepts, or, of the Use of Lully's Spirit of Wine . . .* (London, 1685).

the only irony: the writer also laments the fact that Philalethes's commentaries on the latter six gates of Ripley's *Compound* are now lost, owing to the "Malice or Self-conceitedness" of Starkey, who neglected to share them.[102]

Hortolanus's tone is characteristic of alchemical publications of the later sixteenth and seventeenth centuries, which typically adopt tougher polemics against opposing positions. Although complaints against unsuitable ingredients (particularly organic products) provide a long-standing theme in alchemica, criticism of rival practices was sharpened in the humanist ambience of early modern Europe, exacerbated by medical controversies and increasing concern over alchemical fraud. The escalation of conflict between practitioners of Galenic and chemical medicine flavors alchemical disputations, as the proponents of different principles and techniques sought to distinguish their methods from the general field of alchemical practice.

Ripley's reputation weathered these storms because his intrinsically diplomatic and subtly worded alchemy could sustain multiple interpretations, even for those disillusioned with the approach of dissolving metals in menstruums, known as the *via humida*.[103] At the end of the seventeenth century his works continued to attract variant readings. William Salmon published a new (and rather inaccurate) English translation of the *Medulla* in 1692, apparently derived from the printed Type II Latin version, and thereby adding another layer to the work's complex translation history.[104] The Starkey commentaries circulated in both manuscript and print, contributing to the reception of Ripley's work from the mid-1650s until the present day.[105]

Throughout their metamorphoses, these fifteenth-century sources were viewed as more than empty vehicles for reinterpretation. Just as Ripley himself labored to reconcile puzzling aspects of his pseudo-Lullian authorities, so the marginalia encountered in Ripleian manuscripts, and the commentaries that appeared in print, speak of ongoing, serious attempts by seventeenth-century readers to make practical sense of his writings. Such apples might fall far from the original tree, but their very presence testifies to continued interest in early texts as documentary records of past success: evidence that

102. *Golden Age*, 155.

103. Briefly, the *via humida* (wet path) uses liquid solvents (such as wine-based menstrua), whereas the *via sicca* (dry path) uses mercury; see Principe, *Aspiring Adept*, 153. Both are accommodated by Ripleian alchemy.

104. Salmon, *Medicina Practica*. On the Type II *Medulla*, see CRC, 131.

105. On Newton's use of Philalethes, see William R. Newman, "Starkey's *Clavis* as Newton's Key," *Isis* 78 (1987): 564–74; Newman, *Newton the Alchemist*. In modern times, the Philalethes commentaries have been used to gloss the *Compound* in Linden, *George Ripley's Compound of Alchymy*.

readers sought to square with contemporary preoccupations and theoretical commitments. The mantle of authority slipped forward generation by generation, even as practices were superseded and refined.

THE VEGETABLE STONE REVISED

From the *Secretum secretorum* of pseudo-Aristotle to the verses of Gower, and from Ripley's Lullian synthesis to the patronage suits of Lock, Kelley, and Norton, the trinity of animal, vegetable, and mineral stones shaped how reader-practitioners conceived of and organized alchemical practice over the space of three centuries. Still in vogue at the end of the sixteenth century, the tripartite division of the alchemical work did not disappear suddenly, or for any single reason. By the end of the 1650s, a model grounded in Aristotelian natural philosophy and Galenic medicine may simply have looked outmoded in the face of new theories and practical frameworks for transmutation, from the niter-based chemistry of Sendivogius and the antimonial proselytizing of Basil Valentine to the promise of the universal solvent, or "Alkahest," promoted by Jan Baptist van Helmont, which offered both medicinal and metallurgical applications.[106] Robson had prophetically warned that contradictory methods undermined the possibility of a universal "mineral stone." As the century progressed and methods proliferated, the dream of a universal stone retreated as a practical end, even as its rhetorical value lived on.

In medicine, too, the medieval tradition of the vegetable stone faced pressure from new remedies based on Helmontian theories and practices, which increasingly dominated conversations about the role and efficacy of chemical medicine.[107] For physicians hoping to use alchemy as a route to patronage, the multifunctionality of Helmont's approach, including the diverse applications of the Alkahest, offered an up-to-date take on the medieval model of multiple stones, while conveniently dispensing with its scholastic

106. On Helmontian theory, see Georgiana D. Hedesan, *An Alchemical Quest for Universal Knowledge: The "Christian Philosophy" of Jan Baptist Van Helmont (1579–1644)* (Oxford: Routledge, 2016); on Helmont's attitude toward experiment (and its medieval antecedents), Newman and Principe, *Alchemy Tried in the Fire*, chap. 2.

107. Antonio Clericuzio, "From van Helmont to Boyle: A Study of the Transmission of Helmontian Chemical and Medical Theories in Seventeenth-Century England," *British Journal for the History of Science* 26 (1993): 303–34; Andrew Wear, *Knowledge and Practice in English Medicine, 1550–1680* (Cambridge: Cambridge University Press, 2000), chaps. 8–9; Charles Webster, "English Medical Reformers of the Puritan Revolution: A Background to the 'Society of Chemical Physicians,'" *Ambix* 14 (1967): 16–41. On Helmont's influence on Starkey and Boyle, see Newman and Principe, *Alchemy Tried in the Fire*.

framework. In 1665, a group of English Helmontians called for the founding of a "Society of Chemical Physicians" as a rival to the established College of Physicians.[108] In this setting, pseudo-Lullian alchemy maintained its credibility, but now as a precursor to later models. Helmont himself compared the coagulating power of the stone, as described in the Lullian *Testamentum*, to the Alkahest of Paracelsus.[109]

With the original sericonian pairing of mineral and vegetable elixirs facing competition from new approaches, there was little hope for their awkward third, the animal stone. Neither the manufacture nor the function of this apparently multipurpose stone is consistently expounded in the core writings of Raymond and Ripley. Even their great apologist, Samuel Norton, struggled to explain how an elixir drawn from blood might suffice both to transmute metals and to heal sickness, admitting in the *Key* that anyone with a basic knowledge of the workings of nature would struggle to acknowledge such a possibility: "and vnto myself also at the first veri dificult."[110] By the 1660s, some readers at least viewed the *Medulla*'s animal stone as an encoded recipe for a mineral process. The Boston physician Samuel Lee implied as much when he jotted down the symbol for antimony in the margins of Ripley's discussion of this stone: a fine example of the longevity of alchemical reading techniques and their continued role in the interpretation—and simultaneous transformation—of texts.[111]

If the traditional division of the work into animal, vegetable, and mineral practices was on the wane in philosophical treatises by the 1650s, in Ashmole's *Theatrum* it disappears altogether, to be replaced in the "Prologomena" by a completely different set. Ashmole's iteration of the philosophical stones is one of the most puzzling aspects of his edition. As we have seen, the pseudo-Lullian model of multiple stones was a staple of English alchemical writing from at least the end of the fourteenth century, and the basis

---

108. The enterprise came close to winning a charter from Charles II—an outcome probably capsized by a particularly severe bout of plague in London that year that carried off four of the petitioners, coupled with mounting concern over the intellectual and moral foundations of Helmontianism. See Harold J. Cook, "The Society of Chemical Physicians, the New Philosophy, and the Restoration Court," *Bulletin of the History of Medicine* 61 (1987): 61–77; Wear, *Knowledge and Practice*, 428–33.

109. Hedesan, *An Alchemical Quest*, 181.

110. Ashmole 1421, fol. 191r. He nonetheless cites Ripley's chapter in the *Medulla*, and his eggshell process in the *Bosome Book*, as evidence that a "sulphur" may indeed be drawn from blood, and a white earth from calx of eggs: fols. 192r-v, 193v-94r.

111. Samuel Lee, note in *OOC*, Beinecke Library, Yna31, 649r, 175. On Lee's alchemical books and his antimonial interpretation of Ripley, see Calis et al., "Passing the Book," on 100–101.

for almost every extant English patronage suit prior to 1600. The majority of authorities cited in the *Theatrum*, from Ripley to Blomfild to Kelley, subscribed to an alchemical philosophy and practice grounded on the notion of a multipurpose elixir, comprising a transmuting mineral stone, a medicinal vegetable stone (one that, when mixed with the mineral, also produced a transmuting "compound water"), and a medicinal (and sometimes also transmuting) animal stone.

Yet Ashmole's preface dispenses entirely with this model. In terms of function he preserves only the mineral stone, which retains its conventional role as an agent of transmutation, and the least of the elixirs.[112] His vision of the vegetable stone differs entirely from that presented by his authorities. Rather than a medicinal elixir, it is a magical product, which gives men the ability to understand the nature of plants and animals, and to nurture them to abundance, "yea, in the depth of *Winter*."[113]

Ashmole's set of stones is rounded off by two even more astonishing products. A "*Magicall* or *Prospective Stone*" allows its user to understand the language of birds, to intuit the whereabouts of any person, and "To Convey a *Spirit* into an *Image*" that may be used as an oracle—allegedly without invoking demonic powers.[114] The last is the Angelical stone, of so subtle a substance that it cannot be seen or touched, but only tasted, and so powerful that the devil himself cannot abide its presence:

It hath a *Divine Power, Celestiall*, and *Invisible*, above the rest; and endowes the possessor with *Divine Gifts*. It affords the *Apparition* of *Angells*, and gives a power of conversing with them, by *Dreames* and *Revelations*: nor dare any *Evill Spirit* approach the *Place* where it *lodgeth*.[115]

This distinctly noncanonical set of stones has been lifted directly from another source: the *Epitome of the Treasure of Health*, an English philosophical treatise, dated by colophon to 1561 and signed by an alchemist who calls himself Edwardus Generosus.[116] Although the work draws primarily on ear-

---

112. *TCB*, sig. A4v.

113. Ibid., sig. Br.

114. Ibid., sig. Bv.

115. Ibid.

116. Principe, *Aspiring Adept*, 197–200. See also Kassell, "Reading for the Philosophers' Stone"; William R. Newman, "Newton's Reputation as an Alchemist and the Tradition of Chymiatria," in *Reading Newton in Early Modern Europe*, ed. Elizabethanne A. Boran and Mordechai Feingold (Leiden: Brill, 2017), 313–27, on 324–27; Janacek, *Alchemical Belief*, chap. 5.

lier texts, including Ripley's *Compound* and Norton's *Ordinal*, it also includes Edwardus's own exposition of the philosophical stones. For instance, the vegetable stone, according to Edwardus, was used by Solomon to make "the trees & hearbs to flourish at all times of the yeare," and "bring the birds down to him out of the air to sing chirp & sit by him but also dwell with him."[117] Among other marvels, he describes a shining "lunar stone," which can "congeal" the surface of a mirror or crystal so that besides the normal reflection it will reveal "very merveilous strange things not to be written of"—an apt description of the use of a show stone to scry for spirits. Little wonder that Ashmole, who closely studied Dee and Kelley's angelic conversations and even attempted to cast his own talismans, found much to interest him in Edwardus's seeming conflation of alchemy with natural and angelic magic.[118]

Such a collapsing of the boundaries between alchemy and magic is, as we have seen, distinctly unusual in the English alchemical tradition. On the other hand, the way in which Edwardus describes the virtues of the stone bears a striking similarity to some of the procedures outlined in magical treatises of the fifteenth and sixteenth centuries: the preparation of surfaces suitable for scrying, for instance, or tricks for captivating birds and beasts.[119] A practitioner versed in both alchemy and magic—a Richard Jones, perhaps, or even a William Blomfild—might well have wondered about the outcome of substituting the alchemical quintessence, or the philosophers' stone, into this kind of magical procedure.[120] For such an inquiring reader, alchemical treatises offered the raw material for speculating about the real-world implications of superperfect materials like the stone. Perhaps we can even identify the source of Edwardus's eclectic, bird-catching stone in the medieval quintessence tradition. Take, for instance, his spectacular account of the lunar stone:

117. King's College Library, MS Keynes 22, fol. 13r (hand of Isaac Newton).

118. On Ashmole's interest in magic, see Lauren Kassell, "The Economy of Magic in Early Modern England," in *The Practice of Reform in Health, Medicine, and Science, 1500–2000: Essays for Charles Webster*, ed. Margaret Pelling and Scott Mandelbrote (Aldershot: Ashgate, 2005), 43–57; Vittoria Feola, "Elias Ashmole's Collections and Views about John Dee," *Studies in History and Philosophy of Science* 43 (2012): 530–38.

119. For a representative selection of such practices, see Richard Kieckhefer, *Forbidden Rites: A Necromancer's Manual of the Fifteenth Century* (University Park: Pennsylvania State University Press, 1998). See also Klaassen, *Transformations of Magic*.

120. Given the role of the quintessence in Ficino's conception of the world soul, and the subsequent popularization of that notion by Agrippa, such a splicing of traditions may have provided, at the least, an intriguing thought experiment; see chaps. 4–5, above.

[Birds] will come presently to you into the shining light so as you may presently take them alive in your hand, poor silly fowles! So great is the secret of nature that they have no power to escape or fly away, they are so enamoured with the resplendent ravishment & shining brightnes thereof.[121]

John of Rupescissa also praised the heavenly fragrance of his quintessence of wine, which is such that if the vessel is placed in the corner of the house, its odor will attract all who enter: this is an observable sign, or token, of practical success.[122] This heavenly "token" is elaborated even further in *De secretis naturae*, where Raymond conceives of the quintessence luring the very birds from the sky:

> If a most wonderful odor issues out, such that no worldly fragrance can be compared to it, in so much that the vessel, being set at a corner of the house, by an invisible miracle attracts all those who enter in, or, the vessel being placed upon a turret, attracts all birds whose senses its odor touches, such that it makes them to stand around it; then, my son, you shall have our quintessence, which is otherwise called our "vegetable mercury," according to your desire.[123]

This passage in one of the founding works of the pseudo-Lullian "multiple stones" model, read through the lens of the natural magic tradition, may have suggested a completely different vision of the stone: one that transforms the vegetable stone from an alchemical medicine to a potentially inimical substance capable of luring hapless avians to their doom.

Unlike the switching of litharge for minium, however, or the splicing of texts by Ripley and Ive, the substitution of an alchemical product into a magical procedure takes us far from the practical tradition espoused by most of Ashmole's alchemical authorities. Yet for a reader familiar with such a tradition, even the peculiarities of Edwardus's *Epitome* may not have been cause for dismay. Isaac Newton himself, one of the most rigorous of alchemical

---

121. MS Keynes 22, fol. 12r.

122. John of Rupescissa, *De consideratione*, 27–28: "in tantum quod si fuerit vas positum in angulo domus, ex fragrantia quintae Essentiae (quod mirabile & summe miraculosum est) attrahat ad se vinculo inuisibili vniuersos intrantes."

123. *De secretis naturae*, fols. 23v-24r: "& si odor supra mirabilis exeat, ita quod nulla mundi fragrantia ei comparari ualeat, in tantum quod uas positum ad angulum domus miraculo inuisibili trahit omnes intrantes, aut uase posito supra turrim trahit omnes aues, quibus omnibus odor naris attigerit, ita quod circa se ipsum stare faciat, tunc habebis fili nostram quintam essentiam, quae aliter dicitur mercurius uegetabilis ad tuum libitum."

experimentalists in the closing decades of the seventeenth century, studied both Edwardus and the contents of the *Theatrum* with close attention.[124] Through alchemical reading, almost any term might offer an alchemical *Deckname*—and almost any tale a cover story for chemical operations.

THE ENDS OF ENGLISH ALCHEMY

By the 1650s, the old model of animal, vegetable, and mineral stones was under attack from two directions, and, in each case, by apparent allies. In construing Ripley's writings through the lens of continental chemistry, George Starkey switched the focus not only from lead to antimony, but from medicine to transmutation. Ashmole went farther still, overwriting the storied triad with an entirely different taxonomy of alchemical stones that drew on the literature not of practical alchemy, but of natural and angelic magic. While Ashmole's and Starkey's approaches to the past are in many ways distinct, they have this in common: each served to shift scholarly attention away from the substance of earlier English practices, even while claiming to encapsulate those very traditions. Influential in their own time, and nowadays readily available in modern print and online editions, the writings of both authors offer well-marked paths into the thickets of alchemy, still gratefully trodden by general readers and specialists alike.[125] As with any paths, however, these routes will omit the greater part of the wood, for each is the product of a single time—individual moments in a tradition whose origins lie in the Middle Ages, much of which remains unseen, and unedited, in the sylvan gloom.

To penetrate these thickets requires that we, too, learn how to read our sources like alchemists, alert to the potential multiple meanings of technical terms and allegories alike. But we must also read as historians, taking care to situate the activities and motivations of individual practitioners in

124. On Newton and Edwardus, see Newman, "Newton's Reputation as an Alchemist," 324–27. Newton's copy of the *TCB* is now held in the University of Pennsylvania, Philadelphia: Van Pelt Rare Book and Manuscript Library, E. F. Smith Collection, QD25.A78 1652. He also transcribed individual poems, including Blomfild's *Blossoms* (King's College Library, MS Keynes 15, fols. 1r–4r), and *The Hunting of the Green Lyon* (MS Keynes 20, fols. 1r–3v).

125. Thanks to the publication of a 1966 reprint and availability on *Early English Books Online*, the *TCB* remains the main repository of English alchemical verse for present-day readers; Elias Ashmole, ed., *Theatrum Chemicum Britannicum*, with introduction by Allen G. Debus (New York: Johnson Reprint Corporation, 1966). The writings of Eirenaeus Philalethes are collected in S. Merrow Broddle, comp., *Alchemical Works: Eirenaeus Philalethes Compiled* (Boulder, CO: Cinnabar, 1994), and are available online in a variety of formats.

the context of their own time, place, and available authorities. The status of practical traditions switches around often enough to make a reader dizzy, but broad trends can still be discerned: a wide slate of organic products narrowed down to metals alone; the metallic monopoly exploded into a model of three stones (or more) and back again; the model of multiple stones startlingly conflated with a formally distinct tradition of angelic magic. Frequently these approaches overlapped, and individual practitioners seldom restricted themselves to one alone—even those who, like Ripley or Starkey, later gained distinction for a particular practice.

Practical exegesis—the reinvention of earlier practices through successive cycles of testing and reinterpreting written sources—has made many of these transformations invisible, creating an illusion of stasis through repetition of the same terms and topoi. And what practical exegesis did for alchemical practice, pseudepigraphy has also accomplished for alchemical history. As antiquarians accommodated alchemy within the broader landscape of English knowledge, they sought to impose order on their disparate and frequently overlapping sources—much in the manner of those ecclesiastical historians who struggled to reconstruct the history of the *Ecclesia Anglicana* from a mass of fragmentary, forged, or otherwise unverifiable documentation.[126] Their attempts to devise lineages of adepts no longer carry historical weight, but different kinds of genealogy allow us to trace the changing uses of alchemical texts and practices over time, from the accretion of ownership marks in medieval manuscripts to the "nesting" of old texts within new formulations. Such reiterations of authoritative material reveal how alchemists like Lock and Kelley recycled their sources, often for the purpose of bestowing ancient authority on a signature practice or high-stakes patronage suit. In their hands, the *Accurtations of Raymond* contracted into the *Work of Dunstan* and expanded again into the *Clavis aureae portae*. Ripley's *Medulla* was absorbed into Lock's *Treatise*, which was in turn epitomized in a later treatise that still punned on Lock's name, as the *Picklock to Ripley His Castle*.[127]

These continual adaptations, whether of practices, texts, or histories, provide the strongest evidence for the vitality of early modern alchemy,

---

126. There are clear parallels between early modern attempts to construct alchemical and ecclesiastical history, respectively. Forgery, pseudepigraphy, and the construction of ecclesiastical history by both Catholic and Protestant scholars have been extensively discussed in the work of Anthony Grafton; see, inter alia, Grafton, "Church History in Early Modern Europe."

127. On the permutations of Lock's *Treatise*, see Grund, *Misticall Wordes*, 24–28. Timmermann's methodology for tracing intertextual links between alchemical poems also offers a tool for handling anonymous material; Timmermann, *Verses on the Elixir*.

just as the hardheaded reassessment of pseudepigraphic corpora in modern times is a sure sign that scholars no longer view alchemy as either a fruitful approach to chemistry or a reliable record of the past. As long as alchemy remained a living tradition, it altered from moment to moment, even as its authors strove to present an image of unchanging unity. Diversity of practice was folded and refolded into a timeless narrative of coherence, through the efforts of new generations of reader-practitioners not just to make the stone, but to make a living—and to make sense.

Bibliography

MANUSCRIPTS

Bologna, Biblioteca Universitaria di Bologna
    MS 142 (109), vol. 2
    MS 457, vol. 23, pt. 3

Boston, Massachusetts Historical Society
    Winthrop 20c

Cambridge, Cambridge University Library
    MS DD.3.83
    MS FF.4.12
    MS FF.4.13
    MS KK.1.3

Cambridge, Corpus Christi College Library
    MS 99
    MS 112
    MS 395

Cambridge, King's College Library
    MS Keynes 15
    MS Keynes 17
    MS Keynes 20
    MS Keynes 22

Cambridge, Queens' College Library
    MS 49

Cambridge, Trinity College Library
MS O.2.16
MS O.2.33
MS O.4.39
MS O.5.31
MS O.8.24
MS O.8.25
MS O.8.32
MS O.8.9
MS R.14.14
MS R.14.56
MS R.14.58

Cieszyń, Książnica Cieszyńska
MS SZ DD.vii.33

Copenhagen, Royal Library
GKS 242
GKS 1727
GKS 1746

Edinburgh, Royal College of Physicians of Edinburgh
MS Anonyma 2, vol. 1

Florence, Biblioteca Nazionale Centrale
MS Magliabechiano XVI. 113

Glasgow, Glasgow University Library
MS Ferguson 91
MS Ferguson 102
MS Ferguson 133
MS Hunter 251
MS Hunter 253
MS Hunter 403

The Hague, Royal Library of the Netherlands
MS 46 (Bibliotheca Philosophica Hermetica)

Hatfield, Hatfield House
Cecil Papers 271/1

Kassel, Landesbibliothek und Murhardsche Bibliothek der Stadt Kassel
2° MS chem. 19
4° MS chem. 67
4° MS chem. 68

Kendal, Kendal Archive Centre
MS Hothman 5

Leiden, Leiden Universiteitsbibliotheek
MS Vossianus Chym. F.3

Leipzig, Universitätsbibliothek
MS 0398

London, British Library
MS Add. 10302
MS Add. 11388
MS Add. 15549
MS Add. 36674
MS Cotton Titus B.V.
MS Harley 660
MS Harley 1887
MS Harley 2407
MS Harley 2411
MS Harley 3528
MS Harley 3542
MS Harley 3703
MS Harley 3707
MS Lansdowne 19
MS Lansdowne 21
MS Lansdowne 101
MS Lansdowne 703
Royal MS 7 C XVI
MS Sloane 288
MS Sloane 363
MS Sloane 513
MS Sloane 689
MS Sloane 1091
MS Sloane 1095
MS Sloane 1423
MS Sloane 1744

MS Sloane 1842
MS Sloane 2128
MS Sloane 2170
MS Sloane 2175
MS Sloane 2325
MS Sloane 2413
MS Sloane 3188
MS Sloane 3189
MS Sloane 3191
MS Sloane 3579
MS Sloane 3580A
MS Sloane 3580B
MS Sloane 3604
MS Sloane 3645
MS Sloane 3654
MS Sloane 3667
MS Sloane 3682
MS Sloane 3707
MS Sloane 3744
MS Sloane 3747

London, Lincoln's Inn Library
MS Hale 90

London, National Archives
C 66/475
C 66/490
C 66/506
C66 /522
C66 /538
KB 27/629
SP 1/69
SP 1/73
SP 1/222
SP 4/1
SP 12/36
SP 46/2
SP 46/8
SP 70/80
SP 70/146

SP 81/6
SP 92/32
STAC 2/14/111
STAC 2/14/112
STAC 3/7/85

London, Wellcome Library
MS 239

Los Angeles, Getty Research Institute
MS 18, vol. 10, pt. 2
MS 18, vol. 11

New Haven, Beinecke Rare Book & Manuscript Library
MS Mellon 12
MS Mellon 33
Note of Samuel Lee, Yna31, 649r

Oxford, Bodleian Library
MS Ashmole 208
MS Ashmole 391
MS Ashmole 759
MS Ashmole 971/972
MS Ashmole 1394
MS Ashmole 1406
MS Ashmole 1407
MS Ashmole 1408
MS Ashmole 1415
MS Ashmole 1418
MS Ashmole 1421
MS Ashmole 1423
MS Ashmole 1424
MS Ashmole 1426
MS Ashmole 1441
MS Ashmole 1442
Ms Ashmole 1445
MS Ashmole 1450
MS Ashmole 1451
MS Ashmole 1459
MS Ashmole 1467
MS Ashmole 1478

MS Ashmole 1479
MS Ashmole 1480
MS Ashmole 1485
MS Ashmole 1486
MS Ashmole 1487
MS Ashmole 1490
MS Ashmole 1492
MS Ashmole 1493
MS Ashmole 1507
MS Ashmole 1508
MS Ashmole 1790
MS Digby 133
MS Laud Misc. 708
MS Rawlinson B.306
MS Rawlinson D.241
MS Rawlinson poet 182
MS Savile 18

Oxford, Corpus Christi College Library
MS 118
MS 125
MS 136
MS 172
MS 336

Oxford, Queen's College Library
MS 307

Paris, Bibliothèque Nationale
MS Lat. 12993
MS Lat. 14007

Philadelphia, Van Pelt-Dietrich Library Center, University of Pennsylvania
Codex 111

Vatican City, Biblioteca Apostolica Vaticana
MS Reg. Lat. 1381

Vienna, Österreichische Nationalbibliothek
Sammlung von Handschriften und alten Drücken, Cod. 8964 [Fugger-Zeitungen 1591]

Warminster, Longleat House
The Dudley Papers, MS DUI
MS 178

SOURCES PRINTED BEFORE 1800

Agnello, Giovanni Battista. *Espositione sopra un Libro intitolato Apocalypsis spiritus secreti.* London: John Kingston for Pietro Angeliono, 1566.

——— [John Baptista Lambye]. *A reuelation of the secret spirit: Declaring the most concealed secret of alchymie.* Trans. Richard Napier. London: John Haviland for Henrie Skelton, 1623.

Agrippa von Nettesheim, Heinrich Cornelius. *De incertitudine et vanitate scientiarum et artium atque de excellentia verbi Dei declamatio invectiva.* Cologne, 1527.

———. *De Nobilitate et Praecellentia Foeminei Sexus.* Cologne, 1532.

———. *Three Books of Occult Philosophy.* Trans. J. F. London: R. W. for Gregory Moule, 1651 [i.e., 1650].

———. *The Vanity of Arts and Sciences.* London: Samuel Speed, 1676.

Agrippa von Nettesheim, Heinrich Cornelius [attr. to]. *De arte chimica (The Art of Alchemy): A Critical Edition of the Latin Text with a Seventeenth-Century English Translation.* Ed. Sylvain Matton. Paris: S.É.H.A.; Milan: Archè, 2014.

Ashmole, Elias, ed. *Theatrum Chemicum Britannicum: Containing Severall Poeticall Pieces of Our Famous English Philosophers, Who Have Written the Hermetique Mysteries in Their Owne Ancient Language. Faithfully Collected into One Volume with Annotations Thereon.* London: J. Grismond for Nathanial Brooke, 1652.

Avicenna [pseud.]. *De anima in arte alchimiae.* In *Artis Chemicae Principes, Avicenna atque Geber,* ed. Mino Celsi. Basel: Pietro Perna, 1572.

Bale, John. *Index Britanniae Scriptorum: John Bale's Index of British and Other Writers.* Ed. Reginald Lane Poole and Mary Bateson. 1902. Reprint, Cambridge: D. S. Brewer, 1990.

———. *Scriptorum illustrium maioris Brytanniae . . . Catalogus.* Basel: Johannes Oporinus, 1557.

Barnaud, Nicolas, ed. *Quadriga aurifera (Prima rota: Tractatus de philosophia metallorum, a doctissimo . . . viro anonymo conscriptus; 2a rota: Georgii Riplei . . . Liber duodecim portarum; 3a rota: Liber de mercurio et lapide philosophor. Georgii Riplei; 4a rota: Scriptum . . . docti viri cuius nomen excidit, elixir solis Theophrasti Paracelsi tractans).* Leiden: Christophorus Raphelengius, 1599.

Biringuccio, Vannoccio. *De la pirotechnia. Libri .X.* Venice: Curtio Navò, 1540.

Bonus, Petrus. *Pretiosa margarita novella de thesauro, ac pretiosissimo philosophorum lapide*. Ed. Giovanni Lacinio. Venice: Aldo Manuzio, 1546.

B[ostocke], R[ichard]. *The difference betwene the auncient phisicke, first taught by the godly forefathers, consisting in vnitie peace and concord: and the latter phisicke proceeding from idolaters, ethnickes, and heathen: as Gallen, and such other consisting in dualitie, discorde, and contrarietie*... London: [G. Robinson] for Robert Walley, 1585.

Brunschwig, Hieronymus. *Liber de arte distillandi de simplicibus*. Strassburg: J. Grüninger, 1500.

———. *The vertuose boke of distyllacyon of the waters of all maner of herbes*. Trans. Laurence Andrewe. London: Laurence Andrewe, 1527.

Casaubon, Meric. *A True and Faithful Relation of What Passed for Many Years Between Dr John Dee and Some Spirits*. London: Garthwait, 1659.

Chrysogonus Polydorus, ed. *Alchemiae Gebri Arabis*. Bern: Mathias Apiarius, 1545. Reprint of Chrysogonus Polydorus, ed., *In hoc volumine de alchimia continentur*. Nuremberg: Johannes Petreius, 1541.

Collectanea Chymica: A Collection of Ten Several Treatises in Chymistry, concerning The Liquor Alkahest, the Mercury of Philosophers, and other Curiosities worthy the Perusal... London: for William Cooper, 1684.

Combach, Ludwig, ed. *Tractatus aliquot chemici singulares summum philosophorum arcanum continentes, 1. Liber de principiis naturae, & artis chemicae, incerti authoris. 2. Johannis Belye Angli... tractatulus novus, & alius Bernhardi Comitis Trevirensis, ex Gallico versus. Cum fragmentis Eduardi Kellaei, H. Aquilae Thuringi, & Joh. Isaaci Hollandi*... Geismar: Salomonis Schadewitz for Sebaldi Köhlers, 1647.

Condeesyanus, Hermann [i.e., Johann Grasshof], ed. *Harmoniae imperscrutabilis Chymico-Philosophicae, sive Philosophorum Antiquorum Consentientium*. Frankfurt, 1625.

Conring, Hermann. *De Hermetica Aegyptiorum vetere et Paracelsicorum nova medicina liber unus*. Helmstedt: Henning Müller, 1648.

Cooper, William. *A Catalogue of Chymical Books Which Have Been Written Originally or Translated into English*, printed with W. C. Esquire, *The Philosophical Epitaph*. London: William Cooper, 1673.

———. *A Catalogue of Chymicall Books. In Three Parts*. London: William Cooper, 1675.

———. *The Continuation or Appendix to The Second Part of the Catalogue of Chymical Books*. London: William Cooper, 1688.

De Alchimia opuscula complura veterum philosophorum... Frankfurt am Main: Cyriacus Iacobus, 1550.

Dee, Arthur. *Fasciculus chemicus, or, Chymical collections: expressing the ingress, progress, and egress of the secret Hermetick science, out of the choisest and most famous authors . . . whereunto is added, the Arcanum, or, Grand secret of hermetick philosophy.* Ed. and trans. Elias Ashmole. London: J. Flesher for Richard Mynne, 1650.

Du Chesne, Joseph. *Ad Veritatem Hermeticae Medecinae ex Hippocratis veterumque decretis ac Therapeusi, . . . adversus cujusdam Anonymi phantasmata Responsio.* Paris: Abraham Saugrain, 1604.

Dugdale, Sir William. *The Baronage of England, or, An Historical Account of the Lives and Most Memorable Actions of our English Nobility in the Saxons time to the Norman Conquest . . .* London: Thomas Newcomb, for Abel Roper, John Martin, and Henry Herringman, 1675–76.

Eden, Richard. *The Decades of the newe worlde or west India, Conteyning the nauigations and conquestes of the Spanyardes, with the particular description of the moste ryche and large landes and Ilandes lately founde in the west Ocean perteynyng to the inheritaunce of the kinges of Spayne.* London: Richard Jugge, 1555.

Eirenaeus Philoponus Philalethes. *A Breviary of Alchemy; or a Commentary upon Sir George Ripley's Recapitulation: Being a Paraphrastical Epitome of his Twelve Gates.* London: for William Cooper, 1678.

———. *Introitus apertus ad occlusum Regis palatium.* Amsterdam: Johannes Janssonius van Waesbergen, 1667.

———. *The Marrow of Alchemy, Being an Experimental Treatise, Discovering the Secret and Most Hidden Mystery of the Philosophers Elixer. Divided Into Two Parts.* London: A.M. for Edward Brewster, 1654.

———. *Ripley Reviv'd: or An Exposition Upon Sir George Ripley's Hermetico-Poetical Works.* London: William Cooper, 1677–78.

———. *The Secret of the Immortal Liquor called Alkahest, or Ignis-Aqua.* London: For William Cooper, 1683.

Erbinäus von Brandau, Matthias. *Warhaffte Beschreibung von der Universal-Medicin.* Leipzig: F. Lanckisch, 1689.

Gallus Carniolus, Jacobus [Jacob Handl]. *Quatuor vocum Liber I. Harmoniarum Moralium . . .* Prague: Georgius Nigrinus, 1589.

Geber [pseud.]. *De alchimia. Libri tres.* Strasbourg: Johann Grüninger, 1529.

Gessner, Conrad. *Bibliotheca universalis, sive catalogus omnium scriptorum locupletissimus, in tribus linguis, Latin, Graeca, & Hebraica: extantium & non extantium veterum & recentiorum . . .* Zurich: Christophorus Froschouerus, 1545.

———. *The newe iewell of health wherein is contayned the most excellent secretes of phisicke and philosophie . . .* Trans. George Baker. London: Henrie Denham, 1576.

———. *Pandectarum sive Partitionum universalium libri XXI*. Zurich: Christophorus Froschouerus, 1548.

———. *Thesavrvs Evonymi Philiatri De remediis Secretis* . . . Zurich: Andreas Gesner and Rudolf Wyssenbach, 1552.

Grado [Gradi], Giovanni Matteo Ferrari da. *Consilia . . . cum tabula Consiliorum ecundum viam Avicenne ordinatorum utile repertorium*. [Venice]: [Mandato et impensis heredum Octaviani Scoti & sociorum, impressa per Georgium Arrivabenum], [1514].

Gratarolo, Guglielmo, ed. *Verae alchemiae artisque metallicae, citra aenigmata, doctrina, certusque modus, scriptis tum novis tum veteribus nunc primum & fideliter maiori ex parte editis, comprehensus*. Basel: Heinrich Petri and Peter Perna, 1561.

Guibert, Nicholas. *Alchymia ratione et experientia ita demum viriliter impugnata* . . . Strasbourg: Lazarus Zetzner, 1603.

Hartlib, Samuel. *Chymical, Medicinal, and Chyrurgical Addresses*. London: G. Dawson for Giles Calvert, 1655.

Hortulanus Junior [pseud.]. *The Golden Age: or, the Reign of Saturn Reviewed, Tending to set forth a True and Natural Way, to prepare and fix common Mercury into Silver and Gold . . . An Essay. Written by Hortolanus Junr*. London: J. Mayos for Rich. Harrison, 1698.

Humphrey, Lawrence. *Interpretatio linguarum: seu de ratione conuertendi & explicandi autores tam sacros quam profanos, libri tres*. Basel: Hieronymus Froben, 1559.

John of Rupescissa. *De consideratione Quintae essentiae rerum omnium, opus sanè egregium*. Basel: Conrad Waldkirch, 1597.

Kelley, Edward. *Tractatus duo egregii, de lapide philosophorum*. Hamburg: Schultze, 1676.

Lenglet-Dufresnoy, Nicolas. *Histoire de la Philosophie Hermétique*. Paris: Coustelier, 1742.

Libavius, Andreas. "Analysis Dvodecim Portarvm Georgii Riplaei Angli, Canonici Regularis Britlintonensis." In *Syntagmatis arcanorum chymicorum: tomus [primus] secundus. . .* , 400–436. Frankfurt: Nicolaus Hoffmann, 1613–15.

Maier, Michael. *Tripus Aureus, hoc est, Tres Tractatus Chymici Selectissimi . . .* Frankfurt: Lucas Jennis, 1618.

Manget, Jean-Jacques, ed. *Bibliotheca chemica curiosa*. 2 vols. Geneva: Chouet, 1702.

Martire d'Anghiera, Pietro. *De orbe novo decades*. [Alcalá]: [Arnaldi Guillelmi], [1516].

Münster, Sebastian. *A treatyse of the newe India with other new founde landes and islandes, aswell eastwarde as westwarde, as they are knowen and found in these oure dayes* . . . Trans. Richard Eden. London: S. Mierdman for Edward Sutton, [1553].

Norton, Samuel. *Mercurius Redivivus, seu Modus conficiendi Lapidem Philosophicum* . . . Ed. Edmund Deane. Frankfurt, 1630.

Oviedo, Gonzalo Fernández. *Historia general y natural de las Indias*. Seville, 1530–55.

Pantheo, Giovanni Agostino. *Voarchadumia contra Alchimiam: Ars distincta ab Archimia, et Sophia*. Venice: Giovanni Tacuino, 1530.

Paracelsus, his Archidoxis comprised in ten books : disclosing the genuine way of making quintessences, arcanums, magisteries, elixirs, &c . . . Trans. J. H. London: For W.S., 1660.

Penot, Bernard G., ed. *Dialogus inter Naturam et Filium Philosophiae, Accedunt Abditarum rerum Chemicarum Tractatus Varii scitu dignissimi ut versa pagina indicabit*. Frankfurt, 1595.

Petreius, Johannes. *In hoc volumine de Alchemia continentur haec. Gebri Arabis* Nuremberg: Johannes Petreius, 1451.

Reuchlin, Johannes. *De arte cabalistica libri tres*. Anshelm, 1517.

Richenbourg, Jean Maugin de, ed. *Bibliotheque des philosophes*, 4 vols. Paris: André Cailleau, 1740–54.

Ripley, George. *The Compound of Alchymy* . . . *Divided into Twelue Gates*. Ed. Raph Rabbards. London, 1591.

———. *Opera omnia chemica*. Ed. Ludwig Combach. Kassel, 1649.

———. *Opuscula quaedam Chemica. Georgii Riplei Angli Medvlla Philosophiae Chemicae. Incerti avtoris canones decem, Mysterium artis mira brevitate & perspicuitate comprehendentes* . . . *Omnia partim ex veteribus Manuscriptis eruta, partim restituta*. Frankfurt am Main: Johann Bringer, 1614.

Roth-Scholtz, Friedrich, ed. *Deutsches Theatrum Chemicum, auf welchem der berühmtesten Philosophen und Alchymisten Schrifften* . . . 3 vols. Nürnberg: Felsecker, 1728–30.

Sandys, George. *Anglorum Speculum, or, The Worthies of England in Church and State alphabetically digested into the several Shires and Counties therein* . . . London: for John Wright, Thomas Passinger, and William Thackary, 1684.

Secreta secretorum Raymundi Lulli et hermetis philosophorum in libros tres divisa. Cologne: Goswin Cholinus, 1592.

Suchten, Alexander von. *Antimonii Mysteria Gemina* . . . Leipzig, 1604.

———. *Liber unus de Secretis Antimonii, das ist von der grossen Heimligkeit des Antimonii*. Strassburg, 1570.

Taisner, Jean. *A Very Necessarie and Profitable Booke Concerning Navigation*. Trans. Richard Eden. London: Richard Jugge, [1575].

Tanner, Thomas. *Bibliotheca Britannico-Hibernica: sive, de scriptoribus, qui in Anglia, Scotia, et Hibernia ad saeculi XVII initium floruerunt, literarum ordine juxta familiarum nomina dispositis commentarius.* London, 1748.

Thoroton, Robert. *The Antiquities of Nottinghamshire.* Ed. John Throsby. 2nd ed. Nottingham: G. Burbage, 1790.

———. "Osberton." In *Thoroton's History of Nottinghamshire: Volume 3, Republished With Large Additions By John Throsby*, ed. John Throsby. Nottingham, 1796.

Ulstad, Philipp. *Coelum philosophorum seu de secretis naturae.* Fribourg [Strasbourg]: [Johann Grüninger], 1525.

Usk, Thomas. "Testament of Love." In Geoffrey Chaucer, *The Workes of Geffray Chaucer Newly Printed*, ed. William Thynne. London, 1532.

Valentine, Basil. *Triumph-Wagen Antimonii . . . An Tag geben, durch Johann Thölden . . .* Leipzig, 1604.

Vigenère, Blaise de. *A Discourse of Fire and Salt, Discovering many secret Mysteries, as well Philosophicall, as Theologicall.* London: Richard Cotes, 1649.

———. *Traicté du Feu et du Sel.* Paris: Abel l'Angelier, 1618.

Weidenfeld, Johann Seger. *De secretis adeptorum sive de usu spiritus vini Lulliani libri IV. Opus practicum per concordantias philosophorum inter se discrepantium . . .* London, 1684; re-ed. Hamburg, 1685.

———. *Four Books of Johannes Segerus Weidenfeld Concerning the Secrets of the Adepts, or, of the Use of Lully's Spirit of Wine . . .* London, 1685.

Weston, Elizabeth Jane. *Parthenica.* Vol. I. Prague: Paulus Sessius, [1606].

Zetzner, Lazarus. *Theatrum chemicum, præcipuos selectorum auctorum tractatus de chemiæ et lapidis philosophici antiquitate, veritate, iure, præstantia et operationibus . . .* 6 vols. Ursel and Strasbourg, 1602–61.

SOURCES PRINTED AFTER 1800

Alford, Stephen. *London's Triumph: Merchant Adventurers and the Tudor City.* London: Allen Lane, 2017.

Allen, Martin. *Mints and Money in Medieval England.* Cambridge: Cambridge University Press, 2012.

Allmand, Christopher. *The Hundred Years War: England and France at War, c. 1300–c. 1450.* Rev. ed. 1988. Cambridge: Cambridge University Press, 2001.

Ames-Lewis, Francis. *The Intellectual Life of the Early Renaissance Artist.* New Haven: Yale University Press, 2000.

Andrews, Jonathan. "Richard Napier (1559–1634)." *ODNB.*

Andrews, Kenneth R. *Trade, Plunder, and Settlement: Maritime Enterprise and the Genesis of the British Empire, 1480–1630.* Cambridge: Cambridge University Press, 1984.

Arber, Edward. "The Life and Labors of Richard Eden, Scholar, and Man of Science." In *The First Three English Books on America: [?1511]–1555 A.D.*, ed. Arber, xxxvii-xlviii. Birmingham: [Printed by Turnbull & Spears, Edinburgh], 1885.

Archer, Ian W. "Smith, Sir Thomas (1513–1577)." *ODNB.*

Arnald de Villanova. *Opera medica omnia.* Vol. 5,1: *Tractatus de intentione medicorum*, ed. Michael R. McVaugh. Barcelona: Publicacions I Edicions de la Univ. de Barcelona, 2000.

Ash, Eric H. *Power, Knowledge, and Expertise in Elizabethan England.* Baltimore: Johns Hopkins University Press, 2004.

Ashmole, Elias, ed. *Theatrum Chemicum Britannicum.* With introduction by Allen G. Debus. New York: Johnson Reprint Corporation, 1966.

Aston, Margaret. "English Ruins and English History: The Dissolution and the Sense of the Past." *Journal of the Warburg and Courtauld Institutes* 36 (1973): 231–55.

Backhouse, Janet. "The Royal Library from Edward IV to Henry VII." In *The Cambridge History of the Book in Britain*, vol. 3, *1400–1557*, ed. Lotte Hellinga and J. B. Trapp, 267–73. Cambridge: Cambridge University Press, 1999.

Bäcklund, Jan. "In the Footsteps of Edward Kelley: Some MSS References at the Royal Library in Copenhagen Concerning an Alchemical Circle around John Dee and Edward Kelley." In *John Dee: Interdisciplinary Studies in English Renaissance Thought*, ed. Stephen Clucas, 295–330. Dordrecht: Springer, 2006.

Bacon, Roger. *De universali regimine senum et seniorum.* In *Opera hactenus inedita Rogeri Bacon*, fasc. 9, ed. Andrew G. Little and Edward Withington. Oxford: Clarendon Press, 1928.

———. *Opus majus.* Ed. John Henry Bridges. 2 vols. Oxford: Clarendon Press, 1897.

———. *Opus tertium.* In *Opera quaedam hactenus inedita Rogeri Baconis*, fasc. 1, ed. J. S. Brewer, 3–310. London: Longman, Green, Longman, and Roberts, 1859.

———. *Secretum secretorum cum glossi et notulis, tractatus brevis et utilis ad declarandum quedam obscure dicta Fratris Rogeri.* In *Opera hactenus inedita Rogeri Baconis*, fasc. 5, ed. Robert Steele, 1–175. Oxford: Clarendon Press, 1920.

Baggs, A. P., Ann J. Kettle, S. J. Lander, A. T. Thacker, and David Wardle. "Houses of Cistercian Monks: The Abbey of Combermere." In *A History of the*

County of Chester, vol. 3, ed. C. R. Elrington and B. E. Harris, 150–56. London: Victoria County History, 1980.

Bailey, Michael D. "Diabolic Magic." In *The Cambridge History of Magic and Witchcraft in the West: From Antiquity to the Present*, ed. David J. Collins, 361–92. New York: Cambridge University Press, 2015.

Balsem, Astrid C. "Books from the Library of Andreas Dudith (1533–89) in the Library of Isaac Vossius." In *Books on the Move: Tracking Copies through Collections and the Book Trade*, ed. Robin Myers, Michael Harris, and Giles Mandelbrote, 69–86. London: Oak Knoll Press, 2007.

Baskerville, Geoffrey. *English Monks and the Suppression of the Monasteries.* London: Jonathan Cape, 1937.

Bassnett, Susan. "Absent Presences: Edward Kelley's Family in the Writings of John Dee." In *John Dee: Interdisciplinary Studies in English Renaissance Thought*, ed. Stephen Clucas, 285–94. Dordrecht: Springer, 2006.

Bayer, Penny. "Lady Margaret Clifford's Alchemical Receipt-Book and the John Dee Circle." *Ambix* 52 (2006): 71–84.

Beaujouan, Guy, and Paul Cattin. "Philippe Éléphant (mathématique, alchimie, éthique)." In *Histoire littéraire de la France*, vol. 41, *Suite du quatorzième siècle*, 285–363. Paris: Imprimerie nationale, 1981.

Bell, David N. "A Cistercian at Oxford: Richard Dove of Buckfast and London." *Studia monastica* 31 (1989): 67–87.

———. "Monastic Libraries: 1400–1557." In *The Cambridge History of the Book in Britain*, vol. 3, *1400–1557*, ed. Lotte Hellinga and J. B. Trapp, 229–54. Cambridge: Cambridge University Press, 2008.

Bellamy, J. G. *The Law of Treason in England in the Later Middle Ages.* Cambridge: Cambridge University Press, 1970.

Betteridge, Thomas. *Tudor Histories of the English Reformations, 1530–83.* Aldershot: Ashgate, 1999.

Binksi, Paul, and Stella Panayotova. *The Cambridge Illuminations: Ten Centuries of Book Production in the Medieval West.* London: Harvey Miller, 2005.

Black, William Henry. *A Descriptive, Analytical, and Critical Catalogue of the Manuscripts Bequeathed unto the University of Oxford by Elias Ashmole, Esq., M.D., F.R.S.* Oxford: Oxford University Press, 1845.

Blok, Frans Felix. *Contributions to the History of Isaac Vossius's Library.* Amsterdam: North-Holland, 1974.

Boeren, P. C. *Codices Vossiani chymici.* Leiden: Universitaire pers Leiden, 1975.

Brenner, Robert. *Merchants and Revolution: Commercial Change, Political Conflict, and London's Overseas Traders, 1550–1653.* London: Verso, 2003.

Brewer, J. S. *The Reign of Henry VIII from His Accession to the Death of Wolsey.* Ed. James Gairdner. 2 vols. London: John Murray, 1884.

Broddle, S. Merrow, comp. *Alchemical Works: Eirenaeus Philalethes Compiled.* Boulder, CO: Cinnabar, 1994.

Bryson, Alan. "Whalley, Richard (1498/9–1583)." *ODNB.*

Bueno, Mar Rey. "*La Mayson pour Distiller des Eaües* at El Escorial: Alchemy and Medicine at the Court of Philip II, 1556–1598." *Medical History* 29 (2009): 26–39.

Calendar of Entries in the Papal Registers Relating to Great Britain and Ireland: Papal Letters. Vol. 11, *1455–1464*, prepared by J. A. Twenlow. London: Her Majesty's Stationery Office, 1893.

Calendar of Letter-Books of the City of London. Vol. 1, *1400–1422*, ed. Reginald R Sharpe. London: His Majesty's Stationery Office, 1909.

Calendar of State Papers Domestic: Edward, Mary, and Elizabeth, 1547–80. London: Her Majesty's Stationery Office, 1856.

Calendar of State Papers Domestic: Elizabeth, Addenda, 1566–79. London: Her Majesty's Stationery Office, 1871.

Calis, Richard, Frederic Clark, Christian Flow, Anthony Grafton, Madeline McMahon, and Jennifer M. Rampling. "Passing the Book: Cultures of Reading in the Winthrop Family, 1580–1730." *Past and Present* 241 (2018): 69–141.

Calvet, Antoine. *Les oeuvres alchimiques attribuées à Arnaud de Villeneuve: Grand œuvre, médecine et prophétie au Moyen-Âge.* Paris: S.É.H.A.; Milan: Archè, 2011.

Campbell, Andrew, Lorenza Gianfrancesco, and Neil Tarrant, eds. "Alchemy and the Mendicant Orders of Late Medieval and Early Modern Europe." *Ambix* 65 (2018): 201–9.

Campbell, James Stuart. "The Alchemical Patronage of Sir William Cecil, Lord Burghley." Master's thesis, Victoria University of Wellington, 2009.

Carley, J. P., ed. *The Libraries of King Henry VIII.* London: The British Library in association with The British Academy, 2000.

———. "Monastic Collections and Their Dispersal." In *The Cambridge History of the Book in Britain*, vol. 4, *1557–1695*, ed. John Bernard and D. F. McKenzie, with Maureen Bell, 339–48. Cambridge: Cambridge University Press, 2002.

Carlson, David. "The Writings and Manuscript Collection of the Elizabethan Alchemist, Antiquary, and Herald, Francis Thynne." *Huntington Library Quarterly* 52 (1989): 203–72.

Carpenter, David. "Gold and Gold Coins in England in the Mid-Thirteenth Century." *Numismatic Chronicle* 147 (1987): 106–13.

Carusi, Paola. "*Animalis herbalis naturalis.* Considerazioni parallele sul 'De anima in arte alchimiae' attribuito ad Avicenna e sul '*Miftāh al-hikma*' (Opera di un allievo di Apollonia di Tiana)." *Micrologus* 3 (1995): 45–74.

Challis, C. E. "Mint Officials and Moneyers of the Tudor Period." *British Numismatic Journal* 45 (1975): 51–76.

———. *The Tudor Coinage.* New York: Manchester University Press, 1978.

Charlton, Kenneth. "Holbein's 'Ambassadors' and Sixteenth-Century Education." *Journal of the History of Ideas* 21 (1960): 99–109.

Clark, James G. "Reformation and Reaction at St Albans Abbey, 1530–58." *English Historical Review* 115 (2000): 297–328.

Clark, Stuart. *Thinking with Demons: The Idea of Witchcraft in Early Modern Europe.* Oxford: Oxford University Press, 1997.

Clericuzio, Antonio. *Elements, Principles, and Corpuscles: A Study of Atomism and Chemistry in the Seventeenth Century.* Dordrecht: Kluwer, 2000.

———. "From van Helmont to Boyle: A Study of the Transmission of Helmontian Chemical and Medical Theories in Seventeenth-Century England." *British Journal for the History of Science* 26 (1993): 303–34.

Clucas, Stephen, ed. *John Dee: Interdisciplinary Studies in English Renaissance Thought.* Dordrecht: Springer, 2006.

———. "John Dee's Angelic Conversations and the *Ars notoria*: Renaissance Magic and Mediaeval Theurgy." In Clucas, *John Dee*, 231–73.

Clulee, Nicholas H. "Astronomia inferior: Legacies of Johannes Trithemius and John Dee." In *Secrets of Nature: Astrology and Alchemy in Early Modern Europe*, ed. William R. Newman and Anthony Grafton, 173–233. Cambridge, MA: MIT Press, 2001.

———. *John Dee's Natural Philosophy: Between Science and Religion.* Oxford: Routledge, 1988.

Collette, Carolyn P., and Vincent DiMarco. "The Canon's Yeoman's Tale." In *Sources and Analogues of the Canterbury Tales*, vol. 2, ed. Robert M. Correale and Mary Hamel, 715–47. Cambridge: D. S. Brewer, 2005.

Connolly, Margaret. *Sixteenth-Century Readers, Fifteenth-Century Books: Continuities of Reading in the English Reformation.* Cambridge: Cambridge University Press, 2019.

Cook, Harold J. "The Society of Chemical Physicians, the New Philosophy, and the Restoration Court." *Bulletin of the History of Medicine* 61 (1987): 61–77.

Coote, Lesley A. *Prophecy and Public Affairs in Later Medieval England.* Woodbridge: York Medieval Press, 2000.

Corbett, J. A. *Catalogue des manuscrits alchimiques latins.* 2 vols. Paris: Office International de Labraire, 1939, 1951.

Cox, J. Charles. "The Religious Pension Roll of Derbyshire, temp. Edward VI." *Journal of the Derbyshire Archaeological and Natural History Society* 28 (1906): 10–43.

Crankshaw, David J., and Alexandra Gillespie. "Parker, Matthew (1504–1575)." *ODNB.*

Crisciani, Chiara. "Alchimia e potere: Presenze francescane (secoli XIII–XIV)." In *I Francescani e la politica: Atti del convegno internazionale di studio, Palermo 3–7 dicembre 2002*, ed. Alessandro Musco, 223–35. Palermo: Biblioteca Francescana—Officina di Studi Medievali, 2007.

———. "Opus and sermo: The Relationship between Alchemy and Prophecy (12th–14th Centuries)." *Early Science and Medicine* 13 (2008): 4–24.

Crisciani, Chiara, and Michela Pereira. "Black Death and Golden Remedies: Some Remarks on Alchemy and the Plague." In *The Regulation of Evil: Social and Cultural Attitudes to Epidemics in the Late Middle Ages*, ed. Agostino Paravicini Bagliani and Francesco Santi, 7–39. Florence: SISMEL, 1998.

Daniel, Dane Thor. "Invisible Wombs: Rethinking Paracelsus's Concept of Body and Matter." *Ambix* 53 (2006): 129–42.

Dapsens, Marion. "De la Risālat Maryānus au *De Compositione alchemiae*: Quelques réflexions sur la tradition d'un traité d'alchimie." *Studia graeco-arabica* 6 (2016): 121–40.

Debus, Allen G. *The Chemical Philosophy: Paracelsian Science and Medicine in the Sixteenth and Seventeenth Centuries*. 2 vols. New York: Science History Publications, 1977.

———. *The English Paracelsians*. London: Oldbourne Press, 1965.

Dee, John. *The Diaries of John Dee*. Ed. Edward Fenton. Charlbury: Day Books, 1998.

———. *The Private Diary of Dr. John Dee, and the Catalog of His Library of Manuscripts*. Ed. James Orchard Halliwell. London: Printed for the Camden Society, 1842.

De Schepper, Susanna L. B. "'Foreign' Books for English Readers: Published Translations of Navigation Manuals and Their Audience in the English Renaissance, 1500–1640." PhD diss., University of Warwick, 2012.

DeVun, Leah. *Prophecy, Alchemy, and the End of Time: John of Rupescissa in Medieval Europe*. New York: Columbia University Press, 2009.

Dewar, Mary. *Sir Thomas Smith, Tudor Intellectual in Office*. London: Athlone Press, 1964.

Dickins, Bruce. "The Making of the Parker Library." *Transactions of the Cambridge Bibliographical Society* 6 (1972–76): 19–34.

Dickinson, J. C. *The Origins of the Austin Canons and Their Introduction into England*. London: S.P.C.K., 1950.

Dodds, Madeleine Hope. "Political Prophecies in the Reign of Henry VIII." *Modern Language Review* 11 (1916): 276–84.

Donald, M. B. "Burchard Kranich (c. 1515–1578), Miner and Queen's Physician, Cornish Mining Stamps, Antimony, and Frobisher's Gold." *Annals of Science* 6 (1950): 308–22.

————. *Elizabethan Monopolies: The History of the Company of Mineral and Battery Works from 1565 to 1604*. Edinburgh: Oliver & Boyd, 1961.

————. "A Further Note on Burchard Kranich." *Annals of Science* 7 (1951): 107–8.

Dudík, Beda. *Iter romanum: Im Auftrage des hohen maehrischen Landesausschusses in den Jahren 1852 und 1853*. Pts. 1–2. Vienna: F. Manz, 1855.

Duncan, Edgar H. "The Literature of Alchemy and Chaucer's Canon's Yeoman's Tale: Framework, Theme, and Characters." *Speculum* 43 (1968): 633–56.

Dupré, Sven, ed. *Laboratories of Art: Alchemy and Art Technology from Antiquity to the 18th Century*. Cham: Springer, 2014.

Duwes, Giles. "An introductorie for to lerne to rede, to pronounce and to speke French trewly." In *L'éclaircissement de la langue française . . . la grammaire de Gilles Du Guez*, ed. F. Génin Paris: Imprimerie nationale, 1852.

Eamon, William. "Masters of Fire: Italian Alchemists in the Court of Philip II." In *Chymia: Science and Nature in Medieval and Early Modern Europe*, ed. Miguel López Pérez, Didier Kahn, and Mar Rey Bueno, 138–56. Newcastle-upon-Tyne: Cambridge Scholars, 2010.

————. *Science and the Secrets of Nature: Books of Secrets in Medieval and Early Modern Culture*. Princeton: Princeton University Press, 1994.

Eamon, William, and Gundolf Keil. "*Plebs amat empirica*: Nicholas of Poland and His Critique of the Medieval Medical Establishment." *Sudhoffs Archiv* 71 (1987): 180–96.

Easton, Stewart C. *Roger Bacon and His Search for a Universal Science: A Reconsideration of the Life and Work of Roger Bacon in the Light of His Own Stated Purposes*. Oxford: Blackwell, 1952.

Elton, G. R. *Policy and Police: The Enforcement of the Reformation in the Age of Thomas Cromwell*. Cambridge: Cambridge University Press, 1972.

Elzinga, J. G. "York, Sir John (*d.* 1569)." *ODNB*.

Emden, B. *A Biographical Register of the University of Oxford to A.D. 1500*. Vol. 3. Oxford: Clarendon Press, 1959.

Evans, John. "The First Gold Coins of England." *Numismatic Chronicle and Journal of the Numismatic Society* 20 (1900): 218–51.

Evans, R. J. W. *Rudolf II and His World: A Study in Intellectual History, 1576–1612*. Oxford, 1973. Reprint, London: Thames & Hudson, 1997.

Farmer, David L. "Prices and Wages, 1350–1500." In *The Agrarian History of England and Wales*, vol. 3, *1348–1500*, ed. Edward Miller, 431–525. Cambridge: Cambridge University Press, 1991.

Farmer, S. A., ed. and trans. *Syncretism in the West: Pico's 900 Theses (1486); The Evolution of Traditional Religious and Philosophical Systems*. Tempe, AZ: Medieval & Renaissance Texts & Studies, 1998.

Feingold, Mordechai. "The Occult Tradition in the English Universities of the Renaissance: A Reassessment." In *Occult and Scientific Mentalities in the Renaissance*, ed. Brian Vickers, 73–94. Cambridge: Cambridge University Press, 1984.

Feola, Vittoria. *Elias Ashmole and the Uses of Antiquity*. Paris: Librairie Blanchard, 2012.

———. "Elias Ashmole's Collections and Views about John Dee." *Studies in History and Philosophy of Science* 43 (2012): 530–38.

Ficino, Marsilio. *Three Books on Life: A Critical Edition and Translation*. Ed. and trans. Carol V. Kaske and John R. Clark. Binghamton, NY: Medieval & Renaissance Texts & Studies in Conjunction with the Renaissance Society of America, 1989.

Findlen, Paula. *Possessing Nature: Museums, Collecting, and Scientific Culture in Early Modern Italy*. Berkeley: University of California Press, 1994.

Flynn, Colin George. "The Decline and End of the Lead-Mining Industry in the Northern Pennines, 1865–1914: A Socio-Economic Comparison between Wensleydale, Swaledale, and Teesdale." PhD diss., Durham University, 1999.

Forbes, R. J. *A Short History of the Art of Distillation*. Leiden: Brill, 1970.

Ford, L. L. "Audley, Thomas, Baron Audley of Walden (1487/8–1544)." *ODNB*.

Forshaw, Peter J. "Cabala Chymica or Chemia Cabalistica—Early Modern Alchemists and Cabala." *Ambix* 60 (2013): 361–89.

———. "'Chemistry, that Starry Science': Early Modern Conjunctions of Astrology and Alchemy." In *Sky and Symbol*, ed. Nicholas Campion and Liz Greene, 143–84. Lampeter: Sophia Centre Press, 2013.

Foster, Joseph, ed. *Alumni Oxonienses, 1500–1714*. 4 vols. Oxford: Parker and Co., 1891–92.

Fowler, R. C. "Alchemy in Essex." In *The Essex Review: An Illustrated Quarterly Record of Everything of Permanent Interest in the County*, vol. 16, ed. Edward A. Fitch and C. Fell Smith, 158–59. Colchester: Behnam & Co., 1907.

Foxe, John. *The Unabridged Acts and Monuments Online (1570 edition)*. 1587. Sheffield: HRI Online Publications, 2011. http//www.johnfoxe.org.

French, Peter J. *John Dee: The World of the Elizabethan Magus*. London: Routledge & Kegan Paul, 1972.

French, Roger. *Canonical Medicine: Gentile da Foligno and Scholasticism*. Leiden: Brill, 2001.

———. *Medicine before Science: The Business of Medicine from the Middle Ages to the Enlightenment*. Cambridge: Cambridge University Press, 2003.

French, Roger, and Andrew Cunningham, *Before Science: The Invention of the Friars' Natural Philosophy*. Aldershot: Ashgate, 1996.

Fuller, Thomas. *The History of the Worthies of England: A New Edition.* Ed. P. Austin Nuttall. 2 vols. London: Thomas Tegg, 1840.

Galluzzi, Paolo. "Motivi paracelsiani nella Toscana di Cosimo II e di Don Antonio de' Medici: Alchimia, medicina, 'chimica' e riforma del sapere." In *Scienze, credenze occulte, livelli di cultura,* ed. Paola Zambelli, 189–215. Florence: Olschki, 1982.

Gaskill, Malcolm. "Witchcraft and Evidence in Early Modern England." *Past and Present* 198 (2008): 33–70.

Gentilcore, David. *Medical Charlatanism in Early Modern Italy.* Oxford: Oxford University Press, 2006.

Geoghegan, D. "A Licence of Henry VI to Practise Alchemy." *Ambix* 6 (1957): 10–17.

Getz, Faye. "Kymer, Gilbert (d. 1463)." *ODNB.*

———. *Medicine in the English Middle Ages.* Princeton: Princeton University Press, 1998.

———. "Mirfield, John [Johannes de Mirfeld] (d. 1407)." *ODNB.*

———. "To Prolong Life and Promote Health: Baconian Alchemy and Pharmacy in the English Learned Tradition." In *Health, Disease, and Healing in Medieval Culture,* ed. Sheila Campbell, Bert Hall, and David Klausner, 141–50. New York: Palgrave, 1992.

Gilly, Carlos. "On the Genesis of L. Zetzner's *Theatrum Chemicum* in Strasbourg." In *Magia, alchimia, scienza dal '400 al '700: L'influsso di Ermete Trismegisto,* ed. Carlos Gilly and Cis van Heertum, 1:451–67. Florence: Centro Di, 2002.

Given-Wilson, Chris. *Chronicles: The Writing of History in Medieval England.* London: Hambledon Continuum, 2004.

Goltz, Dietlinde. *Studien zur Geschichte der Mineralnamen in Pharmazie, Chemie und Medizin von den Anfängen bis Paracelsus.* Wiesbaden: Franz Steiner, 1972.

Goodman, David C. *Power and Penury: Government, Technology, and Science in Philip II's Spain.* Cambridge: Cambridge University Press, 1988.

Gould, J. D. *The Great Debasement: Currency and the Economy in Mid-Tudor England.* Oxford: Clarendon Press, 1970.

Gower, John. *Confessio Amantis.* Vol. 2, ed. Russell A. Peck, trans. Andrew Galloway. 2nd ed. Kalamazoo, MI: Medieval Institute Publications, 2013.

Grafton, Anthony. "Church History in Early Modern Europe: Tradition and Innovation." In *Sacred History: Uses of the Christian Past in the Renaissance World,* ed. Katherine Van Liere et al., 3–26. Oxford: Oxford University Press, 2012.

———. *Commerce with the Classics: Ancient Books and Renaissance Readers.* Ann Arbor: University of Michigan Press, 1997.

———. *Defenders of the Text: The Traditions of Humanism in an Age of Science, 1450–1800.* Cambridge, MA: Harvard University Press, 1991.

———. *Joseph Scaliger: A Study in the History of Classical Scholarship*. Vol. 1, *Textual Criticism and Exegesis*. Oxford: Clarendon Press, 1983.

———. "Matthew Parker: The Book as Archive." *History of Humanities* 2 (2017): 15–50.

Grafton, Anthony, and Megan Williams. *Christianity and the Transformation of the Book: Origen, Eusebius, and the Library of Caesarea*. Cambridge, MA: Harvard University Press, 2008.

Grant, Edward. *The Foundations of Modern Science in the Middle Ages: Their Religious, Institutional, and Intellectual Contexts*. Cambridge: Cambridge University Press, 1996.

———. *God and Reason in the Middle Ages*. Cambridge: Cambridge University Press, 2009.

———. "Medieval Natural Philosophy: Empiricism without Observation." In *The Nature of Natural Philosophy in the Late Middle Ages*, 195–224. Washington, DC: Catholic University of America Press, 2010.

Green, Richard Firth. *Poets and Princepleasers: Literature and the English Court in the Late Middle Ages*. Toronto: Toronto University Press, 1980.

Grund, Peter J. "Albertus Magnus and the Queen of the Elves: A 15th-Century English Verse Dialogue on Alchemy." *Anglia: Zeitschrift für englische Philologie* 122 (2004): 640–62.

———. "'Ffor to make Azure as Albert biddes': Medieval English Alchemical Writings in the Pseudo-Albertan Tradition." *Ambix* 53 (2006): 21–42.

———. "The Golden Formulas: Genre Conventions of Alchemical Recipes in the Middle English Period." *Neuphilologische Mitteilungen* 104.4 (2003): 455–75.

———. *"Misticall Wordes and Names Infinite": An Edition and Study of Humfrey Lock's Treatise on Alchemy*. Tempe: Arizona Center for Medieval and Renaissance Studies, 2011.

———. "A Previously Unrecorded Fragment of the Middle English Short Metrical Chronicle in Bibliotheca Philosophica Hermetica M199." *English Studies* 87 (2006): 277–93.

Gwei-Djen, Lu. Joseph Needham, and Dorothy Needham. "The *Coming of Ardent Water*." *Ambix* 19 (1972): 69–112.

Gwyn, David. "Richard Eden, Cosmographer and Alchemist." *Sixteenth Century Journal* 15 (1984): 13–34.

Hadfield, Andrew. "Eden, Richard (*c*.1520–1576)." *ODNB*.

Halleux, Robert. "Les ouvrages alchimiques de Jean de Rupescissa." *Histoire littéraire de la France* 41 (1981): 241–77.

———. *Les textes alchimiques*. Turnhout: Brepols, 1979.

———. "The Reception of Arabic Alchemy in the West." In *Encyclopedia of the*

History of Arabic Science, ed. Roshdi Rashed, 3:886–902. London: Routledge, 1996.

Hamilton, Marie P. "The Clerical Status of Chaucer's Alchemist." *Speculum* 16 (1941): 103–8.

Hammond, Frederick. "Odington, Walter (*fl. c.*1280–1301)." *ODNB*.

Handl, Jacob. *The Moralia of 1596*. Ed. Allen B. Skei. Middleton, WI: Madison, A-R Editions, 1970.

Harkness, Deborah E. *The Jewel House: Elizabethan London and the Scientific Revolution*. New Haven: Yale University Press, 2007.

———. *John Dee's Conversations with Angels: Cabala, Alchemy, and the End of Nature*. Cambridge: Cambridge University Press. 1999.

———. "Managing an Experimental Household: The Dees of Mortlake and the Practice of Natural Philosophy." *Isis* 88 (1997): 247–62.

Harley, David. "Rychard Bostok of Tandridge, Surrey (c. 1530–1605), M.P., Paracelsian Propagandist and Friend of John Dee." *Ambix* 47 (2000): 29–36.

Hatcher, John. "Plague, Population, and the English Economy, 1348–1530." In *British Population History: From the Black Death to the Present Day*, ed. Michael Anderson, 9–94. Cambridge: Cambridge University Press, 1996.

Haye, Thomas. *Das lateinische Lehrgedicht im Mittelalter: Analyse einer Gattung*. Leiden: Brill, 1997.

Heale, Martin. *The Abbots and Priors of Late Medieval and Reformation England*. New York: Oxford University Press, 2016.

Hedesan, Georgiana D. *An Alchemical Quest for Universal Knowledge: The "Christian Philosophy" of Jan Baptist Van Helmont (1579–1644)*. Oxford: Routledge, 2016.

Henry, John. "Occult Qualities and Experimental Philosophy." *History of Science* 24 (1986): 335–81.

Hicks, Michael. "Neville, George (1432–1476)." *ODNB*.

Hinckley, Marlis Ann. "Diagrams and Visual Reasoning in Pseudo-Lullian Alchemy, 1350–1500." MSt thesis, King's College, University of Cambridge, 2017.

Hobbins, Daniel. *Authorship and Publicity before Print: Jean Gerson and the Transformation of Late Medieval Learning*. Philadelphia: University of Pennsylvania Press, 2013.

Hogart, Ron Charles, ed. *Alchemy: A Comprehensive Bibliography of the Manly P. Hall Collection of Books and Manuscripts; Including Related Material on Rosicrucianism and the Writings of Jacob Böhme*. Intro. Manly P. Hall. Los Angeles: The Philosophical Research Society, 1986.

Houlbrooke, Ralph. "Parkhurst, John (1511?–1575)." *ODNB*.

Houston, R. A. *Literacy in Early Modern Europe: Culture and Education, 1500–1800.* London: Longman, 1988.

Hughes, Jonathan. *Arthurian Myths and and Alchemy: The Kingship of Edward IV.* Stroud: Sutton Publishing, 2002.

———. *The Rise of Alchemy in Fourteenth-Century England: Plantagenet Kings and the Search for the Philosopher's Stone.* London: Continuum, 2012.

Hutchison, Keith. "What Happened to Occult Qualities in the Scientific Revolution?" *Isis* 73 (1982): 233–53.

Isidore of Seville. *Isidori Hispalensis episcopi Etymologiarum sive originum libri XX.* Ed. W. M. Lindsay. Oxford: Oxford University Press, 1911.

James, M. R. *A Descriptive Catalogue of the Manuscripts in the Library of Corpus Christi College, Cambridge.* 2 vols. Cambridge: Cambridge University Press, 1912.

———. *The Sources of Archbishop Parker's Collection of MSS at Corpus Christi College, Cambridge.* Cambridge: Cambridge Antiquarian Society, 1899.

———. *The Western Manuscripts in the Library of Trinity College Cambridge: A Descriptive Catalogue.* Vol. 3. Cambridge: Cambridge University Press, 1902.

Janacek, Bruce. *Alchemical Belief: Occultism in the Religious Culture of Early Modern England.* University Park: Pennsylvania State University Press, 2011.

Jardine, Lisa, and Anthony Grafton. "'Studied for Action: How Gabriel Harvey Read His Livy.'" *Past and Present* 129 (1990): 30–78.

Jones, Peter Murray. "Alchemical Remedies in Late Medieval England." In *Alchemy and Medicine from Antiquity to the Eighteenth Century,* ed. Jennifer M. Rampling and Peter M. Jones. London: Routledge, forthcoming.

———. "Complexio and Experimentum: Tensions in Late Medieval English Practice." In *The Body in Balance: Humoral Medicines in Practice,* ed. Peregrine Horden and Elizabeth Hsu, 107–28. New York: Berghahn, 2013.

———. "Four Middle English Translations of John of Arderne." In *Latin and Vernacular: Studies in Late-Medieval Texts and Manuscripts,* ed. A. J. Minnis, 61–89. Cambridge: D. S. Brewer, 1989.

———. "Mediating Collective Experience: The *Tabula Medicine* (1416–1425) as a Handbook for Medical Practice." In *Between Text and Patient: The Medical Enterprise in Medieval and Early Modern Europe,* ed. Florence Eliza Glaze and Brian K. Nance, Micrologus' Library 39, 279–307. Florence: SISMEL, 2011.

———. "The Survival of the *Frater Medicus*? English Friars and Alchemy, ca. 1370–ca. 1425." *Ambix* 65 (2018): 232–49.

Jordan, William Chester. *The Great Famine: Northern Europe in the Early Fourteenth Century.* Princeton: Princeton University Press, 1997.

Josten, C. H., ed. *Elias Ashmole: His Autobiographical and Historical Notes, His Correspondence, and Other Contemporary Sources Relating to His Life and Work.* 5 vols. Oxford: Oxford University Press, 1967.

Kahn, Didier. *Alchimie et Paracelsime en France à la fin de la Renaissance (1567–1625).* Geneva: Librairie Droz, 2007.

———. "Littérature et alchimie au Moyen Age: De quelques textes alchimiques attribués à Arthur et Merlin." *Micrologus* 3 (1995): 227–62.

———. "The *Turba philosophorum* and Its French Version (15th c.)." In *Chymia: Science and Nature in Medieval and Early Modern Europe*, ed. Miguel López Pérez, Didier Kahn, and Mar Rey Bueno, 70–114. Newcastle-upon-Tyne: Cambridge Scholars, 2010.

Kahn, Didier, and Hiro Hirai, eds. "Pseudo-Paracelsus: Forgery and Early Modern Alchemy, Medicine and Natural Philosophy." *Early Science and Medicine* 5–6 (2020): 415–575.

Kaplan, Edward. "Robert Recorde (c. 1510–1558): Studies in the Life and Works of a Tudor Scientist." PhD diss., New York University, 1960.

Karpenko, Vladimír, and Ivo Purš. "Edward Kelly: A Star of the Rudolfine Era." In *Alchemy and Rudolf II: Exploring the Secrets of Nature in Central Europe in the 16th and 17th Centuries*, ed. Ivo Purš and Vladimír Karpenko, 489–534. Prague: Artefactum, 2016.

Kassell, Lauren. "The Economy of Magic in Early Modern England." In *The Practice of Reform in Health, Medicine, and Science, 1500–2000: Essays for Charles Webster*, ed. Margaret Pelling and Scott Mandelbrote, 43–57. Aldershot: Ashgate, 2005.

———. *Medicine and Magic in Elizabethan London: Simon Forman: Astrologer, Alchemist, and Physician.* Oxford: Clarendon Press, 2005.

———. "Reading for the Philosophers' Stone." In *Books and the Sciences in History*, ed. Marina Frasca-Spada and Nick Jardine, 132–50. Cambridge: Cambridge University Press, 2000.

———. "Secrets Revealed: Alchemical Books in Early-Modern England." *History of Science* 48 (2011): 1–27, A1–38.

Keiser, George R. "Preserving the Heritage: Middle English Verse Treatises in Early Modern Manuscripts." In *Mystical Metal of Gold: Essays on Alchemy and Renaissance Culture*, ed. Stanton J. Linden, 189–214. New York: AMS, 2007.

Kelley, Edward. *The Alchemical Writings of Edward Kelly.* Trans. Arthur Edward Waite. London: James Elliott, 1893.

Kempers, Bram. *Painting, Power, and Patronage: The Rise of the Professional Artist in the Italian Renaissance.* Trans. Beverley Jackson. London: Penguin, 1984.

Kendrick, T. D. *British Antiquity*. New York: Barnes & Noble, 1950.

Kieckhefer, Richard. *Forbidden Rites: A Necromancer's Manual of the Fifteenth Century*. University Park: Pennsylvania State University Press, 1998.

————. *Magic in the Middle Ages*. New York: Cambridge University Press, 1989.

Kieffer, Fanny. "The Laboratories of Art and Alchemy at the Uffizi Gallery in Renaissance Florence: Some Material Aspects." In *Laboratories of Art: Alchemy and Art Technology from Antiquity to the 18th Century*, ed. Sven Dupré, 105–27. Cham: Springer, 2014.

Kipling, Gordon. "Duwes [Dewes], Giles [pseud. Aegidius de Vadis] (d. 1535)." *ODNB*.Kitching, Christopher. "Alchemy in the Reign of Edward VI: An Episode in the Careers of Richard Whalley and Richard Eden." *Bulletin of the Institute of Historical Research* 44 (1971): 308–15.

Klaassen, Frank. *The Transformations of Magic: Illicit Learned Magic in the Later Middle Ages and Renaissance*. University Park: Pennsylvania State University Press, 2013.

Knafla, Louis A. "Thynne, Francis (1545?–1608)." *ODNB*.

Knighton, C. S. "Freake, Edmund (*c.* 1516–1591)." *ODNB*.

Knowles, David, and R. Neville Hadcock. *Medieval Religious Houses: England and Wales*. London: Longman, 1971.

Kocher, Paul H. "Paracelsan Medicine in England: The First Thirty Years (ca. 1570–1600)." *Journal of the History of Medicine* 2 (1947): 451–80.

Kraus, Paul. *Jābir b. Ḥayyān: Contribution à l'histoire des idées scientifiques dans l'Islam*. 2 vols. Cairo: Institut français d'archéologie orientale, 1943.

Kühlmann, Wilhelm, and Joachim Telle, eds. *Alchemomedizinische Briefe, 1585 bis 1597*. Stuttgart: Franz Steiner, 1998.

————. *Corpus Paracelsisticum: Dokumente frühneuzeitlicher Naturphilosophie in Deutschland*. Tübingen: Max Niemeyer, 2001–.

Lambley, Kathleen. *The Teaching and Cultivation of the French Language in England during Tudor and Stuart Times*. Manchester: Manchester University Press, 1920.

Langland, William. *The Vision of Piers Plowman: A Critical Edition of the B-Text Based on Trinity College Cambridge MS B.15.17*. Ed. A. V. C. Schmidt. 2nd ed. London: J. M. Dent, 1995.

Lehrich, Christopher I. *The Language of Demons and Angels: Cornelius Agrippa's Occult Philosophy*. Leiden: Brill, 2003.

Lenke, Nils, Nicolas Roudet, and Hereward Tilton. "Michael Maier—Nine Newly Discovered Letters." *Ambix* 61 (2014): 1–47.

Leong, Elaine. "'Herbals she peruseth': Reading Medicine in Early Modern England." *Renaissance Studies* 28 (2014): 556–78.

————. *Recipes and Everyday Knowledge: Medicine, Science, and the Household in Early Modern England.* Chicago: University of Chicago Press, 2019.

Leong, Elaine, and Alisha Rankin, eds. *Secrets and Knowledge in Medicine and Science, 1500–1800.* Farnham: Ashgate, 2011.

Lerer, Seth. *Boethius and Dialogue: Literary Method in "The Consolation of Philosophy."* Princeton: Princeton University Press, 1985.

Letters and Papers, Foreign and Domestic, Henry VIII. Vol. 5, *1531–1532,* ed. James Gairdner. London: Her Majesty's Stationery Office, 1880.

Letters and Papers, Foreign and Domestic, Henry VIII. Vol. 7, *1534 ,* ed. James Gairdner. London: Her Majesty's Stationery Office, 1883.

Letters and Papers, Foreign and Domestic, Henry VIII. Vol. 14, pt. 1, *January–July 1539,* ed. James Gairdner and R. H. Brodie. London: Her Majesty's Stationery Office, 1894.

Letters and Papers, Foreign and Domestic, Henry VIII. Vol. 14, pt. 2, *August– December 1539,* ed. James Gairdner and R. H. Brodie. London: Her Majesty's Stationery Office, 1895.

Letters and Papers, Foreign and Domestic, Henry VIII. Vol. 21, pt. 2, *September 1546–January 1547,* ed. James Gairdner and R. H. Brodie. London: His Majesty's Stationery Office, 1910.

Levitin, Dmitri. *Ancient Wisdom in the Age of the New Science: Histories of Philosophy in England, c. 1640–1700.* Cambridge: Cambridge University Press, 2015.

Linden, Stanton J. *Darke Hieroglyphicks: Alchemy in English Literature from Chaucer to the Restoration.* Lexington: University Press of Kentucky, 1996.

Lippmann, Edmund O. von. "Thaddäus Florentinus [Taddeo Alderotti] über den Weingeist." *Archiv für Geschichte der Medizin* 7 (1913–14): 379–89.

Long, Pamela O. *Artisan/Practitioners and the Rise of the New Sciences, 1400–1600.* Corvallis: Oregon State University Press, 2011.

————. *Openness, Secrecy, Authorship: Technical Arts and the Culture of Knowledge from Antiquity to the Renaissance.* Baltimore: Johns Hopkins University Press, 2001.

Lubac, Henri de. *Exégèse médiévale: Les quatre sens de l'Écriture.* 4 vols. Paris: Aubier, 1959, 1961, 1964.

————. *Medieval Exegesis: The Fourfold Sense of Scripture.* Trans. Mark Sebanc (vol. 1), Edward M. Macierowski (vols. 2 and 3). Grand Rapids, MI: Eerdmans, 1998–2009.

Lüthy, Christoph, John E. Murdoch, and William R. Newman, eds. *Late Medieval and Early Modern Corpuscular Matter Theories.* Leiden: Brill, 2001.

Lyon, Harriet K. "The Afterlives of the Dissolution of the Monasteries, 1536–c. 1700." PhD diss., University of Cambridge, 2018.

Maclean, Ian. *Interpretation and Meaning in the Renaissance: The Case of Law.* Cambridge: Cambridge University Press, 1992.

Maddison, Francis, Margaret Pelling, and Charles Webster. *Essays on the Life and Work of Thomas Linacre, c. 1460–1524.* New York: Oxford University Press, 1977.

Mandelbrote, Scott. "Norton, Samuel (1548–1621)." *ODNB.*

Mandosio, Jean-Marc. "L'alchimie dans les classifications des sciences et des arts à la Renaissance." In *Alchimie et philosophie à la Renaissance,* ed. Jean-Claude Margolin and Sylvain Matton, 11–41. Paris: Vrin, 1993.

Manzalaoui, Mahmoud. "The Pseudo-Aristotelian Kitab Sirr al-asrar: Facts and Problems." *Oriens* 23–24 (1974 [1970–71]): 148–257.

Marshall, Peter. "Forgery and Miracles in the Reign of Henry VIII." *Past and Present* 178 (2003): 39–73.

Martelli, Matteo, ed. and trans. *The Four Books of Pseudo-Democritus.* Sources of Alchemy and Chemistry 1. Leeds: Maney, 2013.

Martels, Zweder von. "Augurello's 'Chrysopoeia' (1515)—A Turning Point in the Literary Tradition of Alchemical Texts." *Early Science and Medicine* 5 (2000): 178–95.

Matton, Sylvain. "Marsile Ficin et l'alchimie: Sa position, son influence." In *Alchimie et philosophie à la Renaissance: Actes du colloque international de Tours (4–7 décembre 1991),* ed. Sylvain Matton and Jean-Claude Margolin, 123–92. Paris: Vrin, 1993.

Matus, Zachary A. *Franciscans and the Elixir of Life: Religion and Science in the Later Middle Ages.* Philadelphia: University of Pennsylvania Press, 2017.

McCallum, R. Ian. *Antimony in Medical History: An Account of the Medical Uses of Antimony and Its Compounds since Early Times to the Present.* Edinburgh: Pentland Press, 1999.

McKisack, Mary. *Medieval History in the Tudor Age.* Oxford: Clarendon Press, 1971.

McLean, Antonia. *Humanism and the Rise of Science in Tudor England.* New York: Heinemann, 1972.

McVaugh, Michael. "The Nature and Limits of Medical Certitude at Early Fourteenth-Century Montpellier." *Osiris,* 2nd ser., 6 (1990): 62–84.

Mitchiner, M. B., and A. Skinner. "Contemporary Forgeries of English Silver Coins and Their Chemical Compositions: Henry III to William III." *Numismatic Chronicle* 145 (1985): 209–36.

Molland, George, "Bacon, Roger (c.1214–1292?)," *ODNB.*

Montford, Angela. *Health, Sickness, Medicine, and the Friars in the Thirteenth and Fourteenth Centuries.* Aldershot: Ashgate, 2004.

Moorat, S. A. J. *Catalogue of Western Manuscripts on Medicine and Science in the Wellcome Historical Medical Library*, 2 vols. London: Wellcome Institute for the History of Medicine, 1962–73.

Moore, Norman. "Moundeford, Thomas (1550–1630)." Rev. Patrick Wallis. *ODNB*.

Moran, Bruce T. *The Alchemical World of the German Court: Occult Philosophy and Chemical Medicine in the Circle of Moritz of Hessen (1572–1632)*. Stuttgart: Franz Steiner Verlag, 1991.

———. *Andreas Libavius and the Transformation of Alchemy: Separating Chemical Cultures with Polemical Fire*. Sagamore Beach, MA: Science History Publications, 2007.

Moureau, Sébastien. "*Elixir Atque Fermentum*: New Investigations about the Link between Pseudo-Avicenna's Alchemical *De anima* and Roger Bacon; Alchemical and Medical Doctrines." *Traditio: Studies in Ancient and Medieval Thought, History, and Religion* 68 (2013): 277–323.

———, ed. *La "De anima" alchimique du pseudo-Avicenne*. 2 vols. Florence: SISMEL, 2016.

———. "Les sources alchimiques de Vincent de Beauvais." *Spicæ: Cahiers de l'Atelier Vincent de Beauvais*, n.s., 2 (2012): 5–118.

———. "*Min Al-Kīmiyāʾ Ad Alchimiam*: The Transmission of Alchemy from the Arab-Muslim World to the Latin West in the Middle Ages." In *The Diffusion of the Islamic Sciences in the Western World*, ed. Agostino Paravicini Bagliani, Micrologus' Library 28, 87–142. Florence: SISMEL, 2020.

———. "Questions of Methodology about Pseudo-Avicenna's *De anima in arte alchemiae*: Identification of a Latin Translation and Method of Edition." In *Chymia: Science and Nature in Medieval and Early Modern Europe*, ed. Miguel López Pérez, Didier Kahn, and Mar Rey Bueno, 1–18. Newcastle-upon-Tyne: Cambridge Scholars, 2010.

———. "Some Considerations Concerning the Alchemy of the *De anima in arte alchemiae* of Pseudo-Avicenna." *Ambix* 56 (2009): 49–56.

Mout, Nicolette. "Books from Prague: The Leiden *Codices Vossiani chymici* and Rudolf II." In *Prag um 1600: Beiträge zur Kunst und Kultur am Hofe Rudolfs II*, ed. E. Fuécíková, 205–10. Freren: Luca Verlag, 1988.

Multhauf, Robert P. "John of Rupescissa and the Origin of Medical Chemistry." *Isis* 45 (1954): 359–67.

Mundill, Robin R. *England's Jewish Solution: Experiment and Expulsion, 1262–1290*. New York: Cambridge University Press, 1998.

Newman, William R. "The Alchemy of Roger Bacon and the *Tres Epistolae* Attributed to Him." In *Comprendre et maîtriser la nature au moyen age:*

Mélanges d'histoire des sciences offerts à Guy Beaujouan, 461–79. Geneva: Librarie Droz, 1994.

———. *Atoms and Alchemy: Chymistry and the Experimental Origins of the Scientific Revolution*. Chicago: University of Chicago Press, 2006.

———. *Gehennical Fire: The Lives of George Starkey, an American Alchemist in the Scientific Revolution*. Cambridge, MA: Harvard University Press, 1994.

———. "The Genesis of the *Summa perfectionis*." *Archives internationales d'histoire des sciences* 35 (1985): 240–302.

———. "New Light on the Identity of Geber." *Sudhoffs Archiv für die Geschichte der Medizin und der Naturwissenschaften* 69 (1985): 76–90.

———. *Newton the Alchemist: Science, Enigma, and the Quest for Nature's "Secret Fire."* Princeton: Princeton University Press, 2018.

———. "Newton's Reputation as an Alchemist and the Tradition of Chymiatria." In *Reading Newton in Early Modern Europe*, ed. Elizabethanne A. Boran and Mordechai Feingold, 313–27. Leiden: Brill, 2017.

———. "The Philosophers' Egg: Theory and Practice in the Alchemy of Roger Bacon." *Micrologus* 3 (1995): 75–101.

———. *Promethean Ambitions: Alchemy and the Quest to Perfect Nature*. Chicago: University of Chicago Press, 2004.

———. "Starkey's *Clavis* as Newton's *Key*." *Isis* 78 (1987): 564–74.

———, ed. *The Summa perfectionis of Pseudo-Geber: A Critical Edition, Translation, and Study*. Leiden: Brill, 1991.

———. "Technology and Alchemical Debate in the Late Middle Ages." *Isis* 80 (1989): 423–45.

Newman, William R., and Anthony Grafton, eds. *Secrets of Nature: Astrology and Alchemy in Early Modern Europe*. Cambridge, MA: MIT Press, 2001.

Newman, William R., and Lawrence M. Principe. *Alchemy Tried in the Fire: Starkey, Boyle, and the Fate of Helmontian Chymistry*. Chicago: University of Chicago Press, 2002.

———. "Alchemy vs. Chemistry: The Etymological Origins of a Historiographic Mistake." *Early Science and Medicine* 3 (1998): 32–65.

Nichols, John Gough. *Narratives of the Days of the Reformation, Chiefly from the Manuscripts of John Foxe, Martyrologist*. London: Camden Society, 1849.

Norrgrén, Hilde. "Interpretation and the Hieroglyphic Monad: John Dee's Reading of Pantheus's Voarchadumia." *Ambix* 52 (2005): 217–45.

Norris, John A. "The Mineral Exhalation Theory of Metallogenesis in Pre-modern Mineral Science." *Ambix* 53 (2006): 43–65.

North, J. D. *The Ambassadors' Secret: Holbein and the World of the Renaissance*. London: Hambledon, 2002.

————. "Chronology and the Age of the World." In *Stars, Minds, and Fate: Essays in Ancient and Medieval Cosmology*, 91–117. London: Hambledon, 1989.

Norton, Thomas. *The Ordinal of Alchemy*. Ed. John Reidy. Oxford: Early English Text Society, 1975.

Nothaft, Carl Philipp Emanuel. "Walter Odington's De etate mundi and the Pursuit of a Scientific Chronology in Medieval England." *Journal of the History of Ideas* 77 (2016): 183–201.

Nummedal, Tara. *Alchemy and Authority in the Holy Roman Empire*. Chicago: University of Chicago Press, 2007.

Obrist, Barbara. "Alchemy and Secret in the Latin Middle Ages." In *D'un principe philosophique à un genre littéraire: Les secrets; Actes du colloque de la Newberry Library de Chicago, 11–14 Septembre 2002*, ed. D. de Courcelles, 57–78. Paris: Champion, 2005.

————. "Art et nature dans l'alchimie médiévale." *Revue d'histoire des sciences* 49 (1996): 215–86.

————. "Nude Nature and the Art of Alchemy in Jean Perréal's Early Sixteenth-Century Miniature." In *Chymists and Chymistry*, ed. Lawrence M. Principe, 113–24. Sagamore Beach, MA: Science History Publications, 2006.

————. "Visualization in Medieval Alchemy." *HYLE—International Journal for Philosophy of Chemistry* 9 (2003): 131–70. www.hyle.org/journal/issues/9-2/obrist.htm.

Oliver, Eugène. "Bernard G[illes] Penot (Du Port), médecin et alchimiste." Ed. Didier Kahn. *Chrysopoeia* 5 (1992–96): 571–667.

Omont, H. "Les manuscrits français des rois d'Angleterre au château de Richmond." In *Études romanes dédiés à Gaston Paris*, 1–13. Paris: É. Bouillon, 1891.

Otten, Willemien. "The Return to Paradise: Role and Function of Early Medieval Allegories of Nature." In *The Book of Nature in Antiquity and the Middle Ages*, ed. A. Vanderjagt and K. VanBerkel, 97–121. Leuven: Peeters, 2005.

Ovid. *Fastorum libri sex: The Fasti of Ovid*. Vol. 3, *Commentary on Books 3 and 4*. Ed. and trans. James George Frazer. Cambridge: Cambridge University Press, 1929.

Oxley, James Edwin. *The Reformation in Essex to the Death of Mary*. Manchester: Manchestesr University Press, 1965.

Page, R. I. *Matthew Parker and His Books*. Kalamazoo, MI: Medieval Institute Publications, 1993.

Page, Sophie. *Magic in the Cloister: Pious Motives, Illicit Interests, and Occult Approaches to the Medieval Universe*. University Park: Pennsylvania State University Press, 2013.

Page, William, ed. *A History of the County of Somerset.* Vol. 2. London: Victoria County History, 1911.

Page, William, and J. Horace Round, eds. *A History of the County of Essex.* Vol. 2. London: Constable, 1907.

Park, Katherine. "Observation in the Margins, 500–1500." In *Histories of Scientific Observation,* ed. Lorraine Daston and Elizabeth Lunbeck, 15–44. Chicago: University of Chicago Press, 2011.

Parry, Glyn. *The Arch-Conjuror of England: John Dee and Magic at the Courts of Renaissance Europe.* New Haven: Yale University Press, 2012.

Parry, Graham. *The Trophies of Time: English Antiquarians of the Seventeenth Century.* Oxford: Oxford University Press, 1995.

Parry, William T., and Edward A. Hacker. *Aristotelian Logic.* Albany: State University of New York Press, 1991.

Parsons, Jotham. *Making Money in Sixteenth-Century France: Currency, Culture, and the State.* Ithaca, NY: Cornell University Press, 2014.

Pastorino, Cesare. "Weighing Experience: Francis Bacon, the Inventions of the Mechanical Arts, and the Emergence of Modern Experiment." PhD diss., Indiana University Bloomington, 2011.

Patten, Jonathan K. van. "Magic, Prophecy, and the Law of Treason in Reformation England." *American Journal of Legal History* 27 (1983): 1–32.

Pejml, Karel. *Dějiny české alchymie.* Prague: Litomyši, 1933.

Pereira, Michela. *The Alchemical Corpus Attributed to Raymond Lull.* London: Warburg Institute, 1989.

———. "Alchemy and the Use of Vernacular Languages in the Late Middle Ages." *Speculum* 74 (1999): 336–56.

———. "Arnaldo da Vilanova e l'alchimia: Un'indagine preliminare." In *Actes de la I Trobada internacional d'estudis sobre Arnau de Vilanova,* vol. 2, ed. Josep Perarnau, 95–174. Barcelona: Institut d'Estudis Catalans, 1995.

———. "Filosofia naturale lulliana e alchimia: Con l'inedito epilogo del *Liber de secretis naturae seu de quinta essentia." Rivista di storia della filosofia* 41 (1986): 747–80.

———. "I francescani e l'alchimia." *Convivium Assisiense* 10 (2008): 117–57.

———. "Le figure alchemiche pseudolulliane: Un indice oltre il testo?" In *Fabula in tabula: Una storia degli indici dal manoscritto al testo elettronico,* ed. Claudio Leonardi, Marcello Morelli, and Francesco Santi, 111–18. Spoleto: Centro italiano di studi sull'alto Medioevo, 1994.

———. "L'elixir alchemico fra *artificium* e *natura." In Artificialia: La dimensione artificiale della natura umana,* ed. Massimo Negrotti, 255–67. Bologna: CLUEB, 1995.

———. *L'oro dei filosofi: Saggio sulle idee di un alchimista del Trecento.* Spoleto: Centro Italiano di Studi sull'Alto Medioevo, 1992.

———. "Mater Medicinarum: English Physicians and the Alchemical Elixir in the Fifteenth Century." In *Medicine from the Black Death to the French Disease,* ed. Roger French, Jon Arrizabalaga, Andrew Cunningham, and Luis Garcia-Ballester, 26–52. Aldershot: Ashgate, 1998.

———. "*Medicina* in the Alchemical Writings Attributed to Raymond Lull (14th–17th Centuries)." In *Alchemy and Chemistry in the Sixteenth and Seventeenth Centuries,* ed. Piyo Rattansi and Antonio Clericuzio, 1–15. Dordrecht: Kluwer, 1994.

———. "Natura naturam vincit." In *De natura: La naturaleza en la Edad Media,* ed. José Luis Fuertes Herreros and Ángel Poncela González, 1:101–20. Porto: Húmus, 2015.

———. "Sulla tradizione testuale del *Liber de secretis naturae seu de quinta essentia* attribuito a Raimondo Lullo: Le due redazioni della *Tertia distinctio*." *Archives internationales des sciences* 36 (1986): 1–16.

———, ed. "Un lapidario alchemico: Il *Liber de investigatione secreti occulti* attribuito a Raimondo Lullo; Studio introduttivo ed edizione." *Documenti e studi sulla tradizione filosofica medievale* 1 (1990): 549–603.

———. "'Vegetare seu transmutare': The Vegetable Soul and Pseudo-Lullian Alchemy." In *Arbor Scientiae: Der Baum des Wissens von Ramon Lull. Akten des Internationalen Kongresses aus Anlaß des 40-jährigen Jubiläums des Raimundus-Lullus-Instituts der Universität Freiburg i. Br.,* ed. Fernando Domínguez Reboiras, Pere Villalba Varneda, and Peter Walter, 93–119. Turnhout: Brepols, 2002.

Pereira, Michela, and Barbara Spaggiari, eds. *Il Testamentum alchemico attribuito a Raimondo Lullo: Edizione del testo latino e catalano dal manoscritto Oxford, Corpus Christi College, 244.* Florence: SISMEL, 1999.

Perifano, Alfredo. *L'alchimie à la cour de Côme Ier de Médicis: Culture scientifique et système politique.* Paris: Honoré Champion, 1997.

Petrina, Alessandra. *Cultural Politics in Fifteenth-Century England: The Case of Humphrey, Duke of Gloucester.* Leiden: Brill, 2004.

Pollard, J. *North-Eastern England during the Wars of the Roses.* Oxford: Clarendon Press, 1990.

Popper, Nicholas. *Walter Ralegh's "History of the World" and the Historical Culture of the Late Renaissance.* Chicago: University of Chicago Press, 2012.

Posset, Franz. *Johann Reuchlin (1455–1522): A Theological Biography.* Berlin: De Gruyter, 2015.

Power, Amanda. *Roger Bacon and the Defence of Christendom*. New York: Cambridge University Press, 2013.

Priesner, Claus. "Johann Thoelde und die Schriften des Basilius Valentinus." In *Die Alchemie in der europäischen Kultur- und Wissenschaftgeschichte*, ed. Christoph Meinel, 107–18. Wiesbaden: Harrassowitz, 1986.

Principe, Lawrence M. "Apparatus and Reproducibility in Alchemy." In *Instruments and Experimentation in the History of Chemistry*, ed. Frederic L. Holmes and Trevor H. Levere, 55–74. Cambridge, MA: MIT Press, 2000.

———. *The Aspiring Adept: Robert Boyle and His Alchemical Quest*. Princeton: Princeton University Press, 1998.

———. "Blomfild, William (fl. 1529–1574)." *ODNB*.

———. "'Chemical Translation' and the Role of Impurities in Alchemy: Examples from Basil Valentine's *Triump-Wagen*." *Ambix* 34 (1987): 21–30.

———. "Chymical Exotica in the Seventeenth Century, or, How to Make the Bologna Stone." *Ambix* 63 (2016): 118–44.

———, ed. *Chymists and Chymistry: Studies in the History of Alchemy and Early Modern Chemistry*. Sagamore Beach, MA: Chemical Heritage Foundation and Science History Publications, 2007.

———. *The Secrets of Alchemy*. Chicago: University of Chicago Press, 2013.

Prinke, Rafał T. "Beyond Patronage: Michael Sendivogius and the Meaning of Success in Alchemy." In *Chymia: Science and Nature in Medieval and Early Modern Europe*, ed. Miguel López Pérez, Didier Kahn, and Mar Rey Bueno, 175–231. Newcastle-upon-Tyne: Cambridge Scholars, 2010.

Pritchard, Alan. "Thomas Charnock's Book Dedicated to Queen Elizabeth." *Ambix* 26 (1979): 56–73.

Pumfrey, Stephen. "John Dee: The Patronage of a Natural Philosopher in Tudor England." *Studies in History and Philosophy of Science* 43 (2012): 449–59.

Pumfrey, Stephen, and Frances Dawbarn. "Science and Patronage in England, 1570–1626: A Preliminary Study." *History of Science* 42 (2004): 137–88.

Purš, Ivo. "Tadeáš Hájek of Hájek and His Alchemical Circle." In *Alchemy and Rudolf II: Exploring the Secrets of Nature in Central Europe in the 16th and 17th Centuries*, ed. Ivo Purš and Vladimír Karpenko, 423–57. Prague: Artefactum, 2016.

Purš, Ivo, and Vladimír Karpenko. "Alchemy at the Aristocratic Courts of the Lands of the Bohemian Crown." In *Alchemy and Rudolf II: Exploring the Secrets of Nature in Central Europe in the 16th and 17th Centuries*, ed. Ivo Purš and Vladimír Karpenko, 47–92. Prague: Artefactum, 2016.

Ralley, Robert. "The Clerical Physician in Late Medieval England." PhD diss., University of Cambridge, 2005.

Rampling, Jennifer M. "The Alchemy of George Ripley, 1470–1700." PhD diss., University of Cambridge, 2010.

———. "Analogy and the Role of the Physician in Medieval and Early Modern Alchemy." In *Alchemy and Medicine from Antiquity to the Enlightenment*, ed. Jennifer M. Rampling and Peter M. Jones. London: Routledge, forthcoming.

———. "The Catalogue of the Ripley Corpus: Alchemical Writings Attributed to George Ripley (d. *ca.* 1490)." *Ambix* 57 (2010): 125–201.

———. "Depicting the Medieval Alchemical Cosmos: George Ripley's Wheel of Inferior Astronomy." *Early Science and Medicine* 18 (2013): 45–86.

———. "English Alchemy before Newton: An Experimental History." *Circumscribere* 18 (2016): 1–11.

———. "Establishing the Canon: George Ripley and His Alchemical Sources." *Ambix* 55 (2008): 189–208.

———. *The Hidden Stone: Alchemy, Art, and the Ripley Scrolls.* Oxford: Oxford University Press, forthcoming.

———. "How to Sublime Mercury: Reading Like a Philosopher in Medieval Europe." *History of Knowledge*, 24 May 2018, https://wp.me/p8bNN8-23p.

———. "John Dee and the Alchemists: Practising and Promoting English Alchemy in the Holy Roman Empire." *Studies in History and Philosophy of Science* 43 (2012): 498–508.

———. "Reading Alchemically: Early Modern Guides to 'Philosophical' Practices." In "Learning by the Book: Manuals and Handbooks in the History of Knowledge," ed. Angela Creager, Elaine Leong, and Matthias Grote, *BJHS Themes* 5 (forthcoming).

———. "Transmission and Transmutation: George Ripley and the Place of English Alchemy in Early Modern Europe." *Early Science and Medicine* 17 (2012): 477–99.

———. "Transmuting Sericon: Alchemy as 'Practical Exegesis' in Early Modern England." *Osiris* 29 (2014): 19–34.

Rankin, Alisha. "How to Cure the Golden Vein: Medical Remedies as *Wissenschaft* in Early Modern Germany." In *Ways of Making and Knowing: The Material Culture of Empirical Knowledge*, ed. Pamela H. Smith, Amy R. W. Mayers, and Harold J. Cook, 113–37. Ann Arbor: University of Michigan Press, 2014.

———. *Panaceia's Daughters: Noblewomen as Healers in Early Modern Germany.* Chicago: University of Chicago Press, 2013.

Raphael, Renee. *Reading Galileo: Scribal Technology and the "Two New Sciences."* Baltimore: Johns Hopkins University Press, 2017.

Raven, James *The Business of Books: Booksellers and the English Book Trade, 1450–1850*. New Haven: Yale University Press, 2007.

Read, John. "Alchemy under James IV of Scotland." *Ambix* 2 (1938): 60–67.

Reynolds, Melissa. "'Here Is a Good Boke to Lerne': Practical Books, the Coming of the Press, and the Search for Knowledge, ca. 1400–1560." *Journal of British Studies* 58 (2019): 259–88.

Rhodes, Neil, Gordon Kendal, and Louise Wilson, eds. *English Renaissance Translation Theory*. London: Modern Human Research Association, 2013.

Richardson, H. G. "Year Books and Plea Rolls as Sources of Historical Information." *Transactions of the Royal Historical Society*, 4th ser., 5 (1922): 28–70.

Riehl, L. D. "John Argentein and Learning in Medieval Cambridge." *Humanistica Lovaniensa* 33 (1984): 71–85.

Ripley, George. *Compound of Alchemy (1591)*. Ed. Stanton J. Linden. Aldershot: Ashgate, 2001.

Roberts, Julian, and Andrew G. Watson, eds. *John Dee's Library Catalogue*. Cambridge: Bibliographical Society, 1990.

Rodríguez-Guerrero, José. "Un repaso a la alquimia del Midi Francés en al siglo XIV (parte I)." *Azogue: Revista electrónica dedicada al estudio histórico crítico de la alquimia* 7 (2010–13): 75–141.

Rotondò, Antonio. "Brocardo, Jacopo." *Dizionario biografico degli italiani 14 (1972)*. http://www.treccani.it/enciclopedia/iacopo-brocardo_(Dizionario -Biografico)/.

Ruska, Julius, ed. *Turba Philosophorum: Ein Beitrag zur Geschichte der Alchemie*. Quellen und Studien zur Geschichte der Naturwissenschaften und der Medizin 1. Berlin: Springer, 1931.

Schuler, Robert M. "An Alchemical Poem: Authorship and Manuscripts." *The Library*, 5th ser., 28 (1973): 240–43.

———. *Alchemical Poetry 1575–1700, from Previously Unpublished Manuscripts*. New York: Garland, 1995.

———. "Charnock, Thomas (1524/6–1581)." *ODNB*.

———. "Hermetic and Alchemical Traditions of the English Renaissance and Seventeenth Century, with an Essay on Their Relation to Alchemical Poetry, as Illustrated by an Edition of 'Blomfild's Blossoms,' 1557." PhD diss., University of Colorado, 1971.

———. "Three Renaissance Scientific Poems." *Studies in Philology* 75 (1978).

———. "William Blomfild, Elizabethan Alchemist." *Ambix* 20 (1973): 75–87.

Secret, François. *Les Kabbalistes chrétiens de la Renaissance*. Paris: Dunod, 1964.

———. "Réforme et alchimie." *Bulletin de la Société de l'histoire du protestantisme français (1903–2015)* 124 (1978): 173–86.

Secretum Secretorum: Nine English Versions. Ed. Mahmoud Manzalaoui. Oxford: Oxford University Press, 1977.

Shackelford, Jole. *A Philosophical Path for Paracelsian Medicine: The Ideas, Intellectual Context, and Influence of Petrus Severinus (1540/2–1602).* Copenhagen: Museum Tusculanum Press, 2004.

Sherman, William H. *John Dee: The Politics of Reading and Writing in the English Renaissance.* Amherst: University of Massachusetts Press, 1995.

Shipman, Joseph C. "Johannes Petreius, Nuremberg Publisher of Scientific Works 1524–1580, with a Short-Title List of His Imprints." In *Homage to a Bookman: Essays on Manuscripts, Books, and Printing Written for Hans P. Kraus on His 60th Birthday Oct. 12, 1967,* ed. H. Lehmann-Haupt, 147–16. Berlin: Mann, 1967.

Shirren, A. J. "'Colonel Zanchy' and Charles Fleetwood." *Notes and Queries* 168 (1953): 431–35.

Simpson, Richard. "Sir Thomas Smith's Stillhouse at Hill Hall: Books, Practice, Antiquity, and Innovation." In *The Intellectual Culture of the English Country House, 1500–1700,* ed. Matthew Dimmock, Andrew Hadfield, and Margaret Healy, 101–16. Manchester: Manchester University Press, 2015.

Singer, Dorothea Waley, and Annie Anderson. *Catalogue of Latin and Vernacular Alchemical Manuscripts in Great Britain and Ireland Dating from before the XVI Century.* 3 vols. Brussels: Maurice Lamertin, 1928, 1930, 1931.

Siraisi, Nancy G. *Taddeo Alderotti and His Pupils: Two Generations of Italian Medical Learning.* Princeton: Princeton University Press, 1981.

Slack, Paul. *The Impact of Plague in Tudor and Stuart England.* Oxford: Clarendon Press, 1990.

Smalley, Beryl. *The Study of the Bible in the Middle Ages.* Oxford: Clarendon Press, 1941.

Smith, Cyril Stanley, and John G. Hawthorne, eds. and trans. *Mappae clavicula: A Little Key to the World of Medieval Techniques.* Philadelphia: AMS, 1974.

Smith, Lesley. *The Glossa Ordinaria: The Making of a Medieval Bible Commentary.* Leiden: Brill, 2009.

Smith, Pamela H. "Alchemy as a Language of Mediation at the Hapsburg Court." *Isis* 85 (1994): 1–25.

———. *The Business of Alchemy: Science and Culture in the Holy Roman Empire.* Princeton: Princeton University Press, 1994.

———. *The Body of the Artisan: Art and Experience in the Scientific Revolution.* Chicago: University of Chicago Press, 2004.

———. "Vermilion, Mercury, Blood, and Lizards: Matter and Meaning in Metalworking." In *Materials and Expertise in Early Modern Europe: Between Market and Laboratory*, ed. Ursula Klein and E. C. Spary, 29–49. Chicago: University of Chicago Press, 2010.

Starkey, George. *Alchemical Notebooks and Correspondence*. Ed. Lawrence M. Principe and William R. Newman. Chicago: University of Chicago Press, 2004.

Stavenhagen, Lee, ed. and trans. *A Testament of Alchemy: Being the Revelations of Morienus to Khālid ibn Yazid*. Hanover, NH: Brandeis University Press, 1974.

Steele, Robert. "Practical Chemistry in the Twelfth Century: Rasis de aluminibus et salibus." *Isis* 12 (1929): 10–46.

Stow, John. *A Survey of London*. Reprinted from the text of 1603 (1908), XLVIII–LXVII. http://www.british-history.ac.uk/report.aspx?compid=60007&strquery=alchemy.

Taape, Tillmann. "Distilling Reliable Remedies: Hieronymus Brunschwig's 'Liber de arte distillandi' (1500) between Alchemical Learning and Craft Practice." *Ambix* 61 (2014): 236–56.

———. "Hieronymus Brunschwig and the Making of Vernacular Knowledge in Early German Print." PhD diss., Pembroke College, University of Cambridge, 2017.

Tatlock, J. S. P. "Geoffrey of Monmouth's Vita Merlini." *Speculum* 18 (1943): 265–87.

Tavormina, M. Teresa. "Roger Bacon: Two Extracts on the Prolongation of Life." In *Sex, Aging, and Death in a Medieval Medical Compendium: Trinity College Cambridge MS R.14.52, Its Texts, Language, and Scribe*, ed. M. Teresa Tavormina, 1:327–72. Tempe: Arizona Center for Medieval and Renaissance Studies, 2006.

Taylor, F. Sherwood. "The Idea of the Quintessence." In *Science, Medicine, and History: Essays on the Evolution of Scientific Thought and Medical Practice Written in Honour of Charles Singer*, ed. Edgar A. Underwood, 1:247–65. Oxford: Oxford University Press, 1953.

———. "Thomas Charnock." *Ambix* 2 (1946): 148–76.

Telle, Joachim. "Astrologie und Alchemie im 16. Jahrhundert: Zu den astroalchemischen Lehrdichtungen von Christoph von Hirschenberg und Basilius Valetinus." In *Die okkulten Wissenschaften in der Renaissance*, ed. August Buck, 227–53. Wiesbaden: O. Harrassowitz, 1992.

Theisen, W. R. "The Attraction of Alchemy for Monks and Friars in the 13th–14th Centuries." *American Benedictine Review* 46 (1995): 239–51.

———. "John Dastin's Letter on the Philosopher's Stone." *Ambix* 33 (1986): 78–87.

Theophilus. *On Divers Arts: The Foremost Medieval Treatise on Painting, Glass-making, and Metalwork*. Ed. and trans. John G. Hawthorne and Cyril Stanley Smith. New York: Dover, 1979.

Thomas, Keith. *Religion and the Decline of Magic*. London: Weidenfeld & Nicolson, 1971. Reprint, London: Penguin, 1991.

Thomas, Phillip D., ed. *David Ragor's Transcription of Walter of Odington's "Icocedron."* Wichita: Wichita State University, 1968.

Thomson, Rodney M. *A Descriptive Catalogue of the Medieval Manuscripts of Corpus Christi College, Oxford*. Oxford: D. S. Brewer, 2011.

———. "John Dunstable and His Books." *Musical Times* 150 (2009): 3–16.

Thorndike, Lynn. *A History of Magic and Experimental Science*. 8 vols. New York: Columbia University Press, 1923–58.

Tillotson, John H., ed. and trans. *Monastery and Society in the Late Middle Ages: Selected Account Rolls from Selby Abbey, Yorkshire, 1398–1537*. Woodbridge: Boydell and Brewer, 1988.

Timmermann, Anke. "Alchemy in Cambridge: An Annotated Catalogue of Alchemical Texts and Illustrations in Cambridge Repositories." *Nuncius* 30 (2015): 345–511.

———. *Verse and Transmutation: A Corpus of Middle English Alchemical Poetry*. Leiden: Brill, 2013.

Tout, T. F. *Chapters in the Administrative History of Mediaeval England: The Wardrobe, the Chamber, and the Small Seals*. Vol. 4. Manchester: Manchester University Press, 1928.

Turner, Wendy J. "The Legal Regulation and Licensing of Alchemy in Late Medieval England." In *Law and Magic: A Collection of Essays*, ed. Christine A. Corcos, 209–25. Durham, NC: Carolina Academic Press, 2010.

Vine, Angus. *In Defiance of Time: Antiquarian Writing in Early Modern England*. Oxford: Oxford University Press, 2010.

———. *Miscellaneous Order: Manuscript Culture and the Early Modern Organization of Knowledge*. Oxford: Oxford University Press, 2019.

Voigts, Linda Ehrsam. "The *Master* of the *King's Stillatories*." In *The Lancastrian Court: Proceedings of the 2001 Harlaxton Symposium*, ed. Jenny Stratford, 233–52. Donington: Shaun Tyas, 2003.

———. "Multitudes of Middle English Medical Manuscripts, or the Englishing of Science and Medicine." In *Manuscript Sources of Medieval Medicine: A Book of Essays*, ed. Margaret R. Schleissner, 183–95. New York: Garland, 1995.

———. "The 'Sloane Group': Related Scientific and Medical Manuscripts from the Fifteenth Century in the Sloane Collection." *British Library Journal* 16 (1990): 26–57.

Voigts, Linda Ehrsam, and Patricia Deery Kurtz, comps. *Scientific and Medical Writings in Old and Middle English: An Electronic Reference*. Ann Arbor: University of Michigan Press, 2000. CD-ROM.

Walker, D. P. *Spiritual and Demonic Magic from Ficino to Campanella*. London: Warburg Institute, 1958. Reprint, University Park: Pennsylvania State University Press, 2000.

Waller, William Chapman. "An Essex Alchemist." *Essex Review* 13 (1904): 19–23.

Walsham, Alexandra. *Providence in Early Modern England*. Oxford: Oxford University Press, 2001.

———. "Providentialism." In *The Oxford Handbook of Holinshed's Chronicles*, ed. Felicity Heal, Ian W. Archer, and Paulina Kewes, 427–42. Oxford: Oxford University Press, 2012.

Warner, George F., and Julius P. Gilson. *Catalogue of Western Manuscripts in the Old Royal and King's Collections*. 2 vols. London: Trustees of the British Museum, 1921.

Watson, Andrew G. "Robert Green of Welby, Alchemist and Count Palatine, c. 1467–c. 1540." *Notes and Queries*, Sept. 1985, 312–13.

Wear, Andrew. *Knowledge and Practice in English Medicine, 1550–1680*. Cambridge: Cambridge University Press, 2000.

Webster, Charles. "Alchemical and Paracelsian Medicine." In *Health, Medicine, and Mortality in the Sixteenth Century*, ed. Charles Webster, 301–34. Cambridge: Cambridge University Press, 1979.

———. "English Medical Reformers of the Puritan Revolution: A Background to the 'Society of Chemical Physicians.'" *Ambix* 14 (1967): 16–41.

Weiss, Roberto. *Humanism in England during the Fifteenth Century*. 3rd ed. Oxford: Blackwell, 1967.

Wiener, Carol Z. "The Beleaguered Isle: A Study of Elizabethan and Early Jacobean Anti-Catholicism." *Past & Present* 51 (1971): 27–62.

Wilding, Michael. "A Biography of Edward Kelly, the English Alchemist and Associate of Dr John Dee." In *Mystical Metal of Gold: Essays on Alchemy and Renaissance Culture*, ed. Stanton J. Linden, 35–89. New York: AMS Press, 2007.

Williams, Steven J. "Esotericism, Marvels, and the Medieval Aristotle." In *Il segreto*, ed. Thalia Brero and Francesco Santi, Micrologus' Library 14, 171–91. Florence: SISMEL, 2006.

———. *The Secret of Secrets: The Scholarly Career of a Pseudo-Aristotelian Text in the Latin Middle Ages*. Ann Arbor: University of Michigan Press, 2003.

Witten, Laurence C., II, and Richard Pachella, comps. *Alchemy and the Occult: A Catalogue of Books and Manuscripts from the Collection of Paul and Mary*

Mellon Given to Yale University Library. Vol. 3, *Manuscripts: 1225–1671.* New Haven: Yale University Library, 1977.

Woolf, D. R. *Reading History in Early Modern England.* Cambridge: Cambridge University Press, 2000.

Young, Francis. *Magic as a Political Crime in Medieval and Early Modern England: A History of Sorcery and Treason.* London: I. B. Tauris, 2018.

Zika, Charles. *Reuchlin und die okkulte Tradition der Renaissance.* Sigmaringen: Thorbecke, 1998.

———. "Reuchlin's *De Verbo Mirifico* and the Magic Debate of the Late Fifteenth Century." *Journal of the Warburg and Courtauld Institutes* 39 (1976): 104–38.

Index

Page numbers in italics refer to figures and tables.

397